FEEDS
AND
FEEDING

FEEDS AND FEEDING

Arthur E. Cullison
University of Georgia
Athens, Georgia

RESTON PUBLISHING COMPANY, INC.
Reston, Virginia
A Prentice-Hall Company

Library of Congress Cataloging in Publication Data

Cullison, Arthur Edison,
   Feeds and feeding.

   Includes bibliographical references and index.
   1. Feeds.   2. Feeding.   I. Title.
SF95.C82   636.08'4   75-6501
ISBN 0-87909-266-1

©1975 by

Reston Publishing Company, Inc.
A Prentice-Hall Company
Reston, Virginia 22090

10   9   8   7   6

Printed in the United States of America.

Contents

Preface   *ix*

Introduction   *1*

1   The Feed Nutrients   *3*

2   General Functions of Feed Nutrients   *10*

3   Proximate Analysis of Feedstuffs   *12*

4   The Van Soest Method of Forage Evaluation   *28*

5   The Digestive Tract   *32*

6   Nutrient Digestion, Absorption, and Transport   *36*

7   Apparent Digestibility   *41*

8   Total Digestible Nutrients or TDN   *46*

9   Energy Utilization   *51*

10   Study Questions and Problems   *55*

11   Protein Nutrition   *70*

12   Mineral Nutrition—General   *81*

13   The Macro Minerals   *84*

14   The Micro Minerals   *101*

15   General Recommendations for Mineral Feeding   *112*

16   Vitamin Nutrition   *115*

17   Vitamins in Livestock Feeding   *119*

18   Physiological Phases of Livestock Production   *130*

19   Study Questions and Problems   *144*

20   Feeds and Feed Groups—General   *159*

21   Air-Dry Energy Feeds   *162*

22   High Protein Feeds   *175*

23   Air-Dry Roughages   *188*

24   The Molasseses   *203*

25   Other High Moisture Feeds   *206*

26   Identifying Feeds from Their Composition   *210*

27   Hay and Hay Making   *216*

28   Silage—Its Production and Use   *224*

29   Processing Feeds   *234*

30   Uniform State Feed Bill   *240*

31   Balancing Rations—General   *255*

32   The Petersen Method of Evaluating Feeds   *260*

33   Calculating a Balanced Daily Ration—General   *267*

34   Balancing a Daily Ration for a Steer Using As Fed Weights and Composition Figures   *270*

35   Balancing a Daily Ration for a Steer Using Dry Feed Weights and Composition Figures   *277*

36   Balancing a Daily Ration for a Steer Using a Restricted Level of Roughage   *282*

37   Balancing a Steer Ration with Minimum Use of the Metric System   *284*

38 The California System for the Net Energy Evaluation of Rations for Growing-Fattening Cattle   *288*

39 Balancing a Ration for a Dairy Cow   *291*

40 Balancing Daily Rations for Horses   *296*

41 Balancing Daily Rations for Sheep   *299*

42 Balancing Daily Rations for Swine—General   *300*

43 Some General Guides for Feeding Livestock   *307*

44 Formulating Balanced Ration Mixtures—General   *310*

45 Formulating a Ration Mixture Based on a Balanced Daily Ration   *313*

46 Formulating Feed Mixtures by the Use of the Square Method   *315*

47 Fortifying Steer Ration Mixtures with Vitamin A   *323*

48 Use of Algebraic Equations in the Formulation of Feed Mixtures   *325*

49 Computerized Least-Cost Rations   *330*

50 Estimating Feed Requirements   *335*

51 Weights, Measures, Volumes, and Capacities   *339*

52 Pastures—General   *342*

53 Some Important Facts About Some of the More Important Pasture Crops   *344*

54 General Use of Pasture in Livestock Feeding   *355*

55 Study Questions and Problems   *362*

56 Tables on Nutrient Requirements   *378*

57 Tables on Feed Composition   *409*

58 Glossary of Terms Frequently Used in Discussing Matters Related to Feeds and Feeding   *456*

Index   *479*

Preface

This handbook on *Feeds and Feeding* has been written primarily as a text for use in the teaching of an undergraduate course in feeds and feeding. Only that information which it is thought should be covered in such a course has been included. While it is believed that this book will answer most basic questions pertaining to feeds and feeding, it does not attempt to deal with the more unusual feeds and is not intended as an all-purpose reference on such materials.

Also, since at many institutions the feeding of poultry is not included in the regular feeds and feeding course, this subject has not been covered in the present text. It is suggested that at those institutions where poultry feeding is included as a part of the regular course in feeds and feeding, the National Research Council publication on the *Nutrient Requirements of Poultry* be used along with this text for teaching purposes.

For the most part *Feeds and Feeding* does not represent a review of original literature but is more of a consolidation of general information on feeds and feeding with facts and figures from several NRC publications and some of the author's personal thoughts and observations. *Feeds and Feeding* by Morrison, *Animal Nutrition* by Maynard and Loosli, and *Applied Animal Nutrition* by Crampton and Harris have been relied on heavily as general information sources. For rather complete bibliographies on subjects related to the feeding of the different species of farm livestock, the reader is referred to these three books as well as the respective NRC publications on nutrient requirements.

Upon making this book available to the general public, the author wishes to express grateful appreciation to Dr. R. S. Lowrey of the University of Georgia Animal Science Department and Dr. James Blakely of the Wharton County (Texas) Junior College for their most helpful critical reviews of the book during its preparation. Appreciation is also expressed to Larry Grimes, formerly of the University of Georgia Animal Science Department, for writing the computer program used in calculating the constants included in Table 17; to Joyce Dilley, the author's secretary, for her patience and cooperation in the typing of the book; to Albert Cunningham, the author's graduate student, whose assistance has contributed significantly to the book's completion; to Vann Cleveland, photographer, University of George College of Agriculture Experiment Stations, who either took or reproduced many of the pictures used in the book; and to all of the other individuals and firms who have contributed toward the book's development.

Also, the author wishes to express special appreciation to the National Academy of Sciences for granting him permission to quote information from certain of its National Research Council publications in the preparation of this book and to all of those who over the years have served as members of the NRC's Committee on Animal Nutrition, and its respective sub-committees, without whose efforts the above-mentioned publications, and in turn this book, would not have been possible.

Arthur E. Cullison

Introduction

Livestock production is one of our nation's largest industries. Sales of livestock and livestock products made up approximately 44%[1] of the nation's gross farm returns in 1973. The United States is currently spending approximately 50%[2] of its food dollar on meat and milk products. Essentially 100% of our production of pasture, hay, and silage must be fed to livestock in order for these materials to be converted into products suitable for human consumption. The same can be said for millions of tons of various by-product materials whose only or primary market is as a livestock feed. Also, a very high percentage of our nation's grain

---

[1] *Farm Income and State Estimates.* Economic Research Service, USDA. September 1974 (FIS 224, supplement).

[2] *Marketing and Transportation Situation* (MIS-194). Economic Research Service, USDA. August 1974.

production is normally used for livestock feeding purposes. From the above the great importance of livestock production in this country's agricultural economy, as well as in its overall economy, becomes apparent.

In this connection it should also be recognized that roughly three-fourths of the costs of producing livestock are feed costs. Furthermore, livestock rations can vary greatly in their efficiency and nutritional adequacy. Also, rations of equal nutritive value can differ greatly in their per unit cost. Consequently, if a livestock operation is to maximize profits, it is most essential that particular attention be given to the feeding program.

A great deal of time and effort has been devoted by animal scientists over the past 50 to 75 years toward developing a better understanding of the principles of animal nutrition and to the application of these principles toward the practical and profitable feeding of farm livestock. Only with a knowledge of these principles and an understanding of their proper application to livestock feeding can one hope to attain maximum efficiency in his livestock production program.

Accordingly, in *Feeds and Feeding* the author deals to a considerable degree with the basic principles of animal nutrition. This is not done with the idea of attempting to make the book an all-inclusive reference on the subject of animal nutrition. It is simply that a modern-day livestock feeding program must be based on nutritional principles if it is to be successful, and a working knowledge of the more basic of these principles is considered to be a must for a student of feeds and feeding under present-day circumstances. Only those principles are discussed, and these at such depths as are considered to be essential for carrying out a practical program of scientific feeding.

1 The Feed Nutrients

I. Of the over 100 known chemical elements, at least 20 enter into the makeup of the various essential feed nutrients. These 20 elements, their symbols, and their atomic weights are as follows:

NAME	SYMBOL	ATOMIC WT	NAME	SYMBOL	ATOMIC WT
Carbon	C	12	Magnesium	Mg	24.3
Hydrogen	H	1	Sodium	Na	23
Oxygen	O	16	Chlorine	Cl	35.5
Phosphorus	P	31	Cobalt	Co	59
Potassium	K	39	Copper	Cu	63.5
Iodine	I	127	Fluorine	F	19
Nitrogen	N	14	Manganese	Mn	55
Sulfur	S	32	Zinc	Zn	65.4
Calcium	Ca	40	Molybdenum	Mo	96
Iron	Fe	55.8	Selenium	Se	79

There is some evidence that chromium (Cr) and possibly others should be included in this group.

II.   These elements, either alone or in various combinations, go to make up what are known as the *feed nutrients*. (The term *feed nutrient* is applied to any feed constituent which may function in the nutritive support of animal life.)

III.   Many different feed nutrients are currently recognized, and new ones are still being found. Those currently recognized are as follows:

    A.   **Carbohydrates.** Carbohydrates contain carbon, hydrogen, and oxygen, with hydrogen and oxygen in the same proportion as in water. They consist largely of hexosans. These are made up of hexose or 6-carbon atom molecules. Pentosans which are made up of pentose or 5-carbon atom molecules are sometimes present. Tetrose, triose, and diose compounds are also sometimes present in small amounts but are generally unimportant.

        1.   **Monosaccharides.** Monosaccharides all have a chemical formula of $C_6H_{12}O_6$. They are formed in plants by the following reaction. This reaction is reversed by animals.

$$6\,CO_2\ +\ 6\,H_2O\ \rightarrow\ C_6H_{12}O_6\ +\ 6\,O_2$$

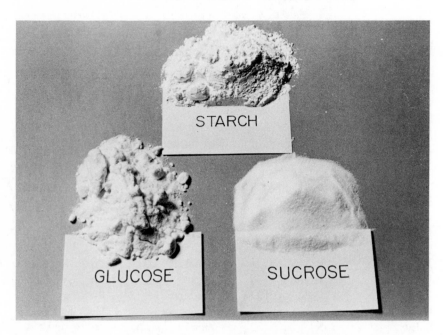

FIGURE 1.   Essentially pure forms of a monosaccharide (glucose), a disaccharide (sucrose), and a polysaccharide (starch). *(Courtesy of the University of Georgia College of Agriculture Experiment Stations)*

The more common monosaccharides are:

    a.   **Glucose.** Glucose (Figure 1) is found in corn syrup and also in blood. It is sometimes referred to as *dextrose* because it rotates the plane of polarized light to the right. It is about three-fourths as sweet as cane sugar.

    b.   **Fructose.** Fructose is found principally in ripe fruits and honey. It is the sweetest of all sugars.

    c.   **Galactose.** Galactose is obtained along with glucose upon the hydrolysis of lactose or milk sugar, a disaccharide listed below.

2.   **Disaccharides.** Disaccharides all have a chemical formula of $C_{12}H_{22}O_{11}$. They are formed from two monosaccharide molecules with the loss of one molecule of water. The more common disaccharides are:

    a.   **Sucrose.** Sucrose (Figure 1) is the same as *cane* and *beet sugar*, commonly used as food sweeteners. It hydrolyzes to glucose and frutcose.

$$1 \text{ sucrose } + 1 H_2O \rightarrow 1 \text{ glucose } + 1 \text{ fructose}$$

    b.   **Maltose.** Maltose is the same as *malt sugar*. It is obtained from the hydrolysis of starch. Maltose is one-fourth as sweet as sucrose. It hydrolyzes entirely to glucose.

$$1 \text{ maltose } + 1 H_2O \rightarrow 2 \text{ glucoses}$$

    c.   **Lactose.** Lactose is commonly referred to as *milk sugar*. It is found principally in milk. Lactose is one-sixth as sweet as sucrose. It hydrolyzes to glucose and galactose.

$$1 \text{ lactose } + 1 H_2O \rightarrow 1 \text{ glucose } + 1 \text{ galactose}$$

3.   **Polysaccharides.** Polysaccharides all have an empirical formula of $(C_6H_{10}O_5)_n$. They are formed by the combination of unknown numbers of hexose molecules. Those polysaccharides usually regarded as important in animal nutrition are:

    a.   **Starch.** Many plants store energy in the form of starch (Figure 1). Starch is a major component of most livestock rations (especially fattening rations) and is highly digestible. Hence, it is a primary energy source for livestock. Starch hydrolyzes as follows:

$$\text{Starch} + H_2O \rightarrow \text{Dextrin}$$
$$\text{Dextrin} + H_2O \rightarrow \text{Maltose}$$
$$\text{Maltose} + H_2O \rightarrow \text{Glucose}$$

b.   **Inulin.** Inulin is similar to starch except it hydrolyzes to fructose rather than glucose. It is not very prevalent. Inulin is found especially in Jerusalem artichokes.

c.   **Glycogen.** Glycogen is sometimes referred to as *animal starch*. Found only in the animal body, it is produced in the liver and is the primary carbohydrate reserve in the animal. It hydrolyzes entirely to glucose.

d.   **Cellulose.** Cellulose (Figure 2) is a principal constituent of the cell wall of plants. It is most abundant in the more fibrous feeds. It is generally low in digestibility. Also, it may reduce the digestibility of other nutrients. Cattle, sheep, and horses digest cellulose fairly effectively; it is only slightly digested by hogs. Cellulose can be hydrolyzed by special processes to glucose.

FIGURE 2.   Two essentially pure forms of cellulose. *(Courtesy of the University of Georgia College of Agriculture Experiment Stations)*

e.   **Lignin.** Lignin is not a true carbohydrate. It contains too much carbon for a carbohydrate; also, the hydrogen and oxygen are not in the right proportion, and some nitrogen is usually present. However, lignin is usually considered with the polysaccharides. It is found largely in overmature hays, straws, and hulls. It is essentially indigestible by all livestock. Also, it may reduce the digestibility of other nutrients, especially cellulose. Lignin is of no known nutritive value except as a bulk factor.

B.   **Fats.** Fats contain carbon, hydrogen, and oxygen with more carbon and hydrogen in proportion to the oxygen than with carbohydrates. Fats contain 2.25 times as much energy per lb or kg as do carbohydrates.

$$\text{Fatty acid} + \text{Glycerol} \rightarrow \text{Fat} + \text{Water}$$

Examples of fats are:

$$\text{Stearic acid} + \text{Glycerol} \rightarrow \text{Stearin} + \text{Water}$$
$$3\,C_{17}H_{35}COOH + C_3H_5(OH)_3 \rightarrow C_{57}H_{110}O_6 + 3\,H_2O$$
$$\text{Palmitic acid} + \text{Glycerol} \rightarrow \text{Palmitin} + \text{Water}$$
$$3\,C_{15}H_{31}COOH + C_3H_5(OH)_3 \rightarrow C_{51}H_{98}O_6 + 3\,H_2O$$
$$\text{Oleic acid} + \text{Glycerol} \rightarrow \text{Olein} + \text{Water}$$
$$3\,C_{17}H_{33}COOH + C_3H_5(OH)_3 \rightarrow C_{57}H_{104}O_6 + 3\,H_2O)$$

1. Stearic acid and palmitic acid are two of many different saturated fatty acids (no double bonds) which combine with glycerol to form two of the more common saturated fats (stearin and palmitin).

2. Oleic and certain other fatty acids (linoleic, linolenic, and arachidonic) are unsaturated (have one or more double bonds) and combine with glycerol to form unsaturated fats. Oleic acid has one double bond, linoleic two, linolenic three, and arachidonic four.

3. Iodine Number of a fat is a measure of its degree of unsaturation. Iodine is taken up in proportion to the degree of unsaturation. The Iodine Number denotes the g of iodine absorbed per 100 g of fat.

4. Saponification:

$$\text{Fat} + \text{Alkali} \rightarrow \text{Soap} + \text{Glycerol}$$

$$
\begin{array}{lcl}
C_{17}H_{35}COOCH_2 & & CH_2OH \\
\;\;\;\;\;\;\;\;\;\;\;| & & \;\;\;\;| \\
C_{17}H_{35}COOCH + 3\,NaOH \rightarrow & 3\,C_{17}H_{35}COONa + & CH\;OH \\
\;\;\;\;\;\;\;\;\;\;\;| & & \;\;\;\;| \\
C_{17}H_{35}COOCH_2 & & CH_2OH
\end{array}
$$

5. Oils are actually fats and differ from other fats only in melting point. Oils have a low melting point and tend to be liquid at room temperature. Generally speaking, the shorter the carbon chain in the fatty acid and the greater the degree of unsaturation, the lower the melting point of the respective fat.

6. Most common fats consist of mixtures of pure fats.

C. **Protein.** Proteins always contain carbon, hydrogen, oxygen and nitrogen and sometimes iron, phosphorus, and/or sulphur. Protein is the only macronutrient which contains nitrogen except for small amounts in lignin. Feed proteins on the average contain 16% nitrogen. Proteins are innumerable in number. They are formed by various combinations of amino acids of which there are some 25+ to be found in proteins. Amino acids are organic acids which carry the amino group (NH_2). (Proteins will be discussed in more detail subsequently.)

D. Minerals. Of the 20 elements that function in animal nutrition, carbon, hydrogen, oxygen, and nitrogen are regarded as the *nonmineral elements*. The other 16 are referred to as the *mineral elements* which function in animal nutrition. Of these, 7 are macro (required in relatively large amounts) and 9 are micro (required in very small or trace amounts). The latter are referred to as the *trace minerals*.

The *macro minerals* are:

Calcium	Sulfur
Phosphorus	Chlorine
Potassium	Magnesium
Sodium	

The *micro* or *trace minerals* are:

Iron	Manganese
Iodine	Zinc
Copper	Molybdenum
Cobalt	Selenium
Fluorine	

(The role of minerals in animal nutrition will be discussed in more detail subsequently.)

E. Vitamins. Vitamins are organic substances required by animals in very small amounts for regulating various body processes toward normal health, growth, production, and reproduction. They are classed as micronutrients. They all contain carbon, hydrogen, and oxygen. In addition, several contain nitrogen. Certain ones also contain one or more of the mineral elements. There are 16 or more that function in animal nutrition. (Vitamins will be discussed in more detail later.)

F. Water. Water contains hydrogen and oxygen. Farm animals will consume from 3 to 8 times as much water as dry matter and will die from lack of water quicker than from lack of any other nutrient. Water is found in all feeds, ranging from around 10% in air-dry feeds to over 80% in fresh green forages. Besides serving as a nutrient, it has many important implications in feeds and feeding.

    1. Functions of water in the animal body.

        a. It enters into many biochemical reactions in the body.

        b. It functions in the transport of other nutrients.

    c.   It helps to maintain normal body temperature.

    d.   It helps to give the body form.

2.   Water is important in feed storage.

    a.   Too much water will cause certain feeds to heat and mold or otherwise lose quality. Some approximate maximum tolerances are as follows:

Ground feeds	11%	Grass hay	20%
Small grains	13%	Molasses	40%
Shelled corn	15%	Silage	75%
Snap corn	18%		

3.   Water is an important factor influencing feed value.

    a.   Water in feed is of no more value to an animal than water from other sources. Hence, appropriate allowances must be made when buying or feeding feeds high in water content.

2 General Functions of Feed Nutrients

I.   There are four general functions which nutrients may serve in the animal body. Three of these may be classed as basic functions, the other as an accessory function.

    A.   **The three basic functions are:**

        1.   **As a structural material for building and maintaining the body structure.** Just as boards, blocks, bricks, mortar, etc., are essential for building a house, so are certain nutrients required for the development of the animal body. Just as a house is a physical structure, so is an animal's body—consisting of bones, muscles, skin, organs, connective tissue, teeth, hair, horn and hoof. Just as a house undergoes constant degeneration and requires more or less constant maintenance, so does an animal's body undergo constant degeneration and require constant maintenance. Essentially the same nutrients are required for body maintenance as for body building. Protein, minerals, fat, and water function in this connection.

        2.   **As a source of energy for heat production, work, and/or fat deposition.** Just as a house needs sources of energy (electricity, gas, coal, oil, wood, etc.) for heat production and for power to run all the gadgets around a home, so must an animal be supplied with energy for keeping the body warm and for the capacity to do work—both work of the vital organs and voluntary work. Also, just as coal is sometimes stored in the basement of a home as a reserve supply of stored energy, so is fat sometimes deposited within the tissues of an animal's body as a reserve supply of stored energy for the animal. Carbohydrates, fats, and proteins function here.

3. **For regulating body processes or in the formation of body produced regulators.** Just as thermostats, switches, faucets, latches, etc., serve to regulate the functioning of various utilities about the home, so are regulators involved in the control of the various functions, processes, and activities in an animal's body. Serving in this capacity are such items as vitamins, enzymes, hormones, minerals, certain amino acids, and certain fatty acids.

B. **The accessory function—milk production.** The production of milk does not actually represent the ultimate use of nutrients but simply the shunting of a portion of the nutrients which an animal has consumed and digested off into the product we know as milk. Not until milk is consumed and utilized by another animal do the nutrients therein actually serve ultimate functions. Since milk contains some of almost all of the essential nutrients, essentially all of the various nutrients function in this connection. With poultry, egg production would fall into this same category.

Table 1
SUMMARY OF THE VARIOUS FUNCTIONS WHICH THE DIFFERENT NUTRIENTS MAY SERVE

	BASIC FUNCTIONS			ACCESSORY FUNCTION
	As a Structural Material for Body Building and Maintenance	As Energy for Heat Production, Work, and Fat Deposition	As or for the Formation of a Body Regulator	As a Source of Nutrients for Milk (or Egg) Production
Protein	Yes	Yes	Certain amino acids	Yes
Carbohydrates	Only as fat formed from carbohydrates enters into makeup of cellular growth	Yes	Yes	Yes
Fats	Only as fat enters into makeup of cellular growth	Yes	Certain fatty acids	Yes
Minerals	Yes	No	Yes	Yes
Vitamins	No	No	Yes	Yes
Water	Yes	No	Yes	Yes

3 Proximate Analysis of Feedstuffs

I.   A system for approximating the value of a feed or material for feeding purposes, without actually using the feed in a feeding trial, was developed at the Weende Experiment Station in Germany over 100 years ago. It is based on the separation of feed components into groups or fractions in accordance with their feeding value. The various fractions are:

Water	Crude fiber
Crude protein	Nitrogen-free extract
Crude fat or ether extract	Mineral matter or ash

II.   Following proper procedures in obtaining and preparing a sample is essential for an accurate analysis on a given lot of feed. No analysis can be any more accurate than the sample used for making it. No information concerning a feed's composition is better than misinformation. Steps to be followed in obtaining and preparing a sample are as follows:

    A.   Obtain a small quantity from several locations within a lot of feed.

    B.   Finely grind or chop the sample and mix until each spoonful is representative of every other spoonful within the sample, as well as of the overall lot of feed.

    C.   Keep the sample in a tightly closed container. Refrigerate if necessary.

III.   The procedures followed in arriving at the amounts of the different fractions ordinarily determined are as follows:

    A.   **Water or moisture.**

        1.   Weigh out a small quantity (usually less than 10 g) of the prepared sample into an appropriate container.

        2.   Dry in an appropriate oven (Figure 3) until there is no further loss in weight.

FIGURE 3.   Aluminum dishes containing feed samples being removed from a 100° C drying oven for weighing. At the left is shown a vacuum oven sometimes used for low temperature drying. *(Courtesy of the University of Georgia College of Agriculture Experiment Stations)*

   a. At 135°C for a short period. (About 2 hours is usually required for air-dry materials.)

   b. At 100°C for a longer period (8-24 hours).

   c. At less than 100°C with forced air or under vacuum.

  3. Weigh sample after drying.

  4. Calculate the percentage of water or moisture by either of two methods.

Method A

$$\frac{\text{Loss of wt during drying}}{\text{Wt of sample before drying}} \times 100 = \% \text{ water}$$

Method B

$$\frac{\text{Wt of sample after drying}}{\text{Wt of sample before drying}} \times 100 = \% \text{ DM}$$

$$100 - \% \text{ DM} = \% \text{ water}$$

Sometimes other methods are used for determining moisture content.

  1. Volumetric distillation using oil or toluene

  2. Electronic—based on conductivity

  3. Freeze drying.

B. **Crude Protein**. This determination is based on the fact that most nitrogen-containing macromaterials in most feeds are proteins, and proteins on the average are approximately 16% nitrogen. Actually, individual proteins will range from about 15% to over 18% nitrogen.

  1. Weigh out a small quantity of the prepared sample (usually from 1-5 g, depending on its nitrogen content) into a piece of nitrogen-free filter paper.

  2. Proceed to determine amount of ammoniacal nitrogen (Figure 4) in sample as follows:

   a. Digest in concentrated sulfuric acid in the presence of potassium or sodium sulfate plus a catalyst to convert all ammoniacal nigrogen to ammonium sulfate—$(NH_4)_2SO_4$.

   b. Add an excess of concentrated sodium hydroxide to make solution strongly alkaline, causing all ammoniacal nitrogen to form ammonium hydroxide—NH_4OH.

   c. Add water and distill ammonia into a known quantity of a standard acid solution, and determine by titration with standard alkali the amount of acid neutralized by the ammonia

FIGURE 4. A Kjeldahl rack for making nitrogen determinations in full operation. At the bottom a set of samples is being subjected to concentrated sulfuric acid digestion while at the top the ammonia is being distilled from an already digested set into beakers containing standard acid just below. *(Courtesy of Russell Research Center, Athens, Georgia)*

formed from the nitrogen in the feed. From this the amount of nitrogen can be calculated.

3. Calculate the amount of protein in the sample by multiplying the amount of nitrogen by 6.25 (since protein is 16% nitrogen, 6.25 times the amount of nitrogen in the sample would equal the total amount or 100% of the protein in the sample).

4. Calculate the percentage of crude protein in the feed as follows:

$$\frac{\text{Amount of protein in sample}}{\text{Wt of sample}} \times 100 = \% \text{ crude protein}$$

Protein so determined is referred to as *crude protein* as contrasted to *true protein* as determined by more involved procedures. It is designated as *crude* since it may contain amounts of certain ammoniacal nitrogen-containing materials which are not true protein such as amino acids, enzymes, certain vitamins, urea, etc. Such materials are referred to as *nonproteins* or *amides*. They are usually present in natural feedstuffs, however, only in very small amounts and so do not ordinarily involve a sizeable error. This is especially true since such materials are usually most prevalent in ruminant feeds, and ruminants can make effective use of nonprotein materials for meeting their protein needs. When the word *protein* is used in this book, it will be used to mean *crude protein* or "N X 6.25" unless otherwise indicated.

C.   Crude fat. Includes all of that portion of a feed soluble in ether.
Hence, crude fat is commonly referred to as *ether extract* or *EE*. While the crude fat in most feeds is usually mostly true fats, it may also embrace varying amounts of other ether-soluble materials such as the fat-soluble vitamins, carotene, chlorophyll, sterols, phospholipids, waxes, etc.—hence, the designation "crude" fat. The amounts of ether-soluble materials in a feed which are not true fats, however, usually represent only a very small percentage of the overall feed. Consequently, no sizeable error is ordinarily involved in assuming that the ether-soluble fraction of a feed is mostly true fat. The steps involved in making this determination (Figure 5) are as follows:

1.   Weigh out a small quantity of the prepared sample (usually less than 5 g) into an extraction thimble.

2.   Remove water from the sample by placing it in a drying oven. This is essential to permit thorough penetration of sample by ether.

3.   Extract the sample with ether in a Soxhlet extractor or some other suitable extractor for several hours.

FIGURE 5.   Feed samples in the process of being analyzed for ether extract. *(Courtesy of the University of Georgia College of Agriculture Experiment Stations)*

4. Evaporate the ether from the extract and weigh what remains. This is crude fat.

5. Calculate the percentage of crude fat:

$$\frac{\text{Wt of crude fat}}{\text{Wt of sample used}} \times 100 = \% \text{ crude fat}$$

D. Crude fiber. This fraction was designed to include those materials in a feed which are of low digestibility. Included here are cellulose, certain hemicelluloses, and some of the lignin, if present. Some of the lignin, however, may be included in the nitrogen-free extract. The steps involved in this determination are as follows:

1. Weigh out a small quantity of the prepared sample (usually less than 5 g).

2. Remove water from the sample by placing it in a drying oven.

3. Extract the sample with ether to remove crude fat. The same sample used for crude fat determination may be used for crude fiber.

4. Boil the remainder of the sample in dilute sulfuric acid (1.25%) for 30 minutes, and filter; then boil in dilute sodium hydroxide (1.25%) for 30 minutes, and filter (Figure 6). These extractions remove the protein, sugars, starches, and the more soluble hemicelluloses and minerals, and also possibly some of the lignin, if present.

5. Dry the residue and weigh. The residue consists of the crude fiber and the more insoluble mineral matter of the feed sample.

6. Ash the residue to oxidize off the crude fiber and weigh ash.

7. Calculate the amount of crude fiber in the sample by subtracting the weight of the ash in step 6 from the weight of the residue in step 5.

8. Calculate the percentage of crude fiber as follows:

$$\frac{\text{Wt of crude fiber}}{\text{Wt of original sample}} \times 100 = \% \text{ crude fiber}$$

E. Mineral matter or ash. This fraction includes for the most part the inorganic or mineral components of a feed. It is determined as follows:

1. Weigh out a small quantity of the prepared sample into a small crucible.

FIGURE 6.   A feed sample having been boiled in dilute acid for 30 minutes is being removed for filtering in a crude fiber determination. *(Courtesy of the University of Georgia College of Agriculture Experiment Stations)*

2.   Ash in a furnace at red heat (600°C) for several hours (Figure 7).

3.   Weigh the ash which includes most of the minerals of the feed in the oxide, chloride, or sulfate forms.

4.   Calculate the percentage of ash or mineral matter as follows:

$$\frac{\text{Wt of ash}}{\text{Wt of original sample}} \times 100 = \% \text{ ash or mineral matter}$$

F.   Nitrogen-free extract. This is commonly referred to as *NFE*. It includes mostly sugars and starches, and also some of the more soluble hemicelluloses and some of the more soluble lignin. Since this fraction was designed to include the more digestible carbohydrates, any lignin which may come out here will tend to distort the meaningfulness of the NFE figure as lignin is essentially indigestible. Nitrogen-free extract is determined by difference—that is, all those fractions discussed above are added together and subtracted from 100, as follows:

% water
% crude protein

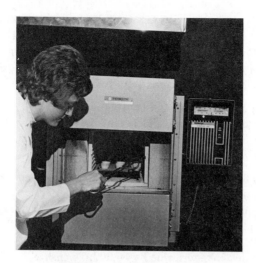

FIGURE 7.   Crucibles containing samples of feed being placed in an ashing furnace for ash determinations. *(Courtesy of the University of Georgia College of Agriculture Experiment Stations)*

% crude fat
% crude fiber
% mineral matter
─────────────────
100 − Total = % nitrogen-free extract

IV.   Vitamins. The amount of total vitamins or of the different individual vitamins is not determined as a part of a routine proximate analysis. This is not to say, however, that the amount of the different vitamins cannot be determined. Procedures have been developed for the determination of the different vitamins when this seems warranted. Vitamins appear as other nutrients in a routine proximate analysis—that is, as crude protein, crude fat or nitrogen-free extract. However, they are quantitatively unimportant insofar as they affect the determination of the amounts of these fractions. The total amount of vitamins occurring in any one of these three fractions will usually amount to less than one-tenth of 1.0%. The amount of choline in the crude protein fraction is a frequent exception to this generalization, but choline has protein value for most animals.

V.   There are different ways of expressing compositions, as outlined below:

    A.   In percent (%). This simply says that a feed contains so many parts (pounds, grams, milligrams, micrograms, etc.) of a particular feed component per 100 parts of the overall feed.

    B.   In parts per million (PPM). This simply says that a feed contains so many parts (lb, g, mg, mcg, etc.) of a particular feed component per

1,000,000 parts of the overall feed. PPM differs from % only in the location of the decimal point. Since one million is 10,000 × 100, to change % to PPM, simply multiply by 10,000 or, in other words, move the decimal point four places to the right. To change PPM to %, simply divide by 10,000 or move the decimal point four places to the left.

C.   **In mg per kilogram (mg/kg).** This says a feed contains so many mg of some component per kilogram of the overall feed. Since a kilogram is equal to 1,000,000 mg, then "mg per kilogram" is the same as "mg per million mg" or "parts per million".

D.   **In milligrams per pound (mg/lb).** This says that a feed contains so many mg of some feed component per lb of the overall feed. Since a pound is equal to 454,000 milligrams, "mg per lb" is the same as "mg per 454,000 mg". Hence, since 454,000 will go into one million 2.2 times, to change mg per lb to PPM, simply multiply by 2.2. To change mg per lb to %, multiply by 2.2 and divide by 10,000. To change PPM to mg/lb, divide by 2.2. To change % to mg/lb, multiply by 10,000 and divide by 2.2.

VI.   The composition of feeds may be expressed on any one or more of three dry matter bases.

A.   **As fed.** Sometimes referred to as the *wet* or *fresh basis*. On this basis dry matter of different feeds may range from 0% to 100%.

B.   **Air-dry.** May be actual or an "assumed dry matter content" basis. The latter is usually 90%. This basis is useful for comparing the composition of feeds having different moisture contents. Most feeds, but not all, are fed in an air-dry state.

C.   **Oven-dry.** Based on a moisture-free or 100% DM state. Also useful for comparing feeds of different moisture contents.

The different bases may be illustrated as follows:

	As fed	Air-dry	Oven-dry
% water	May be any %	Usually 10%	0%
% crude protein	This is dry		
% crude fat	matter—it	Usually 90%	100%
% crude fiber	is always		
% NFE	100% minus		
% ash	the % water		

Composition figures expressed on one basis may be converted to another basis by the use of a simple ratio, as follows:

$$\frac{\% \text{ of any component in a feed on any basis}}{\% \text{ DM in that feed on the same basis}} = \frac{\% \text{ of the component in the feed on another basis}}{\% \text{ of DM in the feed on the same basis}}$$

For example:

If a feed contains 4.0% crude protein on a fresh basis and 75.0% water, the percentage of crude protein on an air-dry basis would be calculated as follows:

100% − 75% = 25% DM in the fresh material

$$\frac{4}{25} = \frac{x}{90}$$

$$25x = 360$$

$$x = 14.4 \text{ (\% crude protein in feed on an air-dry basis)}$$

Table 2
FEEDS LISTED AND GROUPED ACCORDING TO THEIR PERCENTAGE OF DRY MATTER—AS FED BASIS

GROUP A—Mostly Artificially Dried Feeds Along with Several High Fat Feeds, Some Mineral Feeds, and a Few Field Cured Hays

Corn oil	100.0%	Dried whey	93.2%
Ground limestone	99.9	Sunflower meal	93.0
Defluorinated phosphate	99.8	41% cottonseed meal	92.7
Oyster shell flour	99.6	Ground cottonseed	92.7
Dried whole milk	96.2	Oat hulls	92.7
Steamed bone meal	95.7	Distillers dried grains	92.5
Animal fat	95.0	Rice hulls	92.4
Peanut kernels	94.8	Dehydrated alfalfa leaf meal	92.3
Feather meal	94.6	Dried beet pulp/molasses	92.3
Dried skimmed milk	94.3	Redtop hay	92.3
Meat scrap	94.2	Dehydrated alfalfa meal	92.2
Tankage with bone	94.1	Sesame oil meal	92.2
Digester tankage	93.7	36% cottonseed meal	91.8
Meat and bone meal	93.6	Lespedeza hay	91.7
Meadow hay	93.5	Dried bakery product	91.6
Poultry by-product meal	93.4	Soybean hulls	91.6

GROUP B—Primarily Field Cured Grains, Hays, and Straws Along with Several Artificially Dried Products

Kudzu hay	91.5%	Fish meal	91.4%
Peanut hulls	91.5	Potato meal	91.4
Peanut oil meal	91.5	49% soybean oil meal	91.3
Alfalfa hay	91.4	Sweetclover hay	91.3

Table 2 (Continued)

Bermudagrass hay	91.2%	Hominy feed	89.2%
Peanut hay	91.2	Barley grain	89.0
41% solv-extd cottonseed meal	91.1	44% soybean oil meal	89.0
Linseed meal	91.1	Kentucky bluegrass hay	88.9
Brewers dried grains	91.0	Milo grain	88.9
Coastal bermudagrass hay	91.0	Oats grain	88.9
Corn gluten meal	91.0	Soybean hay	88.9
Prairie hay	91.0	Wheat grain	88.9
Dehydrated alfalfa stem meal	90.9	Reed canarygrass hay	88.8
Soybean seed	90.9	Corn bran	88.7
Bahiagrass hay	90.8	Wheat bran	88.7
Sericea hay	90.8	Oat straw	88.6
Dallisgrass hay	90.7	Orchardgrass hay	88.6
Oat groats	90.7	Timothy hay	88.6
Oat hay	90.7	Grain sorghum grain	88.5
Rice bran	90.7	Ground snapped corn	88.5
Dried beet pulp	90.6	Safflower meal without hulls	88.5
Johnsongrass hay	90.6	Copra meal	88.2
Cowpea hay	90.5	Rye grain	88.2
Corn starch	90.4	Wheat straw	87.8
Citrus pulp/molasses, dried	90.3	Alsike clover hay	87.4
Cottonseed hulls	90.2	Wheat middlings	87.4
Dried citrus pulp	90.2	Wheat shorts	87.1
Sweet potato meal	90.2	Dehydrated coastal bermudagrass meal	87.0
Red clover hay	90.0	Barley straw	86.9
Kafir grain	90.0	Tall fescue hay	86.5
Ground corn cob	89.8	Yellow shelled corn	86.0
Bromegrass hay	89.7	Millet hay	85.8
Sudangrass hay	89.6	Ground ear corn	85.4
Hegari grain	89.4	Corn stover	84.4
Blood meal	89.3		

Group C—The Molasseses

Beet molasses	77.5%	Citrus molasses	67.7%
Cane molasses	77.0	Wood molasses	62.4

GROUP D—Mostly the Silages, the Fresh Forages, Wet By-Products, the Root Crops, and the Fresh Milks

Corn ear silage	43.4%	Grain sorghum silage	30.7%
Corn stover silage	35.1	Sweet potatoes	30.6

Table 2 (Continued)

Fresh bromegrass forage	30.0%	Fresh red clover forage	22.7%
Fresh Kentucky bluegrass forage	29.0	Fresh sudangrass forage	20.8
		Fresh rye forage	20.6
Fresh coastal bermudagrass	28.8	Fresh citrus pulp	18.3
Fresh fescue forage	28.5	Fresh ladino clover forage	18.0
Sorgo silage	28.2	Fresh white clover forage	17.7
Corn (fodder) silage	27.8	Fresh crimson clover forage	17.6
Restaurant garbage	26.3	Carrots	12.9
Fresh alfalfa forage	25.9	Fresh cow's milk	12.6
Fresh Dallisgrass forage	25.0	Wet beet pulp	11.3
Fresh orchardgrass forage	24.9	Turnips	9.7
Fresh ryegrass forage	24.1	Cabbage heads	9.6
Wet brewers grains	23.8	Fresh skimmed milk	9.3
White potatoes	23.1		

Table 3

FEEDS LISTED AND GROUPED ACCORDING TO THEIR PERCENTAGE OF ETHER EXTRACT—AS-FED BASIS

GROUP A—The Pure Fats, Dried Whole Milk, and the Oil Seeds

Corn oil	100.0%	Dried whole milk	26.7%
Animal fat, feed grade	95.0	Ground cottonseed	22.9
Peanut kernels	47.7	Soybean seed	17.4

GROUP B—Primarily the Mechanically Extracted Protein Feeds, Other By-Product Concentrates, Some of the Feed Grains, and Certain Other Miscellaneous Products

Rice bran	14.4%	36% cottonseed meal	6.4%
Dried bakery product	13.7	Oat groats	6.0
Poultry by-product meal	13.1	Restaurant garbage	5.9
Tankage with bone	12.4	41% cottonseed meal	5.6
Meat and bone meal	11.1	Linseed meal	5.1
Fish meal	9.8	Peanut hay	5.1
Meat scrap	9.4	Wheat shorts	4.8
Digester tankage	9.0	Corn bran	4.5
Sesame oil meal	8.6	Oats grain	4.5
Copra meal	7.4	Wheat bran	4.4
Distillers dried grains	7.4	Yellow shelled corn	3.8
Hominy feed	6.8	Citrus pulp/molasses, dried	3.6
Safflower meal without hulls	6.7	Fresh cow's milk	3.6
Brewers dried grains	6.6	Dried citrus pulp	3.4

Table 3 (Continued)

Ground ear corn	3.4%	Ground snapped corn	3.0%
Wheat middlings	3.4	Kentucky bluegrass hay	3.0
Dehydrated coastal		Dehydrated alfalfa meal	2.9
bermudagrass meal	3.3	Feather meal	2.9
Dehydrated alfalfa leaf meal	3.1	Kafir grain	2.9
Grain sorghum grain	3.1	Milo grain	2.9

GROUP C—Predominantly the Hays, the Hulls, and the Straws, Some of the Grains, and Certain Solvent-Extracted Materials Along with Several Other Miscellaneous Products

Orchardgrass hay	2.9%	Sericea hay	1.8%
Sunflower meal	2.9	Alfalfa hay	1.7
Lespedeza hay	2.8	Barley grain	1.7
Millet hay	2.8	Barley straw	1.7
Redtop hay	2.8	Bermudagrass hay	1.7
Reed canarygrass hay	2.8	Dehydrated alfalfa stem meal	1.7
Cowpea hay	2.6	Corn ear silage	1.6
Hegari grain	2.6	Fresh orchardgrass forage	1.6
Soybean hulls	2.5	Sudangrass hay	1.6
Alsike clover hay	2.4	Bahiagrass hay	1.5
Corn gluten meal	2.3	Coastal bermudagrass hay	1.5
Kudzu hay	2.3	Rye grain	1.5
Meadow hay	2.3	Wet brewers grains	1.5
Red clover hay	2.3	Blood meal	1.4
Timothy hay	2.3	Cottonseed hulls	1.4
Dallisgrass hay	2.2	Fresh Kentucky bluegrass	
Sweetclover hay	2.2	forage	1.4
Tall fescue hay	2.2	Oat hulls	1.4
Bromegrass hay	2.1	Wheat straw	1.4
41% solv-extd cottonseed meal	2.1	Corn (fodder) silage	1.2
Oat straw	2.1	Fresh bromegrass forage	1.2
Prairie hay	2.1	Fresh fescue forage	1.2
Soybean hay	2.0	Peanut hulls	1.2
Oat hay	1.9	Peanut oil meal	1.2
Wheat grain	1.9	Corn stover	1.1
Steamed bone meal	1.9	Dried whey	1.1
Johnsongrass hay	1.8		

GROUP D—Largely High Moisture Feeds, Mineral Feeds, and Solvent-Extracted Feeds Along with Several Other Miscellaneous Products

Fresh alfalfa forage	1.1%	Fresh coastal bermudagrass	1.1%

Table 3 (Continued)

49% soybean oil meal	1.1%	Ground corn cob	0.6%
Dried skimmed milk	1.0	Dried beet pulp	0.5
Fresh red clover forage	1.0	Sweet potatoes	0.4
Fresh ladino clover forage	0.9	Potato meal	0.3
Fresh rye forage	0.9	Cabbage heads	0.2
Fresh ryegrass forage	0.9	Carrots	0.2
Grain sorghum silage	0.9	Citrus molasses	0.2
44% soybean oil meal	0.9	Corn starch	0.2
Sweet potato meal	0.9	Wet beet pulp	0.2
Corn stover silage	0.8	Beet molasses	0.1
Rice hulls	0.8	Fresh skimmed milk	0.1
Dried beet pulp/molasses	0.7	Turnips	0.1
Fresh sudangrass forage	0.7	White potatoes	0.1
Sorgo silage	0.7	Wood molasses	0.1
Fresh citrus pulp	0.6	Defluorinated phosphate	***
Fresh crimson clover forage	0.6	Ground limestone	***
Fresh Dallisgrass forage	0.6	Oyster shell flour	***
Fresh white clover forage	0.6	Cane molasses	0.0

***Very little, if any.

Table 4
FEEDS LISTED AND GROUPED ACCORDING TO THEIR PERCENTAGE OF CRUDE FIBER—AS-FED BASIS
GROUP A—Extremely Low-Quality Dry Roughages Consisting Primarily of Hulls and Straws

Peanut hulls	59.8%	Wheat straw	38.3%
Rice hulls	41.1	Oat straw	36.3
Cottonseed hulls	40.6	Barley straw	36.2

GROUP B—Made Up Almost Entirely of the Hays and Hay Meals

Dehydrated alfalfa stem meal	33.9%	Kudzu hay	29.9%
Soybean hay	33.2	Oat hulls	29.8
Soybean hulls	33.1	Coastal bermudagrass hay	29.6
Ground corn cob	31.1	Dallisgrass hay	29.1
Prairie hay	30.7	Red clover hay	29.1
Bahiagrass hay	30.5	Bromegrass hay	28.5
Meadow hay	30.5	Redtop hay	28.5
Johnsongrass hay	30.3	Sericea hay	28.4
Orchardgrass hay	30.3	Corn stover	28.2
Timothy hay	30.2	Tall fescue hay	28.1
Reed canarygrass hay	30.0	Alfalfa hay	28.0

Table 4 (Continued)

Lespedeza hay	28.0%	Millet hay	25.0%
Oat hay	27.9	Cowpea hay	24.3
Sudangrass hay	27.5	Dehydrated coastal bermuda-	
Sweetclover hay	27.4	grass meal	24.3
Bermudagrass hay	26.8	Peanut hay	23.7
Kentucky bluegrass hay	26.7	Dehydrated alfalfa meal	22.9
Alsike clover hay	26.3	Dehydrated alfalfa leaf meal	18.5

GROUP C—Consists Largely of a Group of Feeds Generally Referred to as the High Fiber Concentrates

Dried beet pulp	18.2%	Dried citrus pulp	11.6%
Ground cottonseed	16.9	41% solv-extd cottonseed	
Brewers dried grains	14.7	meal	11.4
36% cottonseed meal	14.0	Rice bran	11.2
Peanut oil meal	13.1	41% cottonseed meal	10.9
Dried beet pulp/molasses	13.0	Sunflower meal	10.8
Distillers dried grains	12.8	Oats grain	10.6
Copra meal	12.3	Ground snapped corn	10.4
Corn stover silage	11.8	Citrus pulp/molasses, dried	10.3

GROUP D—Predominantly a Combination of Low-Fiber Air-Dry Concentrates and High Moisture Feeds

Corn bran	9.6%	Soybean seed	5.3%
Linseed meal	8.9	Fresh rye forage	5.2
Wheat bran	8.6	Corn ear silage	5.1
Safflower meal without hulls	8.5	Fresh red clover forage	5.0
Fresh coastal bermudagrass	8.2	Fresh crimson clover forage	4.9
Fresh fescue forage	8.2	Corn gluten meal	4.6
Corn (fodder) silage	7.6	Hominy feed	4.4
Fresh Kentucky bluegrass forage	7.5	Wet brewers grains	3.8
Grain sorghum silage	7.3	Wet beet pulp	3.4
Fresh Dallisgrass forage	7.2	Sweet potato meal	3.3
Ground ear corn	7.1	49% soybean oil meal	2.9
Fresh bromegrass forage	7.0	Fresh white clover forage	2.8
Sorgo silage	6.6	Peanut kernels	2.8
Wheat shorts	6.3	Oat groats	2.6
Fresh orchardgrass forage	6.2	Fresh ladino clover forage	2.5
44% soybean oil meal	6.0	Meat scrap	2.5
Fresh sudangrass forage	5.7	Wheat grain	2.5
Fresh ryegrass forage	5.6	Wheat middlings	2.5
Fresh alfalfa forage	5.5	Tankage with bone	2.4
Sesame oil meal	5.4	Fresh citrus pulp	2.3
Barley grain	5.3	Grain sorghum grain	2.3
		Milo grain	2.3

Table 4 (Continued)

Potato meal	2.1%	Poultry by-product meal	1.6%
Hegari grain	2.0	Sweet potatoes	1.3
Rye grain	2.0	Carrots	1.2
Yellow shelled corn	2.0	Turnips	1.1
Kafir grain	2.0	Cabbage heads	1.0
Steamed bone meal	1.9	Dried bakery product	0.7
Digester tankage	1.8	Restaurant garbage	0.7
Meat and bone meal	1.8	White potatoes	0.6

GROUP E—The Mineral Feeds, the Pure Fats, the Molassesses, the Milk Products, Corn Starch, and Certain Animal By-Products

Blood meal	0.6%	Cane molasses	0.0%
Feather meal	0.6	Citrus molasses	0.0
Fish meal	0.6	Corn oil	0.0
Dried skimmed milk	0.3	Defluorinated phosphate	0.0
Corn starch	0.2	Fresh cow's milk	0.0
Dried whey	0.2	Fresh skimmed milk	0.0
Dried whole milk	0.1	Ground limestone	0.0
Animal fat, feed grade	0.0	Oyster shell flour	0.0
Beet molasses	0.0	Wood molasses	0.0

4 The Van Soest Method of Forage Evaluation

I. While the Weende system of feed analysis has served for many years and continues to serve a very useful purpose in predicting the nutritive value of feeds, this does not mean that the system is without its shortcomings or is not in need of improvement. In fact, there are some definite limitations of the Weende system, especially with respect to both the crude fiber and the nitrogen-free extract fractions.

    A. In the first place, crude fiber as determined is not a chemically uniform substance but a variable mixture, the major components of which are cellulose, hemicellulose, and lignin. While cellulose and hemicellulose are similar in nutritive value, they have much greater feeding values for ruminants than for nonruminants. On the other hand, lignin is essentially indigestible by all livestock. Further complicating the situation is the fact that only a part of the hemicellulose and lignin comes out in the crude fiber fraction, with the remaining portions showing up as NFE, which is ordinarily thought of as consisting largely of highly digestible sugars and starches. Consequently, to the extent that hemicellulose—at best low in digestibility—and lignin—essentially indigestible—appear in the NFE fraction, this fraction would be larger and have a lower average digestibility than would be true if it consisted primarily of sugars and starches. At the same time, the crude fiber value would not reflect all of the more indigestible portion of the feed.

B. Fortunately, the above is a major problem only with the more fibrous feeds from the standpoint of making effective use of proximate analysis figures in predicting nutritive value. However, with fibrous feeds, problems of considerable magnitude are frequently experienced in this connection. As a result, numerous workers over the past several years have tested various procedures which might provide a more definitive separation of feed carbohydrates than does the Weende system of proximate analysis. This has been especially true from the standpoint of forage crop evaluation.

II. A procedure which has received rather wide consideration as a possible substitute for the conventional crude fiber determination was developed by Van Soest and associates[1,2,3,4] working at the USDA's ARS research laboratory in Beltsville, Maryland. This procedure involves the separation of feed dry matter into two fractions—one of high digestibility and the other of low digestibility—by boiling a 0.5-1.0 g sample of the feed in a neutral detergent solution (3% sodium lauryl sulfate buffered to a pH of 7.0) for one hour and filtering.

A. The *neutral detergent solubles (NDS)* consist for the most part of the cell contents. They are composed primarily of lipids, sugars, starches, and protein, and are all high in digestibility, having an average true digestibility of about 98%. Their digestibility does not seem to be

[1] Van Soest, P. J., and L. A. Moore. "New Chemical Methods for Analysis of Forages for the Purpose of Predicting Nutritive Value." *Proceedings of the Ninth International Grassland Congress, Sao Paulo, Brazil*. Sao Paulo, Brazil: International Grassland Conference, 1966.

[2] Van Soest, P. J. "Non-Nutritive Residues: A System of Analysis for the Replacement of Crude Fiber." *Journal of the Association of Official Agricultural Chemists* 49(1966):546.

[3] Van Soest, P. J. "Development of a Comprehensive System of Feed Analyses and Its Application to Forages." *Journal of Animal Science* 26(1967):119.

[4] Goering, H. K., and P. J. Van Soest. "Forage Fiber Analyses." *USDA, ARS Agriculture Handbook No. 379*. Washington, D.C.: Government Printing Office, 1970.

materially influenced by the amount of neutral detergent insolubles present.

The neutral detergent insolubles are usually referred to as *neutral detergent fiber (NDF)*. They embrace for the most part the plant cell wall and are sometimes referred to as the *cell wall components* or *cell wall constituents*. They consist primarily of cellulose, lignin, silica, hemicellulose, and some protein.

In the Van Soest procedure essentially all of the lignin and hemicellulose are included in the NDF fraction, whereas with the Weende method variable amounts of these two components are lost from the crude fiber to the NFE. As a result, NDF as determined by the Van Soest procedure is considerably higher than the conventional crude fiber values for some feeds.

The different NDF (cell wall) components are at best low in digestibility and are entirely dependent on the microorganisms of the digestive tract for any digestion they do undergo. Lignin and silica are essentially indigestible even by microorganisms. Also, lignin has a curvilinear negative influence on cellulose and hemicellulose digestibility.

While, according to Van Soest, NDF corresponds more closely than does conventional crude fiber to the total fiber fraction of a forage feed, it is not a uniform chemical entity but a variable mixture of cell wall components whose overall nutritive availability is influenced to a considerable degree by the proportion of lignin present. Accordingly, Van Soest has proposed that the amount of lignin be determined and appropriate allowance be made for the amount of this component present in attempting to predict NDF digestibility.

B.   For the purpose of determining the lignin in a forage sample, Van Soest has proposed the use of what has come to be known as the *acid detergent lignin procedure*[4]. In this connection the *acid detergent fiber procedure*[4] is used as a preparatory step. This involves the boiling of a 1.0 g sample of air-dry material in an acid detergent solution (49.04 g sulfuric acid and 20 g cetyl trimethylammonium bromide per liter) for one hour and filtering. The insolubles or residue makes up what is known as *acid detergent fiber (ADF)* and consists primarily of cellulose, lignin, and variable amounts of silica. ADF differs from NDF in that NDF contains most of the. feed hemicellulose and a limited amount of protein not present in ADF. The difference in the amount of NDF and ADF is an estimate of the hemicellulose in the feed.

In order to determine the amount of lignin present, the ADF is then digested in 72% H_2SO_4 at 15°C for 3 hours and filtered. The residue remaining after washing and drying is weighed and ashed. The ash remaining approximates the silica present, while the loss in weight during ashing approximates the lignin and is referred to as *acid detergent lignin (ADL)* or more specifically as *acid insoluble lignin*.

C.   An alternative method for determining lignin which has advantages
for certain materials involves the oxidation of the lignin of ADF
with an excess of acetic acid-buffered potassium permanganate solution.
Lignin so determined is referred to as *permanganate lignin*. A variation of
this method may be used to allow for the cutin present in many seed hulls,
which otherwise would be measured as lignin.

D.   Forage processing temperatures of over 50°C tend to increase lignin
yields with either of the above methods largely by the production of
artifact lignin via the nonenzymic browning reaction. The nitrogen
content of the ADF is considered to be[4] a sensitive measure of the extent
of such damage and serves as a basis for estimating artifact lignin.

E.   Once NDS, NDF, ADF, and ADL have been determined for a forage,
the true digestibility of the forage dry matter may be estimated by
application of the following formula:

$$0.98 \text{ NDS} + (1.473 - 0.789 \log_{10} \text{lignin}) \text{ NDF}$$

in which NDS and NDF are expressed as percentages of the forage dry
matter, and lignin is the percentage of acid insoluble lignin in the ADF
fraction.

F.   Apparent digestibility of the forage dry matter may then be
estimated by deducting from the true digestibility figure an
allowance for metabolic dry matter present in the feces, which according
to Van Soest on the average amounts to 12.9% of the dry matter intake.
Should the insoluble ash of a feed exceed 2.0% of the feed dry matter or
should the feed show signs of having undergone heat damage, appropriate
corrections should be made for silica and artifact lignin as outlined by
Goering and Van Soest.[4]

5 The Digestive Tract

I.   The digestive tract, sometimes referred to as the *alimentary tract*, is the passage from the mouth to the anus through which feed passes following consumption as it is subjected to various digestive processes. It consists of the following:

   A.   Mouth and pharynx.

   B.   Esophagus.

   C.   Stomach.

       1.   In animals such as the hog and horse this is a single-compartment organ.

       2.   In animals such as the cow and sheep this is a multiple-compartment structure consisting of the following:

           a.   **Rumen or paunch.**

           b.   **Reticulum or honeycomb or water bag.**

           c.   **Omasum or manyplies.**

           d.   **Abomasum or true stomach.**

   D.   **Small intestine** which is divided into three sections:

       1.   **Duodenum**—upper section.

       2.   **Jejunum**—middle section.

       3.   **Ileum**—lower section.

   E.   **Cecum or caecum.**

   F.   **Large intestine.**

   G.   **Anus.**

   H.   **Associated glands and organs,** including:

       1.   **Salivary glands.**

       2.   **Liver.**

       3.   **Gall bladder.**

       4.   **Pancreas.**

II.   Farm animals are classified according to the nature of their digestive tracts into two general categories:

   A.   **Ruminants.** These are animals such as the cow and sheep which ruminate or, in other words, chew a cud. An animal's cud consists of boluses of feed eaten earlier. Ruminants have a multi-compartment stomach (Figure 9) as outlined above. By means of the microbial fermentation processes of the rumen, such animals are able to make effective use of high fiber feeds and as a result are frequently fed rations containing high levels of fibrous feeds.

   B.   **Nonruminants.** These are animals such as the hog and horse which have a single-compartment stomach (Figure 8) and do not chew a cud. Most nonruminants make very poor use of high fiber feeds. However, the horse by means of the microbial fermentation processes of the cecum and large intestine is able to utilize such feeds effectively, and as a result they are frequently fed rations containing considerable fiber. Hogs, however, are ordinarily fed rations relatively low in fiber.

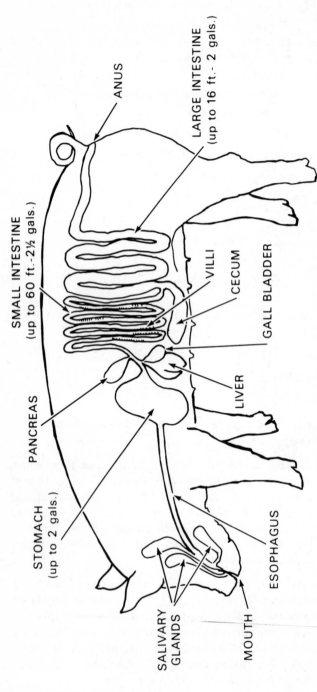

ANUS

LARGE INTESTINE
(up to 16 ft.- 2 gals.)

SMALL INTESTINE
(up to 60 ft.-2½ gals.)

VILLI

CECUM

GALL BLADDER

PANCREAS

LIVER

STOMACH
(up to 2 gals.)

ESOPHAGUS

SALIVARY
GLANDS

MOUTH

Note—The horse is also a nonruminant and has a digestive tract similar to that of a hog, except:

1. It is proportionately larger.
2. The horse has no gall bladder.
3. The horse has a greatly enlarged cecum (up to 3½ ft.—9 gal.) and large intestine
   (up to 21 ft.—25 gal.).

FIGURE 8.   The digestive tract of the hog—a nonruminant.

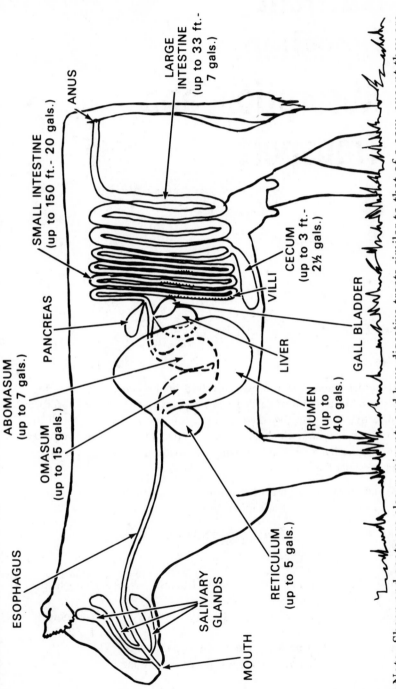

FIGURE 9.   The digestive tract of the cow—a ruminant.

Note—Sheep and goats are also ruminants and have digestive tracts similar to that of a cow, except they are proportionately smaller.

6 Nutrient Digestion, Absorption, and Transport

I. **Nutrient digestion.** The processes involved in the conversion of various feed nutrients into forms (called *end products*) which can be absorbed from the digestive tract. Most of these processes are accomplished through the action of various enzymes which are found in the different digestive juices secreted into the digestive tract. Details of these digestive processes are summarized in Tables 5 and 6.

   A. **End products of digestion.** Forms into which a nutrient must be converted in order for it to be absorbed from the digestive tract.

NUTRIENT	END PRODUCT
Protein	Amino acids
Starch	Glucose
Sucrose	Glucose and fructose (fructose converted to glucose upon absorption)
Lactose	Glucose and galactose (galactose converted to glucose upon absorption)
Cellulose	Organic acids and salts of organic acids
Fats	Primarily fatty acids and glycerol; some soap
Minerals	Any soluble form
Vitamins	Any soluble form

B. **Digestive juices.** Fluid materials secreted into the digestive tract by glands or tissues along the digestive tract. Those usually considered are:

1. Saliva.
2. Gastric juice.
3. Bile.
4. Pancreatic juice.
5. Intestinal juice.

C. **Enzymes.** Enzymes are organic catalysts. Most are protein in makeup. They are usually specific with respect to substrate and medium requirements. Most enzyme names end in "ase". A few early named enzymes end in "in". Today, enzymes are usually named according to the substrate on which they act, and possibly the action which they effect, with an "ase" ending. Those enzymes usually recognized in connection with feed digestion are:

1. Salivary amylase.
2. Salivary maltase.
3. Rennin.
4. Pepsin.
5. Gastric lipase.
6. Pancreatic amylase.
7. Trypsin.
8. Pancreatic lipase.
9. Intestinal peptidases (erepsin).
10. Intestinal maltase.
11. Sucrase.
12. Lactase.

II. **Nutrient absorption.** The movement of the end products of digestion from the digestive tract into the blood and/or lymph system. It is accomplished by the process of osmosis through the semi-permeable membranes which line much of the digestive tract. In nonruminants most nutrient absorption takes place in the small and large intestines. The villi of the small intestine especially facilitate absorption at this location. A large amount of absorption also takes place through the rumen wall in ruminants. All of the end products of digestion, except those of fat digestion and possibly the fat soluble vitamins, are absorbed directly into the blood stream. The end products of fat digestion are absorbed into the lymph system through the lacteals of the villi in the form of chyle. The latter subsequently enters the blood through the thoracic duct and undergoes certain metabolic changes in the liver before being used in the tissues.

III. **Nutrient transport.** The movement of nutrients from the point of absorption to the point of utilization. The blood is the primary basis for the transport of nutrients and other materials in the animal body. However, the lymph serves as the final link between the blood capillaries and the individual cells. All nutrients are transported in solution in the water soluble form. The various nutrients are absorbed from the lymph into the individual cells again by the process of osmosis.

Table 5
SUMMARY OF DIGESTION

NUTRIENT	MOUTH	RUMEN	NUNRUMINANT STOMACH AND RUMINANT ABOMASUM	SMALL INTESTINE	CECUM AND LARGE INTESTINE
Protein	None	Some breakdown of protein to amino acids by microbial fermentation. Also some microbial synthesis of essential amino acids.	Enzyme rennin of gastric juice secreted by wall of stomach curdles milk. Enzyme pepsin of gastric juice acts on protein to form intermediate protein breakdown products (IPBP), such as proteoses, polypeptides, and peptides.	Trypsin and certain other enzymes of the pancreatic juice secreted by the pancreas act on protein and IPBP to produce other IPBP and amino acids. Intestinal peptidases (formerly the enzyme erepsin) of the intestinal juices secreted by the intestinal wall act on IPBP to produce amino acids.	Continued action of trypsin, etc., and the intestinal peptidases.
Starch	In nonruminants enzyme salivary amylase of the saliva secreted by the salivary glands acts on starch to produce maltose. Enzyme salivary maltase acts on maltose to produce glucose.	Some starch undergoes microbial (bacterial and protozoal) fermentation to form primarily acetic, propionic, butyric, and certain other volatile fatty acids (VFA), methane (CH_4), CO_2, and heat.	None	Enzyme pancreatic amylase of the pancreatic juice secreted by the pancreas acts on starch to produce maltose. Enzyme intestinal maltase of the intestinal juices secreted by the intestinal wall acts on maltose to form glucose.	Continued action of pancreatic amylase and intestinal maltase.

NUTRIENT	MOUTH	RUMEN	NONRUMINANT STOMACH AND RUMINANT ABOMASUM	SMALL INTESTINE	CECUM AND LARGE INTESTINE
Sucrose	None	Same as for starch above.	None	Enzyme intestinal sucrase of the intestinal juices secreted by the intestinal wall acts on sucrose to form glucose and fructose.	Continued action of intestinal sucrase if any sucrose still present.
Lactose	None	Lactose ordinarily not found in a functioning rumen.	None	Enzyme intestinal lactase of the intestinal juices secreted by the wall of intestine acts on lactose to form glucose and galactose.	Continued action of intestinal lactase if any lactose is present.
Cellulose	None	Same as for starch above.	None	None	Similar to that in rumen but much less extensive, except in the horse.
Fat	None	Some microbial fermentation of fats to form fatty acids and glycerol. Some fermentation of glycerol to propionic acid.	Enzyme gastric lipase of the gastric juice secreted by wall of stomach acts on some fat to form fatty acids and glycerol.	Certain compounds of bile secreted by the liver react with some fat to form soap and glycerol. Enzyme pancreatic lipase of the pancreatic juice secreted by the pancreas acts on fat to form fatty acids, glycerol, and monoglycerides.	Continued action of pancreatic lipase.

Table 6
ENZYMES AND DIGESTIVE JUICES

ENZYME	ENZYME FOUND IN WHICH DIGESTIVE JUICE?	DIGESTIVE JUICE SECRETED BY?	DIGESTIVE JUICE SECRETED INTO WHAT PART OF DIGESTIVE TRACT?	ENZYME ACTS ON WHAT?	ENZYME ACTIVE IN ACID OR ALKALINE MEDIUM?	ACTION PRODUCES?
Salivary amylase	Saliva	Salivary glands	Mouth	Starch	Neutral to slightly alkaline	Maltose
Salivary maltase	Saliva	Salivary glands	Mouth	Maltose	Neutral to slightly alkaline	Glucose
Rennin	Gastric juice	Wall of stomach	Stomach or abomasum	Milk protein	Acid	Curd
Pepsin	Gastric juice	Wall of stomach	Stomach or abomasum	Proteins	Acid	Proteoses, polypeptides, peptides
Gastric lipase	Gastric juice	Wall of stomach	Stomach or abomasum	Fats	Acid	Fatty acids, glycerol
Pancreatic amylase	Pancreatic juice	Pancreas	Upper small intestine	Starch	Alkaline	Maltose
Trypsin, etc.	Pancreatic juice	Pancreas	Upper small intestine	Proteins, proteoses, polypeptides, peptides	Alkaline	Intermediate protein breakdown products. amino acids
Pancreatic lipase	Pancreatic juice	Pancreas	Upper small intestine	Fats	Alkaline	Fatty acids, glycerol, monoglycerides
Intestinal peptidases (erepsin)	Intestinal juice	Wall of small intestine	Small intestine	Intermediate protein breakdown products	Alkaline	Amino acids
Intestinal maltase	Intestinal juice	Wall of small intestine	Small intestine	Maltose	Alkaline	Glucose
Sucrase	Intestinal juice	Wall of small intestine	Small intestine	Sucrose	Alkaline	Glucose, fructose
Lactose	Intestinal juice	Wall of small intestine	Small intestine	Lactose	Alkaline	Glucose, galactose
	Bile	Liver	Upper small intestine	Bile reacts with fats	Alkaline	Soap, glycerol

7 Apparent Digestibility

I.   Different feeds and nutrients vary greatly in their digestibility. In the evaluation of different feeds for feeding purposes it is helpful to have information on the digestibility of feeds and the nutrients which they contain.

II.   Information on the digestibility of feeds and nutrients is obtained by carrying out a digestion trial. To carry out a digestion trial one simply feeds an animal a known amount of a feed of known composition, and then collects the feces resulting from this feed and determines the composition of the feces (Figure 10). The difference between the nutrients consumed and the nutrients excreted in the feces is considered to be the amount of nutrients digested. Such a procedure gives one *apparent digestibility* rather than *true digestibility*. Apparent digestibility differs from true digestibility in that:

A.   Apparent digestibility considers nutrients lost as methane as having

FIGURE 10.   A metabolism stall such as might be used for conducting digestion trials with calves. Note the pan at the rear for collecting feces. *(Courtesy of the University of Georgia College of Agriculture Experiment Stations, Dairy Nutrition Laboratory)*

been digested and absorbed by the animal. This is especially a factor with ruminants.

B.   Apparent digestibility regards all nutrients remaining in the feces as not digestible, which is not necessarily so. Some may have been digested but not absorbed. Others possibly were not digested but would have been had they remained in the digestive tract for a longer period or were fed again.

C.   Apparent digestibility regards all nutrients in the feces as undigested feed when actually considerable amounts of other materials—especially intestinal mucosa and bacteria—are present in the feces.

III.   For all practical purposes, however, apparent digestibility data provide invaluable information for the evaluation of feeds for feeding purposes, and such data are used extensively in carrying out a scientific feeding program.

IV.   In order to fully understand and appreciate the meaning and significance of *digestibility* data, it behooves a student of feeds and feeding to be familiar with the basic steps of carrying out a digestion trial and the various calculations which are ordinarily made in that connection.

A.   In carrying out a digestion trial it is very difficult, especially in ruminants, to know just which feces resulted from which feed. Consequently, preliminary to carrying out a digestion trial an animal is brought to a constant level of daily feed intake and, hopefully, of daily feces excretion, usually over a period of 7-10 days. Even so, the amount of feces excretion will vary from day to day. Hence, fecal collections are usually made over a period of 7-10 days and an average obtained.

B.   Some of the basic calculations made in connection with carrying out a digestion trial are as follows:

1.   Amount of a nutrient in daily feed − Amount of that nutrient in daily feces = Amount of that nutrient digested daily.

2.   Amount of a nutrient in daily feed = (Amount of feed eaten daily × % of nutrient in feed)/100.

3.   Amount of a nutrient in daily feces = (Average amount of feces excreted daily × % of nutrient in feces)/100.

4.   Coefficient of digestibility of any nutrient = (Amount of that nutrient digested daily/Amount of that nutrient eaten daily) × 100.

5.   % of digestible nutrient in a feed = (% of that nutrient in the feed × Coefficient of digestibility of that nutrient)/100.

or

% of digestible nutrient in a feed = (Amount of that nutrient digested daily/Amount of total feed eaten daily) × 100.

The application of the above calculations is illustrated in Table 7.

Table 7
DIGESTION TRIAL CALCULATIONS ILLUSTRATED

LINE (L.)	ENERGY	DRY MATTER	CRUDE PROTEIN	CRUDE FAT	CRUDE FIBER	N-FREE EXTRACT	MINERAL MATTER*
1. Amount of feed eaten daily—9.11 kg							
2. Composition of feed (as determined by lab analysis)	3839 kcal/kg	91.50%	8.70%	2.01%	27.89%	48.22%	4.68%
3. Amount of each nutrient in daily feed- (L.1 × L.2)/100**	34,973 kcal	8.34kg	0.79kg	0.18kg	2.54kg	4.39kg	0.43kg
4. Average daily feces excretion—20.61 kg							
5. Composition of feces (as determined by lab analysis)	818.9 kcal/kg	20.03%	1.92%	0.41%	8.09%	8.10%	1.51%

LINE (L.)	ENERGY	DRY MATTER	CRUDE PROTEIN	CRUDE FAT	CRUDE FIBER	N-FREE EXTRACT	MINERAL MATTER*
6. Amount of each nutrient in daily feces—(L.4 × L.5)/100**	16,877 kcal	4.13kg	0.40kg	0.08kg	1.67kg	1.67kg	0.31kg
7. Amount of each nutrient digested daily—(L.3 − L.6)	18,096 kcal	4.21kg	0.39kg	0.10kg	0.87kg	2.72kg	0.12kg
8. Coefficient of digestibility of each nutrient—(L.7/L.3) × 100	51.7%	50.5%	49.4%	55.5%	34.3%	62.0%	27.9%
9. kcal/kg or % of digestible nutrient in feed—(L.8 × L.2)/100** or (L.7/L.1) × 100**	1985 kcal/kg	46.2%	4.28%	1.10%	9.55%	29.90%	1.31%

* The digestibility of mineral matter usually is not calculated. It is shown here simply to help accomplish certain instructional objectives.

** Does not apply to the Energy column.

8 Total Digestible Nutrients or TDN

I. TDN is the abbreviation for *total digestible nutrients*. TDN is simply a figure which indicates the relative energy value of a feed to an animal. It is ordinarily expressed in pounds or kilograms or in percent (lb or kg of TDN per 100 lb or kg of feed). It is arrived at by adding together the following:

		AS CALCULATED FOR FEED FROM TABLE 7 (*SEE* LINE 9)
Digestible crude protein	=	4.28%
Digestible crude fiber	=	9.55
Digestible N-free extract	=	29.90
Digestible crude fat		
$\times$ 2.25 (1.10 $\times$ 2.25)	=	2.48
TOTAL		46.21% TDN

Total digestible nutrients or TDN is not an actual total of the digestible nutrients in a feed. In the first place, it does not include the digestible mineral matter. Secondly, the digestible fat is multiplied by 2.25 before being included in the TDN figure. The latter step is necessary to allow for the extra energy value of fats compared to carbohydrates and protein. As a result of this step, feeds high in fat will sometimes exceed 100 in percentage of TDN.

II.   Factors affecting the TDN value of a feed.

A.   **The percentage of dry matter.** Water can in no way contribute in a positive way to the TDN value of a feed. The more water present in a feed, the less there is of other nutrients, and, other things being equal, the lower the TDN value. For example, while milk on a dry basis is quite nutritious, on a fresh basis it is quite low (16%) in TDN because of its high (87%) water content. Silage is low in TDN compared to hay mainly because of a difference in water content. Many other such examples might be cited.

B.   **The digestibility of the dry matter.** Unless the dry matter of a feed is digestible, it can have no TDN value. Only digestible dry matter can contribute TDN. For example, mineral oil has a high gross energy value, but it cannot be digested by the animal and so has no digestible energy or TDN value. Lignin would fall in a similar category. Feeds high in fiber are, in general, low in digestibility and relatively low in TDN. Sand would be another form of dry matter which is indigestible and so would have a 0.0 TDN value.

C.   **The amount of mineral matter in the digestible dry matter.** Mineral compounds in an animal's ration may be digestible, but they contribute no energy to the animal and so have no TDN value. Such materials as salt, limestone, and defluorinated phosphate are all, in effect, digestible by the animal but would have 0.0 TDN values. The more mineral matter a feed contains, other things being equal, the lower will be its TDN value.

D.   **The amount of fat in the digestible dry matter.** As mentioned previously, in calculating TDN the digestible fat is multiplied by 2.25 since fat contributes 2.25 times as much energy per unit of weight as do carbohydrates and protein. Consequently, the more digestible fat a feed contains, other things being equal, the greater will be the TDN value. With feeds exceptionally high in digestible fat, such as peanut kernels or dried whole milk, TDN values may even exceed 100%. In fact, a pure fat which had a coefficient of digestibility of 100% would theoretically have a TDN value of 225% (100% × 2.25).

III.   The following diagram may be helpful in understanding the factors affecting the TDN value of feed.

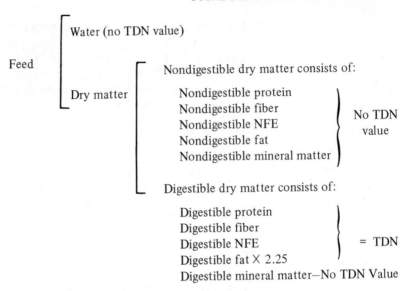

Table 8
FEEDS LISTED AND GROUPED ACCORDING TO THEIR PERCENTAGE OF TDN FOR CATTLE—AS FED BASIS

GROUP A—Primarily the Pure Fats and Other High-Fat Feeds Plus Certain Other Feeds of Generally High Digestibility

Animal fat, feed grade	175.2%	Ground cottonseed	86.6%
Corn oil	172.8	Corn starch	85.8
Peanut kernels	131.1	Hominy feed	84.9
Dried whole milk	110.1	Soybean seed	83.1
Oat groats	90.5	Dried bakery product	82.7

GROUP B—Mostly Low-Fiber Grains and Grain By-Products and Low Fiber- and Low Bone-Containing Protein Feeds, Plus Dried Citrus Pulp and Cane Molasses

Wheat grain	78.3%	Copra meal	73.7%
Wheat middlings	78.3	41% cottonseed meal	73.6
Yellow shelled corn	78.0	Linseed meal	73.6
Dried whey	77.8	Poultry by-product meal	73.0
Dried skimmed milk	77.4	Wheat shorts	72.7
Citrus pulp/molasses, dried	77.0	Sweet potato meal	72.7
Dried citrus pulp	76.6	Ground ear corn	72.6
Distillers dried grains	76.4	Seasame oil meal	72.6
49% soybean oil meal	76.3	Hegari grain	72.0
Corn gluten meal	75.4	44% soybean oil meal	72.0
Cane molasses	73.9	Barley grain	71.9

Table 8 (Continued)

Potato meal	71.7%	Milo grain	71.3%
Rye grain	71.4	Peanut oil meal	70.3
Grain sorghum grain	71.3		

GROUP C—Largely the More Fibrous Grains and By-Product Concentrates and the High Bone-Containing Protein Feeds, Plus Beet Molasses

Safflower meal without hulls	69.5%	Kafir grain	65.0%
41% solv-extd cottonseed meal	68.5	Tankage with bone	62.7
Corn bran	68.1	Wheat bran	62.7
Fish meal	67.8	Feather meal	62.2
Dried beet pulp/molasses	67.7	Meat scrap	62.0
36% cottonseed meal	67.1	Beet molasses	61.1
Ground snapped corn	66.7	Meat and bone meal	60.6
Oats grain	66.3	Sunflower meal	60.6
Digester tankage	65.5	Brewers dried grains	60.3
Dried beet pulp	65.2		

GROUP D—Predominantly the Hays, the Hay Meals, the Straws, and the Hulls (Except Peanut and Rice Hulls), Plus Corn Cobs, Corn Stover, Citrus and Wood Molasses, and Blood Meal

Dehydrated alfalfa leaf meal	59.6%	Citrus molasses	52.5%
Rice bran	58.9	Alsike clover hay	52.4
Blood meal	58.8	Corn stover	51.4
Soybean hulls	58.8	Kudzu hay	51.0
Peanut hay	58.3	Bahiagrass hay	50.8
Redtop hay	57.0	Bromegrass hay	49.8
Dehydrated alfalfa meal	56.7	Alfalfa hay	49.1
Meadow hay	56.1	Coastal bermudagrass hay	49.1
Kentucky bluegrass hay	56.0	Millet hay	49.0
Dehydrated coastal bermuda-		Timothy hay	48.8
grass meal	54.6	Dehydrated alfalfa stem meal	48.2
Oat hay	54.4	Prairie hay	48.1
Johnsongrass hay	53.8	Reed canarygrass hay	46.9
Sweetclover hay	53.5	Cottonseed hulls	46.5
Wood molasses	53.4	Oat straw	46.5
Cowpea hay	53.3	Soybean hay	45.8
Sudangrass hay	53.1	Ground corn cob	44.8
Lespedeza hay	53.0	Bermudagrass hay	44.7
Dallisgrass hay	52.9	Wheat straw	42.8
Orchardgrass hay	52.9	Sericea hay	40.5
Red clover hay	52.7	Barley straw	38.2
Tall fescue hay	52.6	Oat hulls	32.2

Table 8 (Continued)

GROUP E—For the Most Part High Moisture and High Mineral Feeds, Plus Peanut and Rice Hulls

Corn ear silage	31.0%	Fresh Dallisgrass forage	15.2%
Sweet potatoes	25.4	Steamed bone meal	15.0
Restaurant garbage	22.4	Fresh rye forage	14.8
Fresh bromegrass forage	21.4	Fresh ryegrass forage	14.8
Corn stover silage	20.2	Fresh red clover forage	14.6
Corn (fodder) silage	19.7	Fresh sudangrass forage	14.3
Fresh coastal bermudagrass	19.4	Fresh ladino clover forage	13.5
Fresh Kentucky bluegrass		Fresh crimson clover forage	11.8
forage	18.6	Fresh white clover forage	11.8
White potatoes	18.5	Carrots	10.6
Fresh fescue forage	18.4	Rice hulls	10.0
Fresh orchardgrass forage	17.6	Fresh skimmed milk	8.5
Grain sorghum silage	17.6	Cabbage heads	8.3
Fresh alfalfa forage	16.8	Turnips	8.3
Peanut hulls	16.7	Wet beet pulp	7.2
Fresh cow's milk	16.1	Defluorinated phosphate	***
Sorgo silage	16.1	Ground limestone	***
Wet brewers grains	15.9	Oyster shell flour	***
Fresh citrus pulp	15.2		

*** Very little, if any.

9 Energy Utilization

I.   **Definitions** of basic energy terms:

   A.   **Calorie (cal).** The amount of energy as heat required to raise 1 gram of water $1°C$ (precisely from $14.5°C$ to $15.5°C$). Formerly referred to as a small *calorie* and was so designated by being spelled with a lower case c.

   B.   **Kilocalorie (kcal).** The amount of energy as heat required to raise 1 kilogram of water $1°C$ (from $14.5°C$ to $15.5°C$). Equivalent to 1,000 calories. Formerly referred to as a large *Calorie* and was so designated by being spelled with a capital C.

   C.   **Megacalorie (Mcal).** Equivalent to 1,000 kilocalories or 1,000,000 calories. Formerly referred to as a *Therm*.

D.   British Thermal Unit (BTU). The amount of energy as heat required
to raise 1 lb of water 1°F. Equal to 252 calories. Seldom used in
animal nutrition.

E.   Bomb calorimeter. An instrument used for determining the gross
energy content of a material (Figure 11). It consists of an insulated
water container equipped with a combustion bomb, a thermometer, and
certain other accessories.

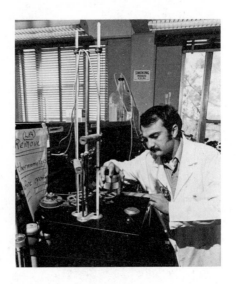

FIGURE 11.   The different parts of a bomb calorimeter are
being explained by a lab supervisor. *(Courtesy of the University of
Georgia College of Agriculture Experiment Stations)*

II.   Energy disposition in the animal may be diagrammed as follows:

Gross energy	Energy of feces		
	Digestible energy	Energy of the urine	
		Energy of methane	
		Metabolizable energy	Heat increment
			Heat of fermentation
			Heat of nutrient metabolism
			Net energy

A. **Gross energy (GE).** This is the total heat of combustion of a material as determined with a bomb calorimeter—ordinarily expressed as kilocalories per kilogram of feed or kcal/kg. The gross energy value of a feed has no relationship to the feed's digestible, metabolizable, or net energy values, except that the latter can never exceed the GE. Certain products such as coal, mineral oil, and lignin have high gross energy values but, because of their indigestibility, are of no energy value to the animal. Roughages have gross energy values comparable to those of concentrates, but the two differ greatly in digestible, metabolizable, and net energy values. Fats, because of their greater proportion of carbon and hydrogen, yield 2.25 times more gross energy per kg than do carbohydrates.

B. **Digestible energy (DE).** This is that portion of the gross energy of a feed which does not appear in the feces. It includes metabolizable energy as well as the energy of the urine and methane. DE differs from TDN in that TDN, as calculated, does not include the energy of the urine—at least that from protein metabolism, which is most of it.

C. **Metabolizable energy (ME).** This is that portion of the gross energy consumed which is utilized by the animal for accomplishing work, growth, fattening, fetal development, milk production, and/or heat production. It is that portion of the gross energy not appearing in the feces, urine, and gases of fermentation (principally CH_4). It is digestible energy minus the energy of the urine and methane. It is comparable to the energy of TDN minus the energy of the fermentation gases.

D. **Net energy (NE).** This is that portion of metabolizable energy which may be used as needed by the animal for work, growth, fattening, fetal development, milk production, and/or heat production. It differs from metabolizable energy in that net energy does not include the heat of fermentation and nutrient metabolism or the heat increment. No net energy is used for heat production unless heat over and above that from other sources is required to keep the animal warm.

E. **Heat increment (HI).** This is the difference between ME and NE. It represents the heat unavoidably produced by an animal incidental with nutrient digestion and metabolism. It has been referred to also as *work of digestion, specific dynamic effect,* and *thermogenic effect.* This heat is useful only for keeping an animal warm during very cold weather. At other times the energy represented by this heat is not only a complete loss but also may actually interfere with production by causing the animal to be too warm.

III. **Comparison of TDN, DE, ME, and NE as measures of a feed's energy value to an animal.** Total digestible nutrients (TDN), digestible energy (DE), metabolizable energy (ME), and net energy (NE) have all been used over the years for expressing the energy value of different feeds and rations for feeding purposes. Each has served and continues to serve a valuable function in ration formulation.

A.   Of these, DE is probably the least precise measure of a feed's energy value to an animal. It embraces all of the energy of a feed which does not appear in the feces but makes no allowances for other energy losses during digestion and utilization.

B.   TDN is superior to DE in this regard since in attributing the same value to digestible protein as is attributed to digestible carbohydrates in calculating TDN, an approximate correction is effected for that part of the protein energy which is excreted in the urine. Whereas digestible carbohydrates have a DE value of approximately 4.15 kcal per gram, digestible protein has a DE value of approximately 5.65 kcal per gram. However, the two are of similar energy value to the animal and are so considered by the procedure followed in calculating TDN.

C.   ME figures, on the other hand, represent a more precise measure of a feed's energy value to an animal than do either DE or TDN figures. In determining the ME of a feed, allowances are made not only for the energy losses in the urine but also for those of the fermentation gases. While the above-mentioned procedure followed in calculating TDN effects an approximate correction for the energy losses of the urine, both TDN and DE values fail to take into account the energy losses of the fermentation gases. ME figures, on the other hand, allow for both. This is especially important from the standpoint of balancing rations for ruminants since such animals normally have significant energy losses of both types, and different feeds and rations vary greatly in the amount of these losses.

D.   In the final analysis, however, to the extent that reliable NE values are available, they represent the most precise measure of an animal's energy needs and the capacity of different feeds to meet these needs. Not only do NE values allow for the energy losses of the urine and fermentation gases, but also they take into consideration the energy losses from the heat of nutrient utilization or the heat increment. Actual NE values, however, have been determined for only a limited number of feeds. Most available NE values are estimates which have been calculated for various feeds based on their composition and digestibility in relation to the composition and digestibility of the feeds on which actual NE values have been determined.

10 Study Questions and Problems

I.   *Give the correct figure for each of the following:*

1.   Number of chemical elements that function in animal nutrition.
2.   Number of mineral elements that function in animal nutrition.
3.   Number of nonmineral elements that function in animal nutrition.
4.   Number of micro mineral elements that function in animal nutrition.
5.   Number of micro elements that function in animal nutrition.
6.   Number of chemical elements found in carbohydrates.
7.   Minimum number of chemical elements found in a protein.
8.   Maximum number of chemical elements found in a protein.
9.   Minimum number of chemical elements found in a true fat.
10. Maximum number of chemical elements found in a true fat.

11. Number of carbon atoms in a fructose molecule.

12. Number of hydrogen atoms in a galactose molecule.

13. Number of oxygen atoms in a glucose molecule.

14. Number of carbon atoms in a lactose molecule.

15. Number of hydrogen atoms in a sucrose molecule.

16. Number of oxygen atoms in a maltose molecule.

17. Number of hexose molecules in a disaccharide.

18. Number of hexose molecules in a polysaccharide.

19. Number of hexose molecules in cellulose.

20. Number of glucose molecules in sucrose.

21. Number of glucose molecules in maltose.

22. Number of fructose molecules in lactose.

23. Number of galactose molecules in lactose.

24. Number of galactose molecules in sucrose.

25. Number of carbon atoms in a pentose molecule.

26. Number of known naturally occurring amino acids.

27. Number of naturally occurring proteins.

28. Percentage of nitrogen in crude protein.

29. What times nitrogen equals crude protein?

30. Fats contain how many times as many cal/lb as does starch?

31. What is the percentage of dry matter in air-dry feed?

32. What times % equals parts per million?

33. What times mg/lb equals parts per million?

34. To convert ppm to % divide by what number?

35. To convert ppm to mg/lb divide by what number?

36. What times % equals mg/lb?

37. Number of pounds in a kilogram.

II.   *Match (using numbers) the items in the right-hand column with the proper chemical formula in the left-hand column:*

1.	$C_6H_{12}O_6$.	1.	Sucrose.
2.	$C_{12}H_{22}O_{11}$.	2.	A pentosan.
3.	$(C_6H_{10}O_5)_n$.	3.	Oleic acid.
4.	$(C_5H_8O_4)_n$.	4.	Stearin.
5.	$C_{17}H_{35}COOH$.	5.	Glucose.
6.	$C_3H_5(OH)_3$.	6.	An amino acid.
7.	$C_{57}H_{110}O_6$.	7.	A soap.
8.	$C_{17}H_{33}COOH$.	8.	Stearic acid.
9.	$C_{17}H_{35}COONa$.	9.	Glycerol.
10.	$C_4H_8(NH_2)COOH$.	10.	Starch.

III. *Indicate by number into which of the following categories each of the items below would fall:*

1. A micro mineral element that functions in animal nutrition.

2. A macro mineral element that functions in animal nutrition.

3. A nonmineral element that functions in animal nutrition.

4. A true fat.

5. A carbohydrate.

6. An enzyme that functions in feed digestion.

7. A part of the digestive tract.

8. None of the above.

1.	Carbon.	18.	Trypsin.	35.	Hydrogen.
2.	Molybdenum.	19.	Sodium.	36.	Lactose.
3.	Glucose.	20.	Peptidase.	37.	Amino acid.
4.	Phosphorus.	21.	Galactose.	38.	Starch.
5.	Olein.	22.	Cadmium.	39.	Chlorine.
6.	Lactase.	23.	Calcium.	40.	Stearin.
7.	Iodine.	24.	Pepsin.	41.	Abomasum.
8.	Sucrose.	25.	Iron.	42.	Cobalt.
9.	Fructose.	26.	Rumen.	43.	Zinc.
10.	Fluorine.	27.	Cellulose.	44.	Bile.
11.	Rennin.	28.	Magnesium.	45.	Maltose.
12.	Palmitin.	29.	Villi.	46.	Reticulum.
13.	$C_{51}H_{98}O_6$.	30.	Selenium.	47.	Manganese.
14.	Oxygen.	31.	Omasum.	48.	Sucrase.
15.	Cecum.	32.	Potassium.	49.	Dextrin.
16.	Nitrogen.	33.	Glycogen.	50.	Inulin.
17.	Sulfur.	34.	Boron.		

IV. *Indicate by number into which of the following categories each of the items below would fall.*

1. A monosaccharide.

2. A disaccharide.

3. A polysaccharide.

4. Other.

1.	Glucose.	5.	$C_{12}H_{22}O_{11}$.	9.	Stearin.
2.	Methane.	6.	Cane sugar.	10.	$(C_6H_{10}O_5)_n$.
3.	Sucrose.	7.	Corn sugar.	11.	Glycogen.
4.	Glycerol.	8.	Cellulose.	12.	Insulin.

13.	Fruit sugar.	19.	Maltose.	25.	Crude fiber.
14.	Maltase.	20.	Inulin.	26.	Starch.
15.	Milk sugar.	21.	Beet sugar.	27.	Lactose.
16.	Fructose.	22.	Urea.	28.	Mannose.
17.	Galactose.	23.	Lignin.	29.	Cotton.
18.	Rennin.	24.	Dextrose.	30.	$C_6H_{12}O_6$.

V. *Correlate by number the various items in the first column with the most appropriate item in the second column.*

1.	Dry roughage.	1.	Olein.
2.	Fructose.	2.	Palmitin.
3.	Low in fiber—high in TDN.	3.	Fruit sugar.
4.	Sucrose.	4.	Ether.
5.	Nonmineral element.	5.	Mineral matter.
6.	Glycogen.	6.	Omasum.
7.	Enzyme.	7.	Stearic acid.
8.	Lactose.	8.	Cotton.
9.	Glycerol.	9.	Honeycomb.
10.	Starch and sugar.	10.	PPM.
11.	Fiber + NFE.	11.	True stomach.
12.	Amid.	12.	Glucose.
13.	A trace mineral.	13.	Cane sugar.
14.	A fat solvent.	14.	Bile.
15.	mg/kg	15.	Milk sugar.
16.	$C_{15}H_{31}COONa$.	16.	Cobalt.
17.	mg/lb.	17.	End product of protein digestion.
18.	Ash.	18.	(mg/kg)/2.2.
19.	CH_4.	19.	Trypsin.
20.	Crude fat.	20.	Animal starch.
21.	Alkaline digestive juice.	21.	Methane.
22.	Percent.	22.	Nitrogen-free extract.
23.	End product of starch digestion.	23.	Carbon.
24.	$C_{51}H_{98}O_6$.	24.	Nonprotein.
25.	Ruminant.	25.	Organic catalyst.
26.	Rumen.	26.	Paunch.
27.	Dry matter.	27.	$C_3H_5(OH)_3$.
28.	Crude protein.	28.	Ether extract.
29.	Oven-dry.	29.	PPM/10,000.
30.	Manyplies.	30.	Soap.
31.	Reticulum.	31.	Carbohydrate.
32.	Unsaturated fat.	32.	Nitrogen × 6¼.
33.	Abomasum.	33.	High in fiber—low in TDN.
34.	Amino acid.	34.	100% dry matter.

35.	Almost pure cellulose.	*35.*	Concentrate.
36.	$C_{17}H_{35}COOH$.	*36.*	Cud-chewing animal.
37.	Proteolytic enzyme.	*37.*	100 minus % H_2O.

VI. *A mixture of the following materials is subjected to a proximate analysis. Indicate by number in which of the following fractions each would appear:*

> *1.* Crude protein.
>
> *2.* Crude fiber.
>
> *3.* Crude fat.
>
> *4.* Nitrogen-free extract.
>
> *5.* Mineral matter.
>
> *6.* None of the above.

1.	True protein.	*13.*	Glycogen.	*25.*	Ether.
2.	Cholesterol.	*14.*	Fatty acids.	*26.*	Sand.
3.	Cotton fiber.	*15.*	Wood ashes.	*27.*	Glucose.
4.	Amids.	*16.*	Dextrin.	*28.*	Hair.
5.	Inulin.	*17.*	$C_5H_{10}O_5$.	*29.*	Amino acids.
6.	Vitamin D.	*18.*	Starch.	*30.*	Wool fiber.
7.	Maltose.	*19.*	Disaccharides.	*31.*	$C_{12}H_{22}O_{11}$.
8.	Salt.	*20.*	Sucrose.	*32.*	Soybean oil.
9.	Calcium oxide.	*21.*	$C_{51}H_{98}O_6N_{11}$.	*33.*	Sucrase.
10.	Stearin.	*22.*	$C_{51}H_{98}O_6$.	*34.*	Lignin.
11.	Mineral oil.	*23.*	Cellulose.		
12.	Chlorophyl.	*24.*	Carotene.		

VII. *Indicate by number which of the following materials on an as fed basis would be among the:*

> *1.* 8 highest in % water.
>
> *2.* 19 highest in % TDN.
>
> *3.* 17 highest in % fiber.
>
> *4.* 14 highest in % fat.

1.	Alfalfa hay.	*11.*	Wheat bran.
2.	Dehydrated alfalfa meal.	*12.*	Corn starch.
3.	Timothy hay.	*13.*	White potatoes.
4.	Grain sorghum grain.	*14.*	Ground corn cob.
5.	Defluorinated phosphate.	*15.*	Red clover hay.
6.	Wheat grain.	*16.*	Cane molasses.
7.	Alfalfa pasture.	*17.*	Lespedeza hay.
8.	Oat straw.	*18.*	Sweet potatoes.
9.	Peanut kernels.	*19.*	Cane sugar.
10.	Corn oil.	*20.*	Steamed bone meal.

21. Soybean seed.
22. Dried skimmed milk.
23. Corn silage.
24. Feather meal.
25. Crimson clover pasture.
26. Barley grain.
27. Sorgo silage.
28. Fish meal.
29. Cottonseed.

30. Yellow shelled corn.
31. Tankage with bone.
32. Dried citrus pulp.
33. 36% cottonseed meal.
34. 44% soybean oil meal.
35. Fresh whole milk.
36. Bermudagrass pasture.
37. Ground limestone.

VIII. *In using the tables on composition of feed, for what does each of the following abbreviations stand.*

1.	blm	6.	IU	11.	res	16.	wo
2.	Ca	7.	kcal	12.	s-c	17.	wt
3.	dehy	8.	mn	13.	sol	18.	DE
4.	fbr	9.	mx	14.	solv-extd	19.	ME
5.	gr	10.	pt	15.	w	20.	mg/kg

IX. *Indicate whether each statement is true or false. In order for a statement to be true, it must be* completely *true.*

1. Bile is produced in the liver and functions in fat digestion.
2. Starch is a carbohydrate and is also a form of nitrogen-free extract.
3. The cell walls of plants and animals are chiefly cellulose.
4. Apparently dry feeds may contain over 10% water.
5. Galactose is a hexose and is a component of milk sugar.
6. The omasum of a cow has a capacity of up to 40-50 gals.
7. Maltase acts on maltose to yield fructose and glucose.
8. Most good hays contain over 60% TDN.
9. Proteins are, on the average, about 6.25% nitrogen.
10. Fat contains a greater proportion of carbon and hydrogen than does starch.
11. The end products of fat digestion are amino acids, glycerol, and soap.
12. Carbohydrates contain carbon, hydrogen, and oxygen.
13. Proteins are soluble in ether and also in dilute H_2SO_4.
14. Another name for crude protein is nitrogen-free extract.
15. Plants get their energy from the soil.
16. Pancreatic lipase is an enzyme which works on fats in the small intestine.
17. Ether extract is another name for crude fat.
18. Fructose is a monosaccharide found in ripe fruits and honey.

19. Fats, on the average, contain 2½ times as much energy per lb as do sugars and starch.
20. The cell walls of animals are high in lignin.
21. Lactose is a disaccharide found in milk.
22. A megacalorie is equal to 1,000 calories.
23. Trypsin is an enzyme which works on protein in the stomach.
24. Sucrose is another name for beet sugar as well as cane sugar.
25. Proteins sometimes contain sulphur, potassium, and/or iron.
26. A pig has no rumen but does have an abomasum.
27. Cane sugar consists of one molecule of glucose and one molecule of fructose.
28. The gross energy value of a feed is a good indication of its TDN value.
29. Pepsin is an enzyme found in the gastric juice and and acts on protein.
30. Of all farm animals the horse has the largest cecum.
31. The horse is a ruminant but does not chew a cud.
32. The pig is classified as a nonruminant.
33. The coefficient of digestibility of a nutrient is the percentage of the total amount of that nutrient present that is digestible.
34. The percentage of digestible protein in a feed may be greater than the percentage of crude protein in the feed.
35. The percentage of crude fiber in a feed may exceed 100% under certain conditions.
36. The percentage of crude fat in a feed may exceed 100% under certain conditions.
37. The percentage of crude protein in a feed may exceed 100% under certain conditions.
38. The percentage of nitrogen-free extract in a feed may exceed 100% under certain conditions.
39. The percentage of TDN in a feed may exceed 100% under certain conditions.
40. The percentage of TDN plus the percentage of mineral matter in a feed should equal the percentage of dry matter in the feed.
41. 100% minus the percentage of water in a feed should equal the percentage of dry matter.
42. The percentage of crude protein plus the percentage of crude fat plus the percentage of crude fiber plus the percentage of nitrogen-free extract should equal the percentage of dry matter in the feed.
43. Metabolizable energy minus the heat increment equals net energy.
44. TDN minus (energy of the urine + energy of methane) would be comparable to metabolizable energy.

45.   A small calorie, a gram-calorie, and a megacalorie are all the same.

46.   A large calorie, a kilo-calorie, and a therm are all the same.

47.   Proteins and carbohydrates are about equal in TDN value per pound.

48.   The sum of digestible protein, digestible carbohydrates, and (digestible fat × 2¼) equals TDN.

49.   TDN may be expressed in lb or kg or percent.

50.   TDN includes all of the digestible nutrients in a feed.

51.   TDN indicates the relative digestible energy value of a feed.

52.   TDN minus the energy of the fermentation gases would be comparable to metabolizable energy.

53.   The heat increment of a feed serves no worthwhile function in livestock production.

54.   Starch is a common component of many feeds.

55.   Net energy plus the heat increment equals metabolizable energy.

56.   Commercially mixed feeds frequently will contain excessive amounts of moisture.

57.   Starch and sugar would be comparable in TDN value per pound.

58.   Cellulose is the primary constituent of cotton.

59.   Reputable feed manufacturing companies welcome rigid feed-mixing controls.

60.   A feed component which is highly digestible could have a low TDN value.

61.   Crude fiber is one of the chief constituents of wool.

62.   Sometimes lignin comes out in the NFE fraction of a feed.

63.   Glycogen is a common component of many feeds.

64.   The percentage of crude protein plus the percentage of crude fat plus the percentage of crude fiber plus the percentage of nitrogen-free extract should equal the percentage of dry matter in the feed.

65.   A feed component which is high in crude fat will always have a high TDN value.

66.   A mega-calorie and a therm are the same.

67.   Inulin is a common component of many feeds.

68.   Digestible protein and digestible nitrogen-free extract are about equal in TDN value per pound.

X.   *A feed has the following composition on an air-dry basis. Calculate its composition on 1. an oven-dry basis; 2. An 80% moisture basis.*

Ash	4.7%
Crude fiber	27.6

Ether extract	1.9
N-free extract	49.4
Crude protein	6.4
TOTAL DM	90.0%

SOLUTIONS:

1.   Composition on an oven-dry basis.

Ash:

$$\frac{4.7}{90.0} = \frac{x}{100}$$

$$90x = 470$$

$$x = 5.22 \ (\% \ ash, oven\text{-}dry \ basis)$$

Calculate the crude fiber, etc.

2.   Composition on an 80% moisture basis.

Ash:

$$\frac{4.7}{90} = \frac{x}{100-80}$$

$$90x = 94$$

$$x = 1.04 \ (\% \ ash, 80\% \ moisture \ basis)$$

Calculate the crude fiber, etc.

XI.   *A feed has the following composition on a fresh basis.*

Ash	2.21%
Crude fiber	6.70%
Ether extract	0.92%
N-free extract	10.00%
Crude protein	4.57%
Calcium	0.40%
Phosphorus	0.06%
Carotene	28.30 mg/lb

1.   What is the % dry matter in the fresh feed?
2.   What is the % crude fiber in the feed on an air-dry basis?
3.   What is the % crude fiber in the feed on an oven-dry basis?
4.   What is the % carotene in the fresh feed?
5.   How many PPM of phosphorus in the fresh feed?

SOLUTIONS:

1.   The % DM in the fresh feed.

2.21 + 6.70 + 0.92 + 10.00 + 4.57 = 24.40

2.   The % crude fiber in the feed on an air-dry basis.

$$\frac{6.70}{24.40} = \frac{x}{90.0}$$

$$24.40x = 603$$

$$x = 24.71$$

3.   The % crude fiber in the feed on an oven-dry basis.

$$\frac{6.70}{24.40} = \frac{x}{100} \quad \text{or} \quad \frac{24.71}{90} = \frac{x}{100}$$

$$24.40x = 670 \qquad\qquad 90x = 24.71$$

$$x = 27.46 \qquad\qquad x = 27.46$$

4.   The % carotene in the fresh feed.

$$\frac{28.3 \times 2.2}{10{,}000} = 0.00623$$

$$or$$

$$\frac{28.3}{453{,}600} \times 100 = 0.00624$$
(mg in a lb)

5.   PPM of phosphorus in the fresh feed.

$$0.06 \times 10{,}000 = 600$$

XII. *A feed has the following composition on a wet basis.*

Ash	3.11%
Crude fiber	8.32%
Ether extract	1.10%
N-free extract	12.44%
Crude protein	5.61%
Calcium	0.32%
Phosphorus	0.16%
Carotene	110.0 mg/kg

1.   What is the % water in the feed on the wet basis?

2.   What is the % NFE in the feed on an air-dry basis?

3. What is the % NFE in the feed on an oven-dry basis?
4. How many mg/lb of phosphorus are in the feed on a wet basis?
5. How many PPM of carotene are in the feed on a wet basis?

ANSWERS:

1. 69.42%.
2. 36.61%.
3. 40.68%.
4. 725.6 mg/lb.
5. 110.0 PPM.

XIII. *A feed contains 1.47% calcium on an air-dry basis. Calculate the following:*

1. % calcium on an oven-dry basis.
2. % calcium on a 75% water basis.
3. Calcium content of air-dry feed in PPM.
4. Calcium content of air-dry feed in mg/lb.
5. Calcium content of air-dry feed in lb per 100 lb feed.

ANSWERS:

1. 1.63%.
2. .408%.
3. 14,700 PPM.
4. 6681.8 mg/lb.
5. 1.47 lb per Cwt.

XIV. *A steer consumes 20 lb of feed daily and excretes 40 lb of feces. The feed and feces had the following compositions:*

	FEED	FECES
Ash	8.0%	1.33%
Crude fiber	28.6%	9.91%
Ether extract	1.9%	0.11%
N-free extract	36.7%	6.20%
Crude protein	15.3%	2.10%

1. What is the % dry matter in the feed?
2. What is the % water in the feces?
3. What is the % digestible DM in the feed, as fed basis?
4. What is the % TDN in the feed, as fed basis?
5. What is the coefficient of digestibility of the DM in the feed?

6.   What is the coefficient of digestibility of the NFE in the feed?

7.   What is the % of digestible NFE in the feed, as fed basis?

8.   What is the % of digestible NFE in the feed, oven-dry basis?

SOLUTIONS:

   1.   The % DM in the feed.

            8.0 + 28.6 + 1.9 + 36.7 + 15.3 = 90.5

   2.   The % water in the feces.

      100 − (1.33 + 9.91 + 0.11 + 6.20 + 2.10) = 80.35

   3.   The % of digestible DM in the feed, as fed.

	ASH	FIBER	EE	NFE	PROTEIN
Nutrients per lb feed (lb)	0.080	0.286	0.019	0.367	0.153
X 20 lb feed	20	20	20	20	20
= Nutrients consumed (lb)	1.60	5.72	0.38	7.34	3.06

	ASH	FIBER	EE	NFE	PROTEIN
Nutrients per lb feces	0.0133	0.0991	0.0011	0.0620	0.0210
X 40 lb feces	40	40	40	40	40
= Nutrients excreted (lb)	0.532	3.964	0.044	2.480	0.840
Nutrients digested (lb)	1.068	1.756	0.336	4.860	2.220

   DM digested per 20 lb feed = 1.068 + 1.756 + 0.336 + 4.860 + 2.220

                                          = 10.24 lb

      % digestible DM in feed $= \dfrac{10.24}{20} \times 100 = 51.2$

   4.   The % TDN in the feed, as fed.

   $\dfrac{1.756 + (0.336 \times 2.25) + 4.860 + 2.220}{20} \times 100 = 47.96$

   5.   The coefficient of digestibility of the DM in the feed.

            $\dfrac{51.2}{90.5} \times 100 = 56.57\%$

   6.   The coefficient of digestibility of the NFE in the feed.

            $\dfrac{4.86}{7.34} \times 100 = 66.21\%$

7. The % of digestible NFE in the feed, as fed.

$$\frac{4.86}{20} \times 100 \quad or \quad \frac{66.21 \times 36.7}{100} = 24.3$$

8. The % of digestible NFE in the feed, oven-dry basis.

$$\frac{24.3}{90.5} = \frac{x}{100}$$

$$90.5x = 2430$$

$$x = 26.85$$

XV. *A steer consumes daily 9.07 kg of a feed containing 11.0% crude protein and excretes daily an average of 14.97 kg of feces containing 2.00% crude protein.*

1. What is the coefficient of digestibility of the crude protein in the feed?
2. What is the % of digestible crude protein in the feed as fed?

ANSWERS:

1. 69.99%.
2. 7.70%.

XVI. *A steer consumes 9.07 kg of feed daily and excretes an average of 6.11 kg of feces. The composition of the feed as fed and the coefficients of digestibility of the respective nutrients are as follows:*

	COMPOSITION	COEFFICIENT OF DIGESTIBILITY
Ash	3.2%	52.0%
Crude fiber	6.2%	55.5%
Ether extract	1.9%	80.0%
N-free extract	66.2%	91.0%
Crude protein	11.8%	77.0%

1. What is the % digestible DM in the feed, as fed basis?
2. What is the % TDN in the feed, as fed basis?
3. What is the % digestible protein in the feed, as fed basis?
4. What was the % DM in the feces?

SOLUTIONS:

1. The % digestible DM in the feed, as fed.

	ASH	FIBER	EE	NFE	PROTEIN
Nutrients per kg feed (kg)	0.032	0.062	0.019	0.662	0.118
× 9.07 kg feed	9.07	9.07	9.07	9.07	9.07
= Nutrients eaten daily (kg)	0.290	0.562	0.172	6.004	1.070
× C of D	0.520	0.555	0.800	0.910	0.770
= Nutrients digested daily (kg)	0.151	0.312	0.138	5.464	0.824

$$\frac{0.151 + 0.312 + 0.138 + 5.464 + 0.824}{9.07} \times 100 = 75.95$$

2.   The % TDN in the feed, as fed.

$$\frac{0.312 + (0.138 \times 2.25) + 5.464 + 0.824}{9.07} \times 100 = 76.20$$

3.   The % digestible protein in the feed, as fed.

$$\frac{0.824}{9.07} \times 100 \quad \text{or} \quad \frac{11.8 \times 77.0}{100} = 9.08$$

4.   The % DM in the feces.

	ASH	FIBER	EE	HFE	PROTEIN
Nutrients eaten daily (kg)	0.290	0.562	0.172	6.004	1.070
− Nutrients digested daily (kg)	0.151	0.312	0.138	5.464	0.824
= Nutrients in daily feces (kg)	0.139	0.250	0.034	0.540	0.246

$$\frac{0.139 + 0.250 + 0.034 + 0.540 + 0.246}{6.11} \times 100 = 19.79 \; (answer)$$

XVII. *A cow consumes 10.0 kg of feed daily and excretes an average of 18.61 kg of feces. The feed had the following composition (as fed basis) and coefficients of digestibility for the respective nutrients.*

	COMPOSITION	COEFFICIENT OF DIGESTIBILITY
Ash	4.7%	44.0%
Crude fiber	27.8%	55.0%
Ether extract	2.1%	50.0%
N-free extract	46.7%	60.0%
Crude protein	9.2%	65.0%

1.   What is the % digestible DM in the feed, as fed basis?

2.   What is the % TDN in the feed, as fed basis?

3. What is the % digestible fiber in the feed, as fed basis?

4. What was the % water in the feces?

ANSWERS:

1. 52.41%.

2. 51.65%.

3. 15.29%.

4. 79.53%.

11 Protein Nutrition

I. **Needs of animals for protein.** Every living animal has a need for protein. It is the basic structural material from which all body tissues are formed. This includes not only the muscles, nerves, skin, connective tissue, and vital organs but also the blood cells, as well as the animal's hair, hoof, and horn. Even the dry matter of bone is over 1/3 protein, protein providing the basic cellular matrix within which the bone mineral matter is deposited. Obviously then protein is essential for an animal's growth and development as well as for fetal development. Also, since all living tissue is in a dynamic state and is undergoing constant degeneration, protein is necessary for its maintenance. Also, protein is required for wool growth and milk production. Furthermore, most body enzymes and hormones are basically protein in composition. Finally, no other nutrient can replace protein in the ration. In view of an animal's many needs for protein and the irreplaceable nature of this nutrient, there is a certain minimum level of dietary protein recommended for each class of animals. This level varies for animals of different classes, depending on the physiological age and type of production, but will usually be somewhere between 8% and 18%.

II.   Proteins are complex organic nitrogenous compounds made up of amino acids. *Amino acids* are organic acids which contain one or more amino groups (NH_2). Some 25 or more different amino acids are present in feed proteins, of which some 20 or more enter into the makeup of animal tissue (Figure 12). Of these, some 10 or 11 are classified as *essential* and the others as *nonessential*.

FIGURE 12.   An animo acid analyzer—used for determining the amino acid make-up of proteins. *(Courtesy of the University of Georgia College of Agriculture Experiment Stations)*

A.   **Essential amino acid.** An essential amino acid is one needed by the animal that cannot be synthesized by the animal in the amounts needed and so must be present in the protein of the feed as such.

B.   **Nonessential amino acid.** A nonessential amino acid is one needed by the animal that can be formed from other amino acids by the animal and so does not have to be present as that particular amino acid in the protein of the feed.

III.   Those amino acids which function in animal nutrition are usually classified on the basis of their essentiality as follows:

ESSENTIAL	NONESSENTIAL
Arginine	Alanine
Histidine	Aspartic acid
Isoleucine	Citrulline
Leucine	Cystine
Lysine	Glutamic acid
Methionine*	Glycine
Phenylalanine	Hydroxyproline
Threonine	Proline
Tryptophan	Serine
Valine	Tyrosine

*May be replaced in part by cystine.

IV.   Limiting essential amino acid. The essential amino acids are required by livestock in definite proportions. While the proportion may vary for different functions, it is always quite definite for any given animal performing any given set of functions. The amino acid which is present in a protein in the least amount in relation to the animal's need for that particular amino acid is referred to as the *limiting amino acid*. Other essential amino acids can be used by the animal toward meeting its essential amino acid requirement only to the extent that the so-called limiting amino acid is present.

The above phenomenon is illustrated by a comparison of the requirements

(a)   (b)

FIGURE 13.   *Lysine deficiency.* (a) This pig gained 11.4 kg in 28 days after lysine was added to the basal diet (2.0 percent DL-lysine). (b) A lysine deficient pig—basal diet only. This pig lost 0.9 kg in 28 days. *(Courtesy of W. M. Beeson, Purdue University)*

of growing pigs for the different essential amino acids with the levels of these amino acids in shelled corn as presented in Table 9. It will be noted that lysine is the limiting essential amino acid of corn.

Table 9
LIMITING ESSENTIAL AMINO ACID OF SHELLED CORN

	PERCENTAGE OF EACH ESSENTIAL AMINO ACID RE-QUIRED IN RATIONS FOR 20-35 KG (44-77 LB) PIGS, AS FED	PERCENTAGE OF EACH ESSENTIAL AMINO ACID PRESENT IN GROUND SHELLED CORN, AS FED	PERCENTAGE OF EACH ESSENTIAL AMINO ACID REQUIREMENT IN CORN
Arginine	0.20%	0.45%	225%
Histidine	0.18	0.18	100
Isoleucine	0.50	0.45	90
Leucine	0.60	0.99	165
Lysine	0.70	0.18	26
Methionine	0.50	0.09	18*
Phenylalanine	0.50	0.45	90
Threonine	0.45	0.36	80
Tryptophan	0.13	0.09	69
Valine	0.50	0.36	72

*At least 50% of the methionine requirement may be met with cystine, and there is about as much cystine as methionine in shelled corn; hence, lysine becomes the limiting essential amino acid of corn protein for growing pigs.

V.   **Biological value**. The biological value of a protein is the percentage of the digestible protein of a feed or feed mixture which is usable as protein by the animal. The biological value of a protein will depend on the amount of the limiting amino acid present in relation to the other essential amino acids, based on the animal's need for each. A protein which has a desirable balance of essential amino acids will have a high biological value and is described as being a protein of *good quality*. A protein which is extremely deficient in one or more of the essential amino acids will have a low biological value and is described as being a protein of *low quality*.

VI.   **Supplementary effect of proteins**. When two proteins have different limiting amino acids and one contains an excess of that amino acid which is limiting in the other, then a *supplementary effect* is realized when the two proteins are mixed together. It is for this reason that more than one source of protein is recommended for nonruminants. Fortunately, the protein of most of the more abundant high protein supplements has a quite favorable supplementary effect with the protein of most farm grains. This phenomenon is well illustrated by the bar graph below.

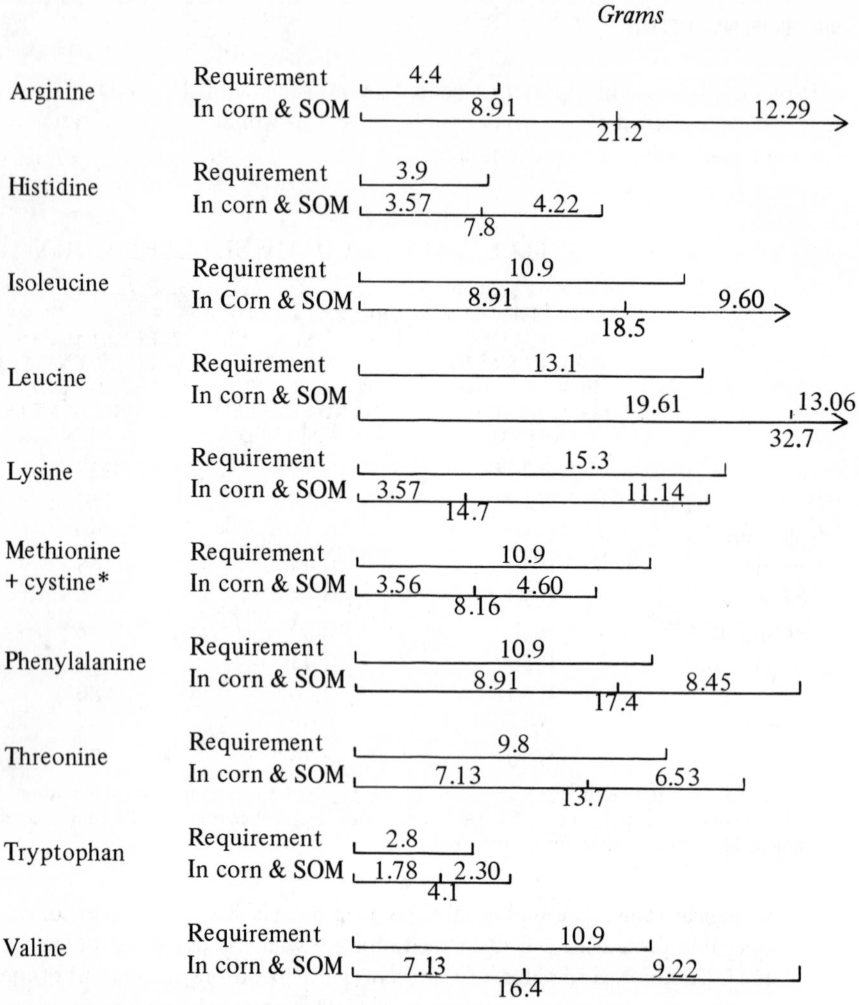

Arginine
Requirement 4.4
In corn & SOM 8.91 12.29
21.2

Histidine
Requirement 3.9
In corn & SOM 3.57 4.22
7.8

Isoleucine
Requirement 10.9
In Corn & SOM 8.91 9.60
18.5

Leucine
Requirement 13.1
In corn & SOM 19.61 13.06
32.7

Lysine
Requirement 15.3
In corn & SOM 3.57 11.14
14.7

Methionine + cystine*
Requirement 10.9
In corn & SOM 3.56 4.60
8.16

Phenylalanine
Requirement 10.9
In corn & SOM 8.91 8.45
17.4

Threonine
Requirement 9.8
In corn & SOM 7.13 6.53
13.7

Tryptophan
Requirement 2.8
In corn & SOM 1.78 2.30
4.1

Valine
Requirement 10.9
In corn & SOM 7.13 9.22
16.4

*At least 50% of the methionine requirement may be met with cystine. Even so, it would appear that the above ration would be deficient in methionine. However, methionine supplementation of corn-soybean oil meal based rations usually produces little if any improvement in performance, especially when choline has been added as a methyl donor to spare the use of methionine by the pig for this purpose. Some swine nutritionists think that the NRC requirement for methionine may be somewhat higher than necessary, especially when supplemental choline is provided.

FIGURE 14. A comparison of the daily requirements of a 47.5 kg growing-finishing pig for the different essential amino acids with the amounts of each of these acids in a daily ration of 1.981 kg corn + 0.384 kg soybean oil meal + minerals and vitamins.

VII. Ruminants vs. nonruminants. While ruminant animals are thought to have just as rigid a physiological requirement for essential amino acids as nonruminants, they do not have a rigid dietary amino acid requirement. Microbes (bacteria and protozoa) in the rumen synthesize, for their own cellular development, essential amino acids from nonessential amino acids and certain other nonprotein-nitrogen-containing materials. The microbes produced in the rumen are then digested in the abomasum and intestinal tract to provide the ruminant animal with most, if not all, of its needs for essential amino acids. Hence, protein quality is not a matter of consideration in formulating rations for ruminant animals.

VIII. Protein as a source of energy. Protein fed in excess of an animal's needs or protein not usable as protein by the animal because of a limiting animo acid is *deaminated* (or *deaminized*) in the liver and used as a source of energy by the animal. In deamination, the NH_2 group is split off of the amino acid molecule and excreted as urea in the urine. While high protein feeds may serve as a source of energy for livestock, they are usually too expensive to be a practical source of feed energy.

IX. Nonprotein nitrogen (NPN) utilization by ruminants. As mentioned previously, ruminants through microbial action in the rumen are able to synthesize protein from nonprotein nitrogen. As a result, nonprotein nitrogen may be used as a substitute for protein in ruminant rations. When so used, however, it must be fed along with a readily fermentable carbohydrate source such as corn, grain sorghum, or some type molasses to supply the energy for the formation of the protein molecule.

X. Urea as an NPN source. While various products have been and are being studied as nonprotein nitrogen sources, the principal NPN source in use today in livestock feeding is urea. It is produced by combining natural gas with water and air. It has a chemical formula of $CO(NH_2)_2$. Consequently, in pure form it contains 46.67% nitrogen. As commercially produced today, however, it commonly carries 45% nitrogen, giving it a crude protein equivalent value of 281% (45.0 × 6.25) (Figure 15). In order for it to be used by rumen microbes, it must first be *hydrolyzed* (chemically combined with water) to form $CO_2 + NH_3$. This hydrolysis is accomplished by the action of an enzyme, urease, secreted by rumen organisms. As NH_3 is liberated in the rumen, it apparently reacts with organic acids of fermentation to form ammonium salts of organic acids, such as ammonium acetate and ammonium propionate. These in turn are metabolized by rumen microbes to form cellular protein.

XI. Urea toxicity. Urea as such is a normal component of body fluids since it is a by-product of protein metabolism in the body. Consequently, in the amounts ordinarily fed, it itself is not toxic to the animal. However, ammonia

UREA (45%N)

FIGURE 15.   Much of today's urea is in a beaded form which helps keep it free flowing for good mixing. *(Courtesy of the University of Georgia College of Agriculture Experiment Stations)*

which is produced upon urea hydrolysis is quite toxic if liberated in excessive amounts—that is, in amounts which exceed the supply of organic acids of fermentation in the rumen with which the urea NH_3 is supposed to combine. Under such circumstances the NH_3 is absorbed into the blood through the rumen wall to produce alkalosis and, in turn, possibly death of the animal. Such trouble is most likely to be experienced when animals do not consume sufficient fermentable carbohydrates to maintain a relatively high level of acidity in the rumen. When rumen acidity is low, not only are there insufficient organic acids to react with the NH_3 present but also urease is most active at a pH of 6 to 8 such as would be found in a rumen low in organic acids of fermentation. Also, the excess NH_3 would tend to form NH_4OH in the rumen to further reduce rumen acidity. In other words, the rumen condition most favorable for urea hydrolysis and NH_3 production is least favorable for effective NH_3 utilization. If urea is to be used effectively for ruminant feeding without toxic results, it must be fed in conjunction with an adequate supply of readily fermentable carbohydrates such as will keep the rumen pH below a level of 6.

XII. Extent of urea feeding. It has been estimated that urea is presently being used for feeding purposes at the rate of about 500,000 tons per year. This amount of urea would have a nitrogen equivalent of 3.2 million tons of 44% soybean oil meal which would be approximately one-fourth of the nation's annual use of SOM (12.6 million tons). SOM makes up about 2/3 of all high protein feeds. While urea in limited amounts is not particularly harmful to nonruminants such as swine, nonruminant animals can make little, if any, effective use of urea since they do not have a rumen. Consequently, urea feeding is restricted to ruminants. Even with ruminants, it is recommended that for effective results it not be used to provide more than one-third of the protein nitrogen in the ration. While more than this amount can be used, the hazards of toxicity increase and the efficiency of utilization decreases at higher urea levels. Whether or not urea should be used for ruminant feeding will depend on the prevailing cost of urea and carbohydrate feeds such as corn, on the one hand, and the cost of feeds high in pre-formed protein, on the other. The following may be helpful in this connection.

$$1 \text{ lb } 45\% \text{ N urea } + 6.5 \text{ lb grnd sh corn } \lessgtr 7.5 \text{ lb } 44\% \text{ SOM}$$

If urea is $200 per ton and corn is $2.80 per bu, then the cost of 1 lb urea (10¢) + 6.5 lb corn (32.5¢) would be 42.5¢. If 7.5 lb of SOM would cost more than 42.5¢ (or 5.67¢ per lb or $113.40 per ton), then it would probably be economical to use urea in the ration.

XIII. **High protein feeds.** High protein feeds are ordinarily named and classified on the basis of their origin and method of processing.

    A.   On the basis of origin, high protein feeds are usually classified into two general categories.

        1.   **Those of animal origin.** Hogs formerly did best with one of these in the ration. This was attributed to a factor known as the *animal protein factor* or *APF*. At the time, this factor was unidentified but has since proved to be what is now known as *vitamin B_{12}*. With the general availability of vitamin B_{12}, high protein feeds of animal origin are not regarded as essential for swine production and so hold little advantage over other high protein feeds at this time.

        2.   **Those of plant origin.** These consist for the most part of oil seed by-products. They vary in composition and feeding value, depending on the seed from which they are produced, and amount of hull and/or seed coat included, and the method of fat extraction used.

    B.   Three methods of extraction have been used over the past several years for extracting fat from oil seeds.

        1.   **Hydraulic or old process.**

            a.   The original process for fat extraction.

            b.   Widely used in the past throughout the Cotton Belt.

            c.   Based on mechanical extraction using a hydraulic press.

            d.   Not a continuous process.

            e.   Leaves considerable oil in meal.

            f.   Few mills use this process today.

        2.   **Expeller or new process.**

            a.   Came in with the soybean industry.

            b.   Also based on mechanical extraction using a screw press.

c.   Preferred over the hydraulic process because it permits continuous operation.

d.   Also leaves considerable oil in the meal.

e.   Still extensively used in the Cotton Belt—almost completely replaced in soybean area with solvent process.

3.   **Solvent process.**

a.   Originally developed in Germany.

b.   Introduced into U.S. following World War II—late 1940s.

c.   Based on extraction of fat in a manner similar to crude fat determination except on much larger scale—also hexane rather than ether used as solvent.

d.   Brings about almost complete fat extraction—some fat must be added.

e.   Most new fat extraction plants are this type—have almost completely taken over in soybean section.

f.   Essentially all SOM is solvent process today.

Table 10
FEEDS LISTED AND GROUPED ACCORDING TO THEIR PERCENTAGE OF CRUDE PROTEIN—AS FED BASIS

GROUP A—Mostly the Meat and Poultry By-Product Meals, Fish Meal, the Oil Seed Meals, the Oil Seeds, and the Dried Milks

Feather meal	87.4%	41% cottonseed meal	41.4%
Blood meal	80.2	Safflower meal without hulls	40.4
Fish meal	60.4	Corn gluten meal	39.0
Digester tankage	59.2	Soybean seed	37.9
Poultry by-product meal	55.4	36% cottonseed meal	35.9
Meat scrap	54.9	Linseed meal	35.9
49% soybean oil meal	50.8	Dried skimmed milk	34.0
Meat and bone meal	49.5	Peanut kernels	28.4
Peanut oil meal	47.4	Distillers dried grains	27.4
Tankage with bone	47.1	Brewers dried grains	25.8
Sunflower meal	46.8	Dried whole milk	25.5
44% soybean oil meal	45.8	Ground cottonseed	23.1
Sesame oil meal	44.3	Copra meal	22.7
41% solv-extd cottonseed meal	41.9		

Table 10 (Continued)

GROUP B—Consists for the Most Part of the Legume Hays and Meals, Barley, Oats, Rye, Wheat, and the Wheat By-Products, and Most of the Grain Sorghums

Dehydrated alfalfa leaf meal	20.8%	Rice bran	13.0%
Dehydrated alfalfa meal	17.9	Lespedeza hay	12.7
Oat groats	16.5	Sericea hay	12.6
Cowpea hay	16.0	Dehydrated alfalfa stem meal	12.5
Wheat shorts	15.8	Alsike clover hay	12.4
Wheat bran	15.7	Wheat grain	11.9
Alfalfa hay	15.5	Kafir grain	11.8
Kudzu hay	15.5	Oats grain	11.7
Wheat middlings	15.5	Barley grain	11.6
Dehydrated coastal bermuda-		Rye grain	11.3
grass meal	15.1	Soybean hulls	11.3
Sweet clover hay	15.0	Milo grain	10.9
Dried whey	14.9	Dried bakery product	10.9
Red clover hay	13.2	Hominy feed	10.6
Soybean hay	13.1	Peanut hay	10.6

GROUP C—Made Up Primarily of the Nonlegume Hays; Snapped, Ear, and Shelled Corn; and Certain Grain Sorghums

Bromegrass hay	10.5%	Ground ear corn	8.0%
Reed canarygrass hay	10.2	Meadow hay	7.8
Orchardgrass hay	9.8	Ground snapped corn	7.7
Potato meal	9.7	Oat hay	7.7
Hegari grain	9.6	Redtop hay	7.4
Kentucky bluegrass hay	9.1	Bermudagrass hay	7.2
Coastal bermudagrass hay	9.0	Steamed bone meal	7.1
Dried beet pulp/molasses	9.0	Johnsongrass hay	6.9
Grain sorghum grain	8.9	Beet molasses	6.6
Yellow shelled corn	8.8	Peanut hulls	6.6
Dried beet pulp	8.7	Dallisgrass hay	6.5
Sudangrass hay	8.7	Dried citrus pulp	6.4
Millet hay	8.5	Timothy hay	6.3
Tall fescue hay	8.2	Citrus pulp/molasses, dried	6.2
Corn bran	8.0		

Table 10 (Continued)

GROUP D—For the Most Part High Moisture Feeds, Mineral Feeds, Certain Low-Quality Roughages, the Pure Fats, Corn Starch, and the Molasseses

Fresh bromegrass forage	6.1%	Fresh skimmed milk	3.3%
Prairie hay	5.8	Wheat straw	3.2
Citrus molasses	5.7	Fresh crimson clover forage	3.0
Corn stover	5.7	Fresh Dallisgrass forage	3.0
Fresh alfalfa forage	5.7	Grain sorghum silage	3.0
Wet brewers grains	5.5	Fresh Sudangrass forage	2.9
Fresh white clover forage	5.0	Ground corn cob	2.8
Sweet potato meal	4.9	Rice hulls	2.8
Cane molasses	4.5	Corn stover silage	2.6
Fresh Kentucky bluegrass		Corn (fodder) silage	2.2
forage	4.5	White potatoes	2.2
Fresh ladino clover forage	4.5	Cabbage heads	2.0
Bahiagrass hay	4.3	Sorgo silage	1.9
Fresh coastal bermudagrass	4.3	Sweet potatoes	1.7
Fresh rye forage	4.3	Carrots	1.3
Fresh red clover forage	4.2	Wet beet pulp	1.3
Restaurant garbage	4.2	Fresh citrus pulp	1.2
Cottonseed hulls	4.0	Turnips	1.1
Fresh ryegrass forage	3.9	Oyster shell flour	1.0
Corn ear silage	3.8	Corn starch	0.6
Fresh orchardgrass forage	3.8	Wood molasses	0.6
Oat straw	3.8	Animal fat, feed grade	***
Barley straw	3.6	Corn oil	***
Oat hulls	3.6	Defluorinated phosphate	***
Fresh cow's milk	3.5	Ground limestone	***
Fresh fescue forage	3.5		

***Very little, if any.

12 Mineral Nutrition— General

I. Needs of animals for minerals.

   A.   In tissue growth and repair.

   1.   Bones and teeth are high in mineral content.

COMPOSITION OF FRESH BONE

Water	45%	Ca	36%
Ash	25%	P	17%
Protein	20%	Mg	0.8%
Fat	10%		

   2.   Some in hair, hoofs, and horns.

   3.   Some in soft tissues.

4.   Some in blood cells.

B.   **As body regulators or for producing body regulators.**

1.   In regulating body processes, minerals function in different forms as listed below and as discussed subsequently.

a.   In the ionic form.

b.   In the molecular form.

c.   As components of vitamins.

d.   In the formation of enzymes, hormones, etc.

2.   Minerals and mineral-containing enzymes, hormones, and vitamins function in the body to regulate:

a.   Various metabolic cycles.

b.   Molecular concentration—make body fluids physiologically compatible with the tissues.

c.   Acid-base balance—help maintain pH of body fluids at about 7.0.

d.   Nerve irritability.

e.   Muscle stimulation and activity.

C.   **In milk production.**

1.   Cow's milk contains 5.8% ash or mineral matter on a dry basis.

2.   Milk normally contains significant amounts of all of the essential minerals—except iron.

3.   The minerals in milk must come either directly or indirectly from the feed and/or water which the animal consumes.

II. **Mineral composition of animal body.** (Average of analyses of 18 steers of varying ages less content of the digestive tract*)

Calcium	1.33%		
Phosphorus	0.74%		
Potassium	0.19%		49% Ca
Sodium	0.16%	2.73%	27% P
Sulfur	0.15%		24% Other
Chlorine	0.11%		
Magnesium	0.04%		
Iron	0.01%		

*Hogan, Albert G., and John L. Nierman. *Studies in Animal Nutrition—VI The Distribution of Mineral Elements in the Animal Body as Influenced by Age and Condition.* Missouri Agricultural Experiment Station Research Bulletin 107. 1927.

Cobalt
Copper
Fluorine
Iodine   Present
Manganese   in
Molybdenum   trace
Selenium   amounts
Zinc   only
Others

13 The Macro Minerals

I. Sodium and chlorine.

    A. Functions.

        1. Formation of digestive juices.

        2. Control of body fluid concentration.

        3. Control of body fluid pH.

        4. Nerve and muscle activity.

    B. Requirements.

        1. Specific requirements for sodium and chlorine have not been worked out for most livestock classes.

        2. For young pigs weighing 13-35 kg, the requirement would

appear to be:

Sodium = 0.08 to 0.10% of the ration

Chlorine = 0.12 to 0.13% of the ration

3. For livestock in general, the requirement for salt is considered to be:

0.25 to 0.50% of the ration

or

0.005 to 0.010% of the body weight daily

4. Requirements of different livestock classes will vary depending on:
    a. Class of animal.
    b. Type of feed fed.
    c. Activity of animal.
    d. Production of animal.

C. Deficiency.

1. Rations consisting of farm-produced feeds are usually deficient in these two minerals, and animals not receiving supplemental salt will tend to develop sodium and/or chlorine deficiency.
    a. Such a deficiency is slow to develop. Salt is reused or recycled (not excreted) by animals on low salt intake. Development of a deficiency may take several weeks.
    b. Even with the development of a sodium and/or chlorine deficiency, under ordinary feedlot conditions, there are no specific deficiency symptoms—just unthrifty appearance and impaired performance.

2. With heavily perspiring animals, such as hard-working horses, on low salt intake, an acute salt deficiency may develop resulting in disrupted nerve and muscle function and possible nervous prostration.

D. Supplementation. Farm livestock should be provided with supplemental salt under almost all circumstances. This may be accomplished in any one of several ways.

1. **As block salt.**
    a. **Types of block salt.**
        - *Plain*—contains only NaCl.
        - *Yellow*—contains NaCl plus sulfur.
        - *Red, brown, or purple*—usually contains mostly NaCl plus the critical trace minerals.

      b.   **Advantages of block salt.**
- Easy to provide—no protection required.
- Stimulates salivation.
- No danger of overconsumption.

      c.   **Disadvantages of block salt.**
- Animal sometimes has difficulty obtaining sufficient salt.

2.   **As loose salt.**

      a.   **Types of loose salt.**
- Plain.
- Trace-mineralized.

      b.   **Advantages of loose salt.**
- Easy for animal to consume.

      c.   **Disadvantages of loose salt.**
- Must have protected mineral box.
- Must have adequate water available with salt-hungry animals to avoid possible deaths.

3.   **As part of a mineral mix.** From 20% to 50% of either plain or trace-mineralized salt may be included in an overall mineral mix.

      a.   **Advantages of a mineral mix.**
- Easy for animal to consume adequate amount of salt.
- Induces animal to consume other less palatable minerals.

      b.   **Disadvantages of a mineral mix.**
- Must have protected mineral box.
- Must have adequate water available with salt-hungry animals to avoid possible deaths.
- Forces animals to consume minerals they may not need.

4.   **As a component of the overall ration mix.**

      a.   Usually added at 0.25-0.5% of ration.

      b.   Ensures adequate salt consumption.

      c.   It may improve ration palatability.

      d.   Free choice salt may or may not be provided in addition.

      e.   High levels of salt in the ration may kill an animal if plenty of water is not readily available.

II.   **Calcium.**

    A.   **Functions.**

        1.   Bone and teeth formation—99% of body calcium in the bones and teeth.

        2.   Nerve and muscle function.

        3.   Acid base balance.

        4.   Milk production—also egg production.

    B.   **Requirements.**

| | | % OF AIR-DRY RATION | | |
	Beef cattle	Dairy cattle	Horses	Sheep	Swine
Growing and fattening	0.54-0.20%	0.31%	0.74-0.26%	0.23-0.18%	0.65-0.50%
Pregnant females	0.15	0.31	0.34	0.23	0.75
Lactating females	0.26	0.39-0.48	0.43	0.28	0.75

    C.   **Deficiency.**

        1.   Calcium may be deficient if a good feed source is not provided. Some good feed sources are:

Legume roughages	Meal and bone meal
Grass roughages from calcium rich soils	Fish meal
Tankage	Milk
Tankage with bone	Skimmed milk
Poultry by-product meal	Citrus pulp
Meat scrap	Citrus molasses

Table 11
FEEDS LISTED AND GROUPED ACCORDING TO THEIR PERCENTAGE OF CALCIUM—AS FED BASIS

GROUP A—The Ca and the Ca + P Supplements

Oyster shell flour	37.95%	Defluorinated phosphate	33.00%
Ground limestone	35.85%	Steamed bone meal	30.92

GROUP B—The Bone-Containing High Protein Feeds

Tankage with bone	11.47%	Digester tankage	6.43%
Meat and bone meal	11.42	Fish meal	5.14
Meat scrap	8.49	Poultry by-product meal	3.00

Table 11 (Continued)

GROUP C—Primarily Dry Legume Roughages, Dried Milk Products, Dried Citrus Pulp, and the Molasseses (Except Beet)

Kudzu hay	2.15%	Dried skimmed milk	1.27%
Dried citrus pulp	2.00	Citrus molasses	1.20
Sesame oil meal	1.99	Peanut hay	1.16
Citrus pulp/molasses, dried	1.66	Alsike clover hay	1.13
Dehydrated alfalfa leaf meal	1.64	Lespedeza hay	1.09
Wood molasses	1.45	Soybean hay	1.08
Cowpea hay	1.37	Dehydrated alfalfa stem meal	1.00
Dehydrated alfalfa meal	1.36	Sericea hay	0.94
Sweetclover hay	1.31	Dried whey	0.91
Red clover hay	1.30	Dried whole milk	0.90
Alfalfa hay	1.29	Cane molasses	0.81

GROUP D—Predominantly the Nonlegume Hays, Fresh Legume Forages, Certain Air-Dry Low-Quality Roughages, and Some of the High Protein Feeds

Johnsongrass hay	0.80%	Bromegrass hay	0.32%
Dried beet pulp	0.68	Prairie hay	0.32
Dried beet pulp/molasses	0.56	Safflower meal without hulls	0.32
Meadow hay	0.55	44% soybean oil meal	0.32
Soybean hulls	0.54	Barley straw	0.31
Sunflower meal	0.53	Coastal bermudagrass hay	0.31
Corn stover	0.50	Blood meal	0.30
Bahiagrass hay	0.45	Dehydrated coastal bermuda-	
Fresh alfalfa forage	0.44	grass meal	0.29
Tall fescue hay	0.43	49% soybean oil meal	0.29
Feather meal	0.42	Millet hay	0.28
Fresh red clover forage	0.41	Orchardgrass hay	0.28
Fresh citrus pulp	0.40	Brewers dried grains	0.27
Kentucky bluegrass hay	0.40	Fresh white clover forage	0.25
Dallisgrass hay	0.39	Peanut hulls	0.25
Linseed meal	0.39	Fresh crimson clover forage	0.24
Redtop hay	0.39	Oat straw	0.24
Bermudagrass hay	0.37	Soybean seed	0.24
Reed canarygrass hay	0.37	36% cottonseed meal	0.23
Sudangrass hay	0.36	Fresh ladino clover forage	0.23
Timothy hay	0.36	Oat hay	0.22

Table 11 (Continued)

GROUP E–A Combination of Fresh Nonlegume Forages, the Silages, Wet By-Products, the Root Crops, the Fresh Milks, the Feed Grains and Grain By-Products, Certain Air-Dry Low-Quality Roughages, the Pure Fats, Corn Starch, and Certain Other Miscellaneous Products

Peanut oil meal	0.20%	Sorgo silage	0.09%
41% cottonseed meal	0.19	Wheat shorts	0.09
Copra meal	0.17	Corn (fodder) silage	0.08
Fresh bromegrass forage	0.17	Grain sorghum silage	0.08
Distillers dried grains	0.16	Oat groats	0.08
41% solv-extd cottonseed meal	0.16	Rice hulls	0.08
Fresh Kentucky bluegrass		Wheat grain	0.08
forage	0.15	Wheat middlings	0.08
Fresh ryegrass forage	0.15	Barley grain	0.07
Oat hulls	0.15	Potato meal	0.07
Sweet potato meal	0.15	Rice bran	0.07
Corn gluten meal	0.14	Rye grain	0.07
Fresh coastal bermudagrass	0.14	Wet brewers grains	0.07
Fresh Dallisgrass forage	0.14	Cabbage heads	0.06
Ground cottonseed	0.14	Dried bakery product	0.06
Wheat bran	0.14	Ground snapped corn	0.06
Wheat straw	0.14	Peanut kernels	0.06
Corn stover silage	0.13	Hominy feed	0.05
Cottonseed hulls	0.13	Turnips	0.05
Fresh orchardgrass forage	0.13	Carrots	0.04
Fresh skimmed milk	0.13	Ground ear corn	0.04
Beet molasses	0.12	Kafir grain	0.04
Fresh cow's milk	0.12	Corn bran	0.03
Fresh fescue forage	0.12	Corn ear silage	0.03
Fresh rye forage	0.11	Grain sorghum grain	0.03
Ground corn cob	0.11	Milo grain	0.03
Restaurant garbage	0.11	Sweet potatoes	0.03
Fresh Sudangrass forage	0.10	Yellow shelled corn	0.03
Hegari grain	0.10	White potatoes	0.01
Wet beet pulp	0.10	Corn starch	***
Oats grain	0.09	Animal fat, feed grade	0.00
Sorgo silage	0.09	Corn oil	0.00

***Very little, if any.

2.   Deficiency symptoms are:

   a.   Rickets in young animals. Joints become enlarged. Bones
        become soft and deformed. Condition may be corrected
   in early stages with calcium feeding.

   b.   Osteomalacia or osteoporosis in older animals. Bones
        become porous and weak. Condition may be corrected
   with calcium feeding if bone does not break. Rear-quarter
   paralysis in swine sometimes due to a calcium deficiency
   resulting in a crushed vertebra and pinched spinal cord.

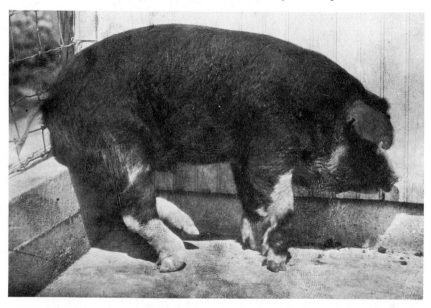

FIGURE 16.   *Calcium deficiency.* Note the abnormal bone
development and the rachitic condition in advanced stage of
deficiency. Lack of calcium retards normal skeletal development but
does not usually depress total gain. *(Courtesy of N. R. Ellis, U.S.
Department of Agriculture)*

D.   **Supplementation.**

   1.   The need for supplementation will depend on the ration. If
        needed, it may be accomplished by using:

      a.   **Calcium-only supplements.**
         • Ground limestone.
         • Oystershell flour.
         • Marble dust.
   All three are primarily $CaCO_3$, containing 33-40% calcium.
   Pure $CaCO_3$ is 40% calcium. All are about equal in nutritive

value. Use whichever is cheapest. The cheapest will probably be the one produced nearest by since transportation is a major factor in determining selling price.

    b.   **Calcium with phosphorus.**
- Steamed bone meal.
- Defluorinated phosphates.

These usually contain around 30% calcium along with 14% to 20% phosphorus. They are usually regarded about equal in feeding value per unit of calcium and phosphorus. They are usually more expensive than the calcium-only supplements—hence, they are used as calcium sources only when phosphorus is also needed.

   2.   Any of the above calcium sources may be either included in a mineral mix provided free choice or added to the ration at the level required.

III. Phosphorus.

   A.   **Functions.**

     1.   Bone and teeth formation—about 80% of body phosphorus in the bones and teeth.

     2.   As a component of protein in the soft tissues.

     3.   Milk production—also egg production.

     4.   In various metabolic processes.

   B.   **Requirements.**

	% OF AIR-DRY RATION				
	Beef cattle	Dairy cattle	Horses	Sheep	Swine
Growing and fattening	0.39-0.20%	0.23%	0.46-0.22%	0.21-0.16%	0.50-0.40%
Pregnant females	0.15	0.23	0.26	0.18	0.50
Lactating females	0.21	0.30-0.35	0.35	0.21	0.50

   C.   **Deficiency.**

     1.   Phosphorus may be deficient if one or more of the following good feed sources of phosphorus are not in the ration.

Wheat bran and middlings   Whole or skimmed milk
All high protein feeds   Grains in general are fair sources
All bone-containing feeds

Table 12
FEEDS LISTED AND GROUPED ACCORDING TO THEIR PERCENTAGE OF PHOSPHORUS—AS FED BASIS

GROUP A—The Ca + P Supplements

Defluorinated phosphate	18.00%	Steamed bone meal	14.01%

GROUP B—The Bone-Containing High Protein Feeds

Meat and bone meal	5.69%	Digester tankage	3.39%
Tankage with bone	5.25	Fish meal	2.91
Meat scrap	4.18	Poultry by-product meal	1.70

GROUP C—Mostly Nonbone-Containing High Protein Feeds and the Wheat By-Products, Along with a Limited Number of Miscellaneous Concentrates

Rice bran	1.59%	Peanut oil meal	0.65%
Sesame oil meal	1.33	49% soybean oil meal	0.65
Wheat bran	1.16	Copra meal	0.59
41% cottonseed meal	1.09	Safflower meal without hulls	0.59
Distillers dried grains	1.06	Soybean seed	0.58
41% solv-extd cottonseed meal	1.06	Hominy feed	0.54
Dried skimmed milk	1.03	Feather meal	0.51
36% cottonseed meal	0.92	Sunflower meal	0.50
Linseed meal	0.87	Wheat middlings	0.50
Wheat shorts	0.81	Brewers dried grains	0.48
Dried whey	0.76	Dried bakery product	0.47
Dried whole milk	0.72	Corn gluten meal	0.46
Ground cottonseed	0.68	Oat groats	0.45
44% soybean oil meal	0.67	Peanut kernels	0.43

GROUP D—Predominantly the Hays and the Feed Grains, Along with Just a Few Other Miscellaneous Products

Barley grain	0.40%	Johnsongrass hay	0.27%
Cowpea hay	0.34	Kentucky bluegrass hay	0.27
Ground snapped corn	0.34	Sudangrass hay	0.27
Rye grain	0.34	Yellow shelled corn	0.27
Wheat grain	0.34	Lespedeza hay	0.24
Kafir grain	0.33	Sweetclover hay	0.24
Oats grain	0.33	Alsike clover hay	0.23
Kudzu hay	0.32	Blood meal	0.23
Tall fescue hay	0.31	Dehydrated alfalfa leaf meal	0.23
Grain sorghum grain	0.29	Dehydrated coastal bermuda-	
Soybean hay	0.29	grass meal	0.23
Dehydrated alfalfa meal	0.28	Ground ear corn	0.23
Milo grain	0.28	Orchardgrass hay	0.23
Hegari grain	0.27	Reed canarygrass hay	0.23

Table 12 (Continued)

Sericea hay	0.22%	Corn bran	0.16%
Peanut hay	0.21	Millet hay	0.16
Bahiagrass hay	0.20	Soybean hulls	0.16
Oat hay	0.20	Dallisgrass hay	0.15
Potato meal	0.20	Meadow hay	0.15
Red clover hay	0.20	Timothy hay	0.15
Redtop hay	0.20	Coastal bermudagrass hay	0.14
Alfalfa hay	0.19	Dried citrus pulp	0.14
Bermudagrass hay	0.19	Sweet potato meal	0.14
Bromegrass hay	0.17		

GROUP E—For the Most Part High Moisture Feeds, Low-Quality Air-Dry Roughages, the Ca-Only Supplements, the Pure Fats, and Corn Starch, Plus a Very Limited Number of Other Materials

Fresh bromegrass forage	0.13%	Restaurant garbage	0.07%
Citrus molasses	0.12	Rice hulls	0.07
Corn ear silage	0.12	Wheat straw	0.07
Fresh Kentucky bluegrass		Corn stover silage	0.07
forage	0.12	Corn (fodder) silage	0.06
Fresh orchardgrass forage	0.12	Cottonseed hulls	0.06
Prairie hay	0.12	Fresh red clover forage	0.06
Wet brewers grains	0.12	Grain sorghum silage	0.06
Citrus pulp/molasses, dried	0.11	Peanut hulls	0.06
Fresh ryegrass forage	0.10	Sorgo silage	0.06
Fresh skimmed milk	0.10	Fresh crimson clover forage	0.05
Oat hulls	0.10	Fresh Dallisgrass forage	0.05
Barley straw	0.09	Sweet potatoes	0.05
Dried beet pulp	0.09	White potatoes	0.05
Fresh cow's milk	0.09	Carrots	0.04
Fresh rye forage	0.09	Ground corn cob	0.04
Fresh Sudangrass forage	0.09	Beet molasses	0.03
Fresh white clover forage	0.09	Cabbage heads	0.03
Oat straw	0.09	Turnips	0.03
Cane molasses	0.08	Wood molasses	0.03
Corn stover	0.08	Dehydrated alfalfa stem meal	0.02
Dried beet pulp/molasses	0.08	Fresh citrus pulp	0.02
Fresh coastal bermudagrass	0.08	Ground limestone	0.02
Fresh fescue forage	0.08	Wet beet pulp	0.01
Fresh ladino clover forage	0.08	Corn starch	***
Fresh alfalfa forage	0.07	Animal fat, feed grade	0.00
Oyster shell flour	0.07	Corn oil	0.00

***Very little, if any.

2. Symptoms of a phosphorus deficiency are:

   a. Rickets in young animals similar to that of a calcium deficiency. May be corrected in the early stage with phosphorus feeding if due to phosphorus shortage.

   b. Osteomalacia or osteoporosis in older animals similar to calcium deficiency. Seldom results in broken bones. May be corrected with phosphorus feeding if due to phosphorus shortage.

   c. Phosphorus-deficient animals usually show poor appetite, slow gain, lowered milk production, low blood phosphorus, and general unthriftiness. Animals will eat soil and chew on nonfeed objects, but this is not specific for phosphorus deficiency.

FIGURE 17.   *Phosphorus deficiency.* On the left is a typical phosphorus-deficient pig in an advanced stage of deficiency. Leg bones are weak and crooked. The pig on the right received the same ration as the one on the left, except that the ration was adequate in available phosphorus. *(Courtesy of M. P. Plumlee and W. M. Beeson, Purdue University)*

D. **Supplementation.**

   1. The need for supplements will depend on the ration but may be accomplished by using either of the following:

      a. **Steamed bone meal.**
         • Around 14% phosphorus.
         • Excellent as source of supplemental phosphorus.
         • Frequently in short supply.

(a)   (b)

FIGURE 18.   *Phosphorus deficiency in feedlot cattle.* (a) This
steer was fed a ration consisting of wet beet pulp, alfalfa hay, and
beet molasses containing 0.12 percent of phosphorus. (b) This steer
received the same ration plus 0.1 lb of steamed bone meal daily,
which brought the phosphorus content up to 0.18 percent and
provided an average of 17 g of phosphorus daily. *(Courtesy of W. M.
Beeson; taken at Idaho Agricultural Experiment Station)*

    b.   **Defluorinated phosphates.**
- Are of several types.
- Vary in phosphorus content from 14% to 20%.
- Most natural deposits of phosphate contain fluorine at levels that would be toxic if used in their natural state as a primary source of phosphorus. Hence, most must be defluorinated before being fed to livestock.
- Mineral supplements sold in most states shall not legally contain fluorine in excess of:

      0.30% for cattle   0.45% for swine
      0.35% for sheep   0.60% for poultry

- AAFC officials recommend not more than 1 part of fluorine per 100 parts of phosphorus.
- Most phosphorus supplements offered for sale by reputable companies are satisfactory sources of phosphorus from the standpoint of fluorine content, phosphorus content, and phosphorus availability.
- Such supplements are usually in adequate supply.

  2.   Either of the above phosphorus sources may be either included in a mineral mix for free choice or *ad libitum* feeding or added directly to the ration at the level needed.

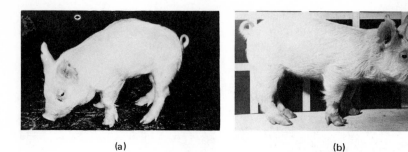

(a)   (b)

(c)

FIGURE 19. *(a) Calcium, (b) phosphorus, and (c) vitamin D deficiencies.* Note the similarity of deficiency symptoms in these 5-week-old pigs. *(Courtesy of E. R. Miller and D. E. Ullrey, Michigan State University)*

IV.   **Magnesium.**

   A.   **Functions.**

   1.   Necessary for many enzyme systems.

   2.   Plays a role in carbohydrate metabolism.

   3.   Necessary for the proper functioning of the nervous system.

   B.   **Requirements.**

   1.   Requirements under normal conditions seem to be:

   For cattle and sheep = 12-30 mg/kg body wt/day

   or

   = 600-1000 mg/kg feed

   For young pigs = 300 mg/kg feed

   2.   Blood serum normally contains about 2.5 mg of magnesium per 100 ml.

3. Most rations under most conditions contain more than adequate levels of magnesium to meet the requirements—that is, more than 0.06% magnesium for cattle and sheep and 0.03% for young pigs.

C. **Deficiency.**

1. Cattle and sheep grazing on certain growing plants under certain conditions such as prevail during the late winter and early spring because of either a low magnesium level in the plant or the presence of some substance in the plant which renders the magnesium unavailable tend to develop a *hypomagnesemia* associated with severe toxic symptoms and frequently death.

2. The condition is commonly referred to as *grass tetany*—also as *grass staggers* and *wheat poisoning.*

3. In such instances, blood magnesium levels usually fall below 1 mg/100 ml with the development of *hypomagnesemic tetany*, a hyperirritability of the neuromuscular system producing hyperexcitability, incoordination, and frequently death.

4. The condition is most common with cattle and sheep grazing on small grains, highly fertilized fescue, and certain other green growing crops in the late winter and early spring. It is most common with older lactating animals.

D. **Supplementation.**

1. Most cattle and sheep feeds contain adequate levels of

FIGURE 20. *Magnesium deficiency.* Five-week-old pig showing stepping syndrome; the pig keeps stepping almost continuously while standing. Weakness of pasterns is apparent. *(Courtesy of E. R. Miller and D. E. Ullrey, Michigan State University)*

magnesium to meet the animal's requirement under most conditions.

2.   Cattle and sheep grazed on small grains and certain other crops during the late winter or early spring probably should be supplied with supplemental sources of magnesium.

3.   Under such conditions it is recommended that cattle be provided with 1+ oz of supplemental magnesium per head daily. Presumably proportionately lesser amounts would suffice for sheep.

4.   While both $MgSO_4$ and MgO have been used for this purpose, the latter is less purgative and probably to to be preferred for feeding purposes.

5.   Consumption of supplemental magnesium may be induced by mixing it with an animal's salt or with a small amount of supplemental feed.

6.   A mixture containing 2 parts MgO to 1 part salt used as the only source of salt seems to be a practical method of magnesium administration.

7.   A mixture of 30% magnesium oxide, 32% defluorinated phosphate, 30% trace-mineralized salt, and 8% cottonseed meal fed free choice has also been used with good results for minimizing the incidence of grass tetany.

V.   **Potassium.**

A.   **Functions.**

1.   Potassium is required by livestock for a variety of body functions, such as osmotic relations, acid-base balance, and digestion.

B.   **Requirements.**

1.   The requirements for potassium have not been critically assessed for most classes of livestock.

2.   A level of 0.6% to 0.8% of the ration dry matter has been reported as optimum for growing-finishing steers.

3.   The potassium requirement of the baby pig has been reported to be 0.26% of the diet.

C.   **Deficiency.**

1.   A potassium deficiency is most unlikely under ordinary conditions.

2.   Symptoms of a potassium deficiency are rather nonspecific,

such as decreased feed consumption, lowered feed efficiency, slow growth, stiffness, and emaciation.

D. **Supplementation.**

1. Potassium supplementation should not be needed with most practical type livestock rations.

2. However, grains are, in general, much lower in potassium than are roughages, and while high-grain rations will usually meet the potassium requirement of hogs, such rations may be marginal or even deficient in potassium for cattle and sheep. However, no general recommendations have as yet been developed for providing cattle and/or sheep on high- or all-grain rations with supplemental sources of potassium.

VI. **Sulfur.**

A. **Functions.**

1. In the synthesis of sulfur-containing amino acids in the rumen.

2. In the formation of various body compounds.

B. **Requirements.**

1. The sulfur requirement of cattle and sheep appears to be around 0.1% of ration dry matter.

2. In nonruminants, sulfur, at least for the most part, should be in the form of sulfur-containing proteins. In ruminants and also probably horses, it may be supplied as protein, as elemental sulfur, or as sulfate sulfur.

C. **Deficiency.**

1. Most livestock rations provide more than the required level of sulfur—hence, a deficiency is seldom experienced under ordinary conditions.

2. A deficiency of sulfur will express itself as a protein deficiency—a general unthrifty condition and poor performance.

D. **Supplementation.**

1. Livestock seldom benefit from sulfur supplementation.

2. Heavy wool producing sheep being fed on mature grass hay or other low sulfur feeds may give a positive response to sulfur supplementation.

3. Animals fed urea as a source of protein nitrogen may benefit from supplemental sulfur.

4.   Supplemental sulfur may be provided as either elemental
       sulfur or in the sulfate form at levels not to exceed about
0.05% of the ration dry matter.

14 The Micro Minerals

I. Iron.

    A.   **Functions.**

        1.   Necessary for hemoglobin formation.

    B.   **Requirements.**

        1.   The precise minimum requirements of various classes of livestock have not been determined.

        2.   As little as 80 mg of iron per kg of diet has proved to be more than adequate for most animals.

    C.   **Deficiency.**

        1.   Most livestock feeds contain more than 80 mg of iron per kg—hence, most livestock rations are more than adequate in

iron content, and an iron deficiency seldom occurs with older animals.

2.   Milk is low in iron content—especially sow's milk (20-25 mg iron per kg milk dry matter).

3.   Newborn animals normally have sufficient iron reserves in the liver and spleen to carry them through the early nursing period.

4.   However, newborn pigs kept on concrete during early life frequently do not possess sufficient iron reserves and will develop an iron deficiency.

5.   An iron deficiency in the young pig is characterized by:

Low blood hemoglobin   Pale eyelids, ears, and nose
Labored breathing ("The Thumps")   Flabby, wrinkled skin
Listlessness   Edema of head and shoulders

D.   **Supplementation.**

1.   Young pigs raised on concrete are usually supplied with supplemental iron in the form of:

a.   Concentrated ferrous sulfate or other iron solution administered orally a few drops daily during the first 3-4 weeks or as a weekly drench of 1/3 to 1 teaspoonful.

b.   Iron dextran injection:

100 mg at 3 days of age
50 mg at 21 days of age

2.   Many livestock producers provide their animals with supplemental iron as a safety measure against a possible deficiency through the use of trace-mineralized salt, which ordinarily contains iron.

II.   **Iodine.**

A.   **Functions.**

1.   In the production of thyroxin by the thyroid gland.

B.   **Requirements.**

1.   A level of 0.2 mg per kg of air-dry diet is considered to be more than adequate for most classes of livestock.

2.   The requirement per kg body weight daily has been estimated to be:

Cattle = 1.0 – 2.0 micrograms

Swine = 4.4 micrograms

C. **Deficiency.**

1. The natural feeds and water of most areas of the U.S. contain sufficient iodine to meet livestock needs.

2. Exceptions to this are the Great Lakes region, the far Northwest, and possibly isolated areas in other sections.

3. In iodine deficient areas, iodine deficiency symptoms are most frequently observed in young animals in the form of:

Goiter at birth or soon thereafter	Hairlessness at birth
Dead or weak at birth	Infected navels—especially in foals

D. **Supplementation.**

1. Where there is the slightest reason to suspect a possible iodine deficiency, the supplying of livestock with supplemental iodine is recommended.

2. Supplementation of livestock with iodine is easily accomplished by the feeding of iodized salt—salt which contains 0.01% potassium iodide or 0.0076% iodine. The iodine should be in stabilized form to prevent its loss into the air.

III. **Cobalt.**

A. **Functions.**

1. As a component of the vitamin B_{12} molecule.

2. In the rumen synthesis of vitamin B_{12}.

B. **Requirements.**

1. For cattle and sheep, feed which contains from 0.05 to 0.10 mg of cobalt per kg feed prevents any symptoms of a cobalt deficiency.

2. For horses, the requirement appears to be less than that for ruminants.

3. For hogs, only as vitamin B_{12}.

C. **Deficiency.**

1. Cobalt deficiencies have been noted in livestock grazing the natural forages of Florida, Massachusetts, New Hampshire,

Pennsylvania, New York, Michigan, and Alberta, Canada.

2.   The cobalt content in the leaves of catalpa trees is regarded as a good indicator of the adequacy of cobalt in an area.

3.   Symptoms of a cobalt deficiency are simply those of general malnutrition—poor appetite, unthriftiness, weakness, anemia, decreased fertility, slow growth, and decreased milk and wool production.

D.   **Supplementation.**

1.   Increasing numbers of livestock producers, even outside the cobalt deficient areas, are providing their animals with supplemental cobalt as insurance against a possible cobalt deficiency.

2.   Supplemental cobalt is easily provided to livestock simply by the use of trace-mineralized salt, all brands being fortified with this mineral usually in the form of cobalt chloride, cobalt sulfate, or cobalt carbonate.

3.   A level of 12.5 g of cobalt in any of the above forms per 100 kg of salt would be about right for preparing trace-mineralized salt.

4.   While cobalt in the mineral form is not known to be beneficial to hogs, it is still included in mineral mixes for this class of animals.

IV.   **Copper**

A.   **Functions.**

1.   In iron absorption.

2.   In hemoglobin formation.

3.   In various enzyme systems.

4.   In synthesis of keratin for hair and wool growth.

B.   **Requirements.**

1.   The following levels of copper per kg of diet have been found adequate for:

Cattle and sheep   —   4-5 mg/kg

Swine   —   6 mg/kg

Horses   —   5-8 mg/kg

2.   High levels of molybdenum and/or sulfate may increase the copper requirement two- to threefold.

C.   **Deficiency.**

1.   The copper content of feedstuffs varies considerably but usually is three to four times the requirement—hence, copper deficiencies are seldom experienced in the U.S. outside of Florida and certain other sections of the Southeast. Copper deficiencies are common in Australia.

2.   Copper deficiency symptoms are not too specific and may include any of the following:

Low blood and liver copper   Muscular incoordination
Bleaching of hair in cattle   Weakness at birth
Abnormal wool growth in   Anemia
    sheep
Abnormal bone metabolism

D.   **Supplementation.**

1.   Supplementation of livestock with copper is essential in copper deficient areas.

2.   Supplementation of livestock with copper in other areas as insurance against a possible marginal supply is becoming a common practice.

3.   Supplementation is easily accomplished through the use of trace-mineralized salt containing 0.25% to 0.50% copper sulfate ($CuSO_4 \cdot 5\ H_2O$).

V.   **Fluorine.**

A.   **Functions.**

1.   Reduces incidence of dental caries in humans and possibly other animals.

2.   Possibly retards osteoporosis in adults.

B.   **Requirements.**

1.   About 1 PPM in drinking water or less.

C.   **Deficiency.**

1.   Excesses of fluorine are more of a concern than are deficiencies in livestock production because of its presence at high levels in the drinking water and forages of certain areas; also because of its presence at high (3-4%) levels in most natural phosphate sources.

2.   The only reported symptoms of fluorine deficiency have been noted in children in the form of excessive dental caries.

D.   Supplementation.

   1.   No need has ever been demonstrated for supplementing
        livestock with fluorine since most, if not all, livestock rations
   seem to contain more than an adequate amount of this mineral.

   2.   Should fluorine supplementation ever become necessary, it
        would appear that the addition of 1 PPM to the drinking water
   should suffice.

   3.   No feeding or supplementation practice should be followed
        which would provide more than 30 mg of fluorine per kg of
   diet for breeding cattle (and probably breeding sheep) or over 100
   mg per kg diet for fattening steers and lambs.

VI.   Manganese.

A.   Functions.

   1.   Probably in enzyme systems influencing estrus, ovulation, fetal
        development, udder development, milk production, growth,
   and skeletal development.

B.   Requirements.

Beef cattle and sheep   —   1 to 25 mg per kg of diet

Dairy cattle   —   10 to 20 mg per kg of diet

Swine   —   20 mg per kg of diet

High levels of calcium and phosphorous in the ration may increase
the above requirement levels.

C.   Deficiency.

   1.   Most livestock rations will contain adequate manganese. Most
        roughages contain 40 to 140 mg/kg; grains other than corn,
   15-45 mg/kg; corn, around 5 mg/kg.

   2.   Symptoms which have been shown to be the result of a
        manganese deficiency have been noted in cattle in certain
   areas.

   3.   Manganese deficiency symptoms have been produced in swine
        on a low manganese diet.

   4.   Manganese deficiency symptoms take on the form of:

Delayed estrus	Deformed young
Reduced ovulation	Poor growth
Reduced fertility	Lowered serum alkaline phosphatase

FIGURE 21.   *Manganese deficiency* in a newborn calf; legs are weak and deformed. *(Courtesy of I. A. Dyer, Washington State University)*

Abortions	Lowered tissue manganese
Resorptions	"Knuckling over" in calves

  D.   Supplementation.

   1.   Seldom necessary for beef cattle except on all-concentrate diets based on corn and nonprotein nitrogen supplements—also in certain isolated areas where manganese is unusually low.

   2.   Conventional swine rations based on corn and soybean oil meal may be improved with manganese supplementation.

   3.   Manganese supplementation of livestock is easily accomplished through the use of trace-mineralized salt containing 0.25% manganese (higher levels used for poultry all right for livestock).

VII. Molybdenum.

  A.   Functions.

   1.   As a component of the enzyme xanthine oxidase—especially important to poultry for uric acid formation.

   2.   Stimulates action of rumen organisms.

  B.   Requirements.

   1.   Probably 0.01 mg or less per kg of dry diet for beef cattle and other farm livestock.

2.   Toxicity symptoms have been noted at levels of 5-10 mg per kg of dry diet. Prevented with copper supplementation.

C.   **Deficiency.**

1.   The essentiality of molybdenum has been demonstrated in poultry.

2.   Only evidence of a positive nutritional role of molybdenum with farm livestock has been with lambs, in which it improved growth rate.

D.   **Supplementation.**

1.   Since extremely low levels of molybdenum are sometimes toxic to livestock and since most all normal rations are adequate in this mineral, it is not recommended that farm livestock be supplemented with molybdenum.

VIII. **Selenium.**

A.   **Functions.**

1.   In vitamin E absorption and utilization.

2.   Possibly others—prevents degeneration and fibrosis of the pancreas in chicks.

B.   **Requirements.**

1.   About 0.1 mg or less per kg of dry diet.

2.   Over about 5 mg per kg of dry feed may produce toxic symptoms.

C.   **Deficiency.**

1.   Forages in the following sections of the country are low to very low in selenium and may result in a selenium deficiency:

   Southeastern coastal area   States adjoining the
                                                        Great Lakes
   New England states   Coastal Northwest

2.   Selenium deficiency symptoms are similar to those of vitamin E deficiency:

   Nutritional muscular dystrophy   Poor growth
       in lambs and cattle
   Heart failure   Low fertility
   Paralysis   Liver necrosis

D.   Supplementation.

1.   The general inclusion of supplemental selenium in rations or supplements for livestock has not as yet been authorized by the Food and Drug Administration.

2.   In the January 8, 1974 *Federal Register,* the FDA ruled that selenium in the form of either sodium selenite or sodium selenate could be added at the rate of 0.1 PPM to complete rations for swine, for growing chickens up to 16 weeks of age, for breeder hens producing eggs only for hatching purposes, and for nonfood animals (zoo animals, pets, and horses not intended for human consumption); and at the rate of 0.2 PPM in complete rations for turkeys. The selenium must be added to the final ration in the form of a pre-mix containing not more than 90.3 mg of selenium per lb. In this connection the selenium is regarded as a nutrient and not as a drug.

3.   It is reasonable to assume that the addition of selenium to rations for other classes of livestock and poultry will receive FDA approval as soon as sufficient data have been submitted to substantiate that such use is not a hazard to human health.

IX.   Zinc.

A.   Functions.

1.   Exact functions not understood.

2.   It prevents parakeratosis.

3.   It promotes general thriftiness and growth.

4.   It promotes wound healing.

5.   Related to hair and wool growth and health.

B.   Requirements.

Beef cattle	= 20-30 mg per kg air-dry feed
Sheep	= About the same as for cattle
Hogs	= 50 mg per kg air-dry feed

C.   Deficiency.

1.   Zinc deficiencies seldom occur in cattle and sheep on normal rations.

2.   Zinc deficiencies are frequently experienced in growing and fattening swine being fed on concrete with rations containing recommended levels of calcium. What would otherwise be adequate

FIGURE 22.   *Zinc deficiency.* The pig on the left received 17 ppm of zinc and gained 1.4 kg in 74 days; note the severe dermatosis or parakeratosis. The pig on the right received the same diet as the pig on the left, except that the diet contained 67 ppm of zinc. This pig gained 50.3 kg. *(Courtesy of J. H. Conrad and W. M. Beeson, Purdue University)*

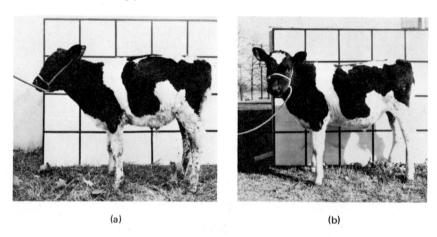

(a)   (b)

FIGURE 23.   (a) Calf showing loss of hair on legs and severe scaliness, cracking, and thickening of the skin as the result of a zinc deficiency. (b) The same calf after receiving supplemental zinc. *(Courtesy of W.J. Miller of the University of Georgia College of Agriculture Experiment Stations)*

zinc is sometimes rendered unavailable and inadequate with an adequacy of calcium.

3.   Deficiency symptoms include:

Parakeratosis	Unhealthy looking hair or
General unthriftiness	wool
Poor growth	Slow wound healing

D.   Supplementation.

1.   Modern-day swine are usually supplemented with about 50 mg of zinc per kg of air-dry feed in the form of $ZnCO_3$ or $ZnSO_4$ added through the supplement or as a component of trace-mineralized salt.

2.   Cattle and sheep rations generally do not require zinc supplementation. However, many are provided with trace-mineralized salt containing zinc as a precautionary measure.

15 General Recommendations for Mineral Feeding

I. When livestock are being fed totally or in part on a mixed feed, those minerals considered necessary or possibly necessary are usually included in the mixture.

A. Salt is included at 0.25-0.50% of the total ration. If the lower level is used, salt may also be provided free choice.

B. Calcium and phosphorus are added as needed to balance the ration—ground limestone or oystershell flour for calcium, bonemeal or defluorinated phosphate for phosphorus, or for calcium and phosphorus.

C. If it is thought that any of the trace minerals might be in short supply, trace-mineralized salt will be used above. It is cheap insurance, not harmful, and good for the soil.

FIGURE 24. Four mineral products widely used in livestock feeding. *(Courtesy of the University of Georgia College of Agriculture Experiment Stations)*

    D. Other minerals will not ordinarily be added unless special circumstances indicate a need for them.

II. When livestock are being fed on unmixed rations or are being run on pasture, minerals may be supplied on a free choice basis in either of two ways:

    A. Cafeteria style. Involves use of a multi-compartment feeder with a separate compartment for each of several different minerals or mineral combinations. This method is based on the theory that an animal will consume as much of each mineral as it requires on a free choice basis. This has never been definitely established. Such feeders are more expensive and are more trouble to keep filled with minerals than are feeders for single mixtures.

    B. Single mixtures. Of two general types:

        1. For livestock primarily on heavy grain feeding:

            1 part salt (usually trace mineralized)

FIGURE 25. Livestock frequently exhibit a craving for some nutritive factor by chewing on wooden gates, etc. *(Courtesy of Ralston Purina Co., St. Louis, Mo.)*

1-2 parts defluorinated phosphate or steamed bone meal
1-2 parts ground limestone or oystershell flour ($CaCO_3$)

Trace-mineralized salt is usually recommended because it is cheap insurance against a possible trace mineral deficiency, no harmful effects have ever been experienced from the proper use of properly formulated trace-mineralized salt, and the use of trace-mineralized salt will tend to maintain the trace-mineral level of the soil.

2.   For livestock primarily on pasture, hay, and/or pasture

1 part salt (usually trace mineralized for same reasons as mentioned above)
1-3 parts defluorinated phosphate or steamed bone meal

Since forages are much more apt to be deficient in phosphorus than in calcium and since the commonly used phosphorus supplements (defluorinated phosphate and steamed bone meal) contain calcium as well as phosphorus, no limestone or oystershell flour (sources of calcium only) is needed in this mix.

16 Vitamin Nutrition

I.   **Definition of vitamin.** Vitamins are organic substances required by animals in very small amounts for regulating various body processes toward normal health, growth, production, and reproduction.

II. **Brief history of vitamin nutrition.** The existence of nutritive factors, such as vitamins, was not recognized until about the start of the 20th century. There was no such word as "vitamin" prior to that time. The existence of vitaminlike factors was first recognized in the Orient when prisoners fed on unpolished rice seemed to be freer of beri beri than were nonprisoners consuming polished rice. It was theorized that there was a nutritive factor in rice polishings which prevented this disease. Tests with poultry confirmed this theory. It was subsequently determined that the factor was water soluble and was an amine (nitrogen-containing). Since the factor seemed to be essential for life, and apparently contained nitrogen, it was named *vitamine*.

Soon thereafter, workers in the United States recognized the presence in the butter fat of milk of a factor which prevented nightblindness in calves. However, the factor seemed to be different from that in rice polishings since the milk fat factor was fat soluble rather than water soluble and also did not contain nitrogen. As a result it was concluded that there were two factors, both of which were essential for life. However, since only one contained nitrogen, it was agreed that such nutrients should be referred to as *vitamins* rather than *vitamines*—also that the anti-nightblindness factor should be called vitamin A and the anti-beri beri factor vitamin B.

Subsequently, the observation was made that there must be a third such factor since there seemed to be something in fresh fruits and vegetables that prevented scurvy in man. It was believed to be different from either of those discovered earlier since it was water soluble but did not contain nitrogen. Consequently, it was designated as vitamin C.

Soon thereafter, it was proposed that vitamin B, which was initially thought to be a single factor, consisted of two different factors, both of which were water soluble amines. The two factors were differentiated by calling one vitamin B_1 and the other vitamin B_2. As vitamin research was continued, it was determined that instead of consisting of two factors, vitamin B actually consisted of several factors, the total number of which has risen to 11 to date. This overall group has come to be known as the *B complex*. The different members of this group are referred to as vitamins B_1, B_2, etc., or in some instances by their actual chemical names.

A total of 16 different vitamins which function in animal nutrition has been discovered to date. In addition to those already mentioned, there are vitamins D, E, and K. The last vitamin to be discovered was vitamin B_{12}, the existence of which was demonstrated in the late 1940s. It is believed that most, if not all, of the vitamins which function in animal nutrition have been identified as of this time.

III. Vitamin designations. Initially, in the absence of a more specific identification, the different vitamins were distinguished from each other by the use of letters of the alphabet such as A, B, C, etc. As pointed out above, it subsequently became necessary to use subscript numbers with these letters in order to distinguish between different vitamins of a particular letter group or to distinguish between different forms of a particular vitamin.

In many instances, vitamins have also been referred to on the basis of the function which they performed or the symptoms which they prevented. As information concerning the chemical makeup of the different vitamins has been developed, an increasing number of the vitamins are being referred to by their actual chemical names, and today most of the vitamins are referred to in this manner.

IV. Chemical formula. In dealing with the different vitamins, it is sometimes helpful to have information concerning their elemental composition and molecular structure. In the early stages of the vitamin story such information was quite limited. However, today, as the result of the diligent efforts of chemists, not only is the elemental composition of each vitamin known but also the structural formula of each vitamin molecule.

As may be noted from the empirical formulas for the different vitamins shown in Table 13, all of the vitamins contain carbon, hydrogen and oxygen. In addition, all of the B vitamins except one (inositol) also contain nitrogen. Certain of the B vitamins also contain one or more of the mineral elements in their molecular structure.

V.   Solubility. Most organic materials are soluble either in water or in fats and fat solvents, or else they are not soluble in either. Very few are soluble in both water, on the one hand, and fat and fat solvents, on the other.

All of the B vitamins and vitamin C are soluble in water and so are said to be *water soluble*. Vitamins A, D, E and K, on the other hand, are soluble in fats and fat solvents and so are said to be *fat soluble*. Many phenomena of vitamin nutrition are related to a vitamin's solubility. Consequently, it behooves every student of animal nutrition to be constantly aware of the solubility differences in vitamins and to make use of such differences whenever possible in developing practices, processes, or procedures.

VI.   Color. Not too much of significance from the standpoint of vitamin nutrition is related to vitamin color. In some instances, however, the presence or absence of a particular vitamin may be indicated by the color of the material in question. Several of the vitamins, however, are colorless and consequently would not lend themselves to any such evaluations.

VII.   Rumen synthesis. Under ordinary circumstances most, if not all, of the B vitamins are synthesized in the rumen in amounts which will more than meet the requirements for these vitamins by ruminant animals. These vitamins are formed in the rumen as metabolic by-products of microbial fermentations. Different strains of microorganisms produce different metabolic by-products. The metabolic by-products of one strain contribute to the nutrition of other strains, and vice versa, in a complicated symbiotic relationship. In turn, the combined metabolic by-products of the various strains of rumen microorganisms represent a significant contribution toward the nutritive needs of the ruminant animal.

VIII.   Body reserves. An animal tends to store reserves of certain vitamins in its body so that a daily intake is not required. This is more true for the fat soluble vitamins than for the water soluble vitamins. The large amount of water which passes through most animals daily tends to carry out and thereby deplete the water soluble vitamins of the body. Fat soluble vitamins, on the other hand, are more inclined to remain in the body. This is especially true of vitamin A and/or carotene which may be stored by an animal in its liver and fatty tissue in sufficient quantities to meet its requirements for vitamin A for periods of up to 6 months or even longer.

IX.   Commercial synthesis. Initially vitamin nutrition was dependent on the feeding of feeds and products which were known to have a high natural content of the respective vitamins. Over the past 30 years, however, industry has developed methods for the laboratory synthesis of the various vitamins. As a result, all of the different vitamins are today obtainable in pure crystalline forms at prices which make them economical for use in the vitamin supplementation of livestock.

X.   Basic functions and deficiency symptoms. Each of the various vitamins performs one or more basic functions in the regulation of various metabolic processes within the body. As a deficiency of a certain vitamin develops, there usually appears a certain set of deficiency symptoms. Originally it was thought that vitamins functioned simply to prevent these symptoms. However, it is realized today that deficiency symptoms are simply outward manifestations of disturbances of basic metabolic processes which occur as the result of a vitamin shortage. The nature of these functions and the symptoms are covered in Table 13.

17 Vitamins in Livestock Feeding

I.   **Vitamins in ruminant feeding.**

    A.   While all of the different vitamins are apparently metabolic essentials for all of the various classes of farm livestock, under most conditions only vitamin A needs to be given attention from the standpoint of meeting the dietary needs of ruminants.

        Under ordinary circumstances all of the various B vitamins are apparently synthesized in the rumen in sufficient quantities to overcome any dietary shortage of these vitamins in ruminant rations. Also, vitamin C, while apparently a metabolic essential for ruminants, is synthesized within the body tissues in sufficient quantities to meet the animal's needs. Ruminants, certainly under most conditions, will usually receive sufficient exposure to direct sunlight to meet their needs for vitamin D. Most ruminant rations are considered to be more than adequate in vitamin E, assuming that sufficient selenium is present to bring about its effective utilization. Most ruminant rations also are more than adequate in vitamin K. In addition, at least one form of vitamin K (K_2) is synthesized in the rumen. Finally, with the large amounts of pasture, hay, and/or silage included in most ruminant rations and the large intake of carotene which these feeds provide, the vitamin A needs of most ruminants are met without supplementation. Especially is this so since reserves of carotene and/or vitamin A can be stored in the liver and body tissues during periods of high intake for use during periods of low intake.

B.   Consequently, it would seem that in most instances ruminant animals would have no need for vitamin supplementation. However, it has been observed that under certain feeding conditions, there is sometimes a very poor conversion of carotene to vitamin A in the ruminant animal. Especially has this condition been noted with feedlot steers, which, in spite of the fact that they are receiving more than the recommended allowance of carotene, frequently develop extremely low blood vitamin A levels and display classical symptoms of vitamin A deficiency. Consequently, it is generally recommended that feedlot steers be supplemented with 20,000-30,000 IU of actual vitamin A per head per day, with the higher amount being fed during hot weather.

    1 IU vitamin A = 0.3 microgram vitamin A alcohol

    1 IU carotene   = 0.6 microgram beta-carotene

(The above holds only for rats and chicks which convert carotene to vitamin A with an efficiency of 50%. Cattle convert carotene to vitamin A with an efficiency of about 12.0% or less.)

    Other classes of ruminants may be supplemented with vitamin A if the ration is low in carotene content or if the animals should exhibit symptoms of a vitamin A deficiency.

    Vitamin A deficiency symptoms in cattle include reduced feed intake; slow gains; nightblindness; swollen hocks, knees, and brisket; excess lacrimation; total blindness; diarrhea; muscular incoordination; staggering gait; and reduced sexual activity; and low fertility in bulls and poor conception, resorptions, and abortions in cows.

II.   **Vitamins in swine nutrition.** Swine are dependent on their diet for vitamins to a much greater degree than are ruminants. Like ruminants, swine can synthesize vitamin C in their tissues in adequate amounts to meet their needs. Also, swine, like other livestock, can synthesize their own vitamin D, provided they are exposed to sufficient amounts of direct sunlight. Synthesis of varying amounts of other vitamins may also take place in the large intestine of swine through microbial action. The extent of this synthesis, however, is generally quite limited, and this is not regarded as an adequate source of any of the vitamins. On the other hand, conventional swine feeds are good sources of several of the vitamins and so in themselves provide adequate amounts of these vitamins. It is generally recognized at this time that under most conditions of practical swine production, special consideration should be given to the matter of the adequacy of the following vitamins from the standpoint of the need for supplementation.

A. **Vitamin A.** Rations based primarily on yellow corn, soybean oil meal, and minerals are at best barely adequate in vitamin A value. Where grain sorghum or barley is substituted for a major portion or all of the corn, vitamin A will definitely be in short supply. Consequently, unless at least 2.5% good quality alfalfa meal is included in such rations or unless such rations are fed on pasture, which is not too frequently done in modern-day swine production, then a supplemental source of vitamin A should be used.

B. **Vitamin D.** Since so many hogs today are fed out in confinement where they receive little exposure to direct sunlight, supplementation of modern-day swine rations with vitamin D is generally recommended.

C. **Riboflavin.** Swine rations which contain little or no alfalfa or milk products are likely to be deficient in riboflavin if not supplemented with this vitamin (Figure 26). Since today milk products are seldom used in swine rations and alfalfa meal is used in only limited amounts, if at all, it is generally recommended that riboflavin be included in present-day swine supplements.

(a)   (b)

FIGURE 26.   *Riboflavin deficiency.* (a) Pig that received no riboflavin. (b) Pig that received adequate riboflavin. *(Courtesy of R. W. Luecke, Michigan Agricultural Experiment Station)*

D.   **Niacin.** Since many corn-based rations for swine are marginal in their content of available niacin without niacin supplementation, it is generally recommended that swine rations be supplemented with this vitamin as a safety measure.

E.   **Pantothenic acid.** Corn is a poor source of pantothenic acid, and other commonly used swine feeds are not sufficiently high in this vitamin to provide adequate levels of it in most ordinary swine rations without supplementation. Consequently, supplementation of swine rations with pantothenic acid is generally recommended.

F.   **Vitamin B_{12}.** Swine rations which do not contain liberal amounts of protein supplements of animal origin will probably be deficient in vitamin B_{12} unless supplemented with this vitamin. Since present-day swine rations for the most part involve the use of soybean oil meal as the protein source, it is generally desirable to provide a supplemental source of vitamin B_{12}.

FIGURE 27.   Seven of the ten pigs farrowed by this sow were born dead. The sow received a riboflavin deficient ration. *(Courtesy of T. J. Cunha and J. P. Bowland, Washington State University)*

G.   **Choline.** While choline is usually adequate in most swine rations, extra choline has a sparing effect on the methionine requirement.

FIGURE 28.   *Pantothenic acid deficiency.* Locomotor in-coordination (goose-stepping) was produced by feeding a ration (corn-soybean meal) low in pantothenic acid. *(Courtesy of R. W. Luecke, Michigan Agricultural Experiment Station)*

Since choline can be provided more cheaply than methionine, choline is frequently included in supplements for swine.

H.   Vitamin E. Most swine rations normally contain adequate levels of vitamin E. However, effective vitamin E utilization seems to be dependent on the presence of adequate selenium, and selenium is sometimes deficient in feeds from certain areas. Before selenium supplementation of swine rations received FDA approval, some swine producers were supplementing their rations with vitamin E in an effort to offset a possible selenium shortage and a resulting poor vitamin E utilization through providing extra amounts of this vitamin. However, with the FDA's recent approval of the use of supplemental selenium in swine rations, the use of supplemental vitamin E in such rations should not be necessary in the future if adequate selenium is provided.

In supplementing swine rations with vitamins, the customary practice is simply to include in the supplement the levels of those critical vitamins which will supply the animal's minimum needs of these vitamins. Any of these vitamins which are present in the feed are over and above the animal's needs and simply serve as a margin of safety. While excesses of vitamins serve no useful purpose, they are not harmful in reasonable amounts. Large excesses, however, may be toxic. The requirements of swine for the different vitamins may be found in Tables 32, 33, 34, and 35.

Table 13

VITAMINS IN LIVESTOCK NUTRITION

NAME(S) AND MOLECULAR FORMULA(S)	FAT OR WATER SOLUBLE	SYNTHE- SIZED IN RUMEN?	PRODUCED COMMER- CIALLY?	BASIC FUNCTIONS	DEFICIENCY SYMPTOMS	GOOD FARM SOURCES	ADDITIONAL INFORMATION
Vitamin A $C_{20}H_{30}O$	Fat	No	Yes	Essential for health of epithelial cells. Functions in eyesight—also in bone formation.	Slow growth. Nightblindness. Reproductive disorders. Rough coat. Stiff and/or swollen joints. Total blindness. Low level of liver vitamin A.	Whole milk (carotene).	Not found in plants. Formed in animals from carotene. Formed in wall of intestine, liver, and certain other tissues. Rapidly destroyed by oxygen in heat and light. Reserves stored in liver and body fat. Colorless. May be deficient.
(Carotene) $C_{40}H_{56}$	Fat	No	Yes	As a precursor of vitamin A.	Possible vitamin A deficiency.	Green pasture, good hay, alfalfa meal, silage, yellow corn, whole milk.	Formed in plants. Used by animals to form vitamin A. Rapidly destroyed by oxygen in heat and light. Reserves stored in body fat. Yellow in color. Vitamin A + Carotene = Vitamin A value. May be deficient.

NAME(S) AND MOLECULAR FORMULA(S)	FAT OR WATER SOLUBLE	SYNTHE-SIZED IN RUMEN?	PRODUCED COMMER-CIALLY?	BASIC FUNCTIONS	DEFICIENCY SYMPTOMS	GOOD FARM SOURCES	ADDITIONAL INFORMATION
Vitamin D (Anti-rachitic factor) (Vitamin D_2) $C_{28}H_{44}O$ (Vitamin D_3) $C_{27}H_{44}O$	Fat	No	Yes	Functions in Ca absorption and in Ca and P metabolism.	Poor growth, rickets, osteomalacia, osteoporosis.	Sun-cured hays, whole milk, (sunlight).	Formed by irradiation of sterols with ultra violet light. Ergosterol + UVL → D_2. 7-dehydrocholesterol + UVL → D_3. Quite stable. Some storage. Sometimes deficient.
Vitamin E (Tocopherols) Alpha $C_{29}H_{50}O_2$ Beta $C_{28}H_{48}O_2$ Gamma $C_{28}H_{48}O_2$ Others	Fat	No	Yes	As an antioxidant. As a metabolic regulator of the cell nucleus. Probably others.	Poor growth. Muscular dystrophy—"White muscle" and "stiff-lamb" disease. Reproductive failures?	Cereal grains, (especially the germ), green forage, good hay, oil seeds.	Alpha tocopherol most active. Utilization dependent on adequate selenium. Some storage in liver and other tissues. Fairly stable but rapidly destroyed in presence of rancid fat. Seldom deficient if selenium is adequate.
Vitamin K K_1—$C_{31}H_{46}O_2$ K_2—$C_{41}H_{56}O_2$ K_3—$C_{11}H_8O_2$	Fat Fat Depends on form	No Yes No	Yes	Essential for prothrombin formation and blood clotting.	Prolonged clotting time of blood. Multiple hemorrhages.	Green forage, good hay, most seeds.	Fairly stable under normal conditions. Considerable intestinal synthesis. Seldom deficient for livestock.

NAME(S) AND MOLECULAR FORMULA(S)	FAT OR WATER SOLUBLE	SYNTHE-SIZED IN RUMEN?	PRODUCED COMMER-CIALLY?	BASIC FUNCTIONS	DEFICIENCY SYMPTOMS	GOOD FARM SOURCES	ADDITIONAL INFORMATION
Thiamine (Vitamin B_1) (Antineuritic factor) (Thiamine hydrochloride) $C_{12}H_{18}ON_4SCI_2$	Water	Yes	Yes	As a coenzyme in energy metabolism.	Poor appetite. Slow growth. Weakness. Hyperirri-tability.	Whole grains, germ meals, brans, green forage, good hay, milk.	Fairly stable under normal conditions. Limited intestinal synthesis. Seldom deficient for livestock.
Riboflavin (Vitamin B_2) (Vitamin G) $C_{17}H_{20}N_4O_6$	Water	Yes	Yes	In several enzyme systems related to energy and protein metabolism.	Slow growth. Dermatitis. Eye abnormalities. Diarrhea and leg troubles in pigs.	Milk, skimmed milk, green forages, good hay—especially alfalfa.	Fairly stable under normal conditions. Most swine rations not containing milk and/or alfalfa meal will be low in this vitamin.
Niacin (Nicotinamide) (Niacinamide) $C_6H_6N_2O$ (Nicotinic acid) $C_6H_5O_2N$	Water	Yes	Yes	In enzyme systems related to glycolysis and tissue respiration.	Digestive disorders, dermatitis, and retarded growth in pigs.	Some present in most feeds, but that in corn, wheat, and grain sorghum is largely unavailable.	Quite stable under normal conditions. Can by synthesized in body tissues from surplus tryptophan. Swine rations sometimes deficient in available niacin.

NAME(S) AND MOLECULAR FORMULA(S)	FAT OR WATER SOLUBLE	SYNTHESIZED IN RUMEN?	PRODUCED COMMERCIALLY?	BASIC FUNCTIONS	DEFICIENCY SYMPTOMS	GOOD FARM SOURCES	ADDITIONAL INFORMATION
Vitamin B_6 (Pyridoxine) $C_8H_{11}O_3N$ (Pyridoxal) $C_8H_9NO_2$ (Pyridoxamine) $C_8H_{11}N_2O$	Water	Yes	Yes	As a coenzyme in protein metabolism. In EFA metabolism. In antibody production.	Poor growth, anemia, and convulsions in pigs.	Most common feeds are fair to good sources.	Rations for livestock normally require no B_6 supplementation.
Pantothenic acid $C_9H_{17}O_5N$	Water	Yes	Yes	As a part of coenzyme A. In other metabolic reactions.	"Goose-stepping" in pigs—also digestive disorders and unhealthy appearance.	Widely distributed but it is low in corn.	Quite stable under normal conditions. Can be deficient for swine.
Biotin $C_{10}H_{16}O_3N_2S$	Water	Yes	Yes	In enzyme systems related to carbon dioxide fixation and decarboxylation.	Dermatitis, loss of hair, and retarded growth in all species.	Widely distributed.	Quite stable. Considerable intestinal synthesis. Not likely to be deficient for livestock. Rendered unavailable by raw egg white.

NAME(S) AND MOLECULAR FORMULA(S)	FAT OR WATER SOLUBLE	SYNTHE-SIZED IN RUMEN?	PRODUCED COMMER-CIALLY?	BASIC FUNCTIONS	DEFICIENCY SYMPTOMS	GOOD FARM SOURCES	ADDITIONAL INFORMATION
Choline $C_5H_{15}O_2N$	Water	Probably	Yes	In transmethylation. In fat metabolism in the liver. In cell structure (not a true vitamin function).	Unthriftiness, incoordination, fatty livers, and poor reproduction in swine.	Most commonly used feeds are fair to good sources.	Some synthesis in body tissues. Not likely to be deficient. Surpluses may be used as a methyl donor to spare methionine.
Folic acid (Folacin) $C_{19}H_{19}N_7O_6$	Water	Yes	Yes	It catalyzes the transfer of single carbon units in various biochemical reactions.	Poor growth and various blood disorders in some species.	Most commonly used feeds are fair to good sources.	Some intestinal synthesis. Not likely to be deficient for livestock.
Vitamin B$_{12}$ (Cyanocobalamin) $C_{63}H_{90}O_{14}N_{14}PCo$	Water	Yes	Yes	As a coenzyme in several biochemical reactions.	Slow growth, incoordination, and poor reproduction.	Protein feeds of animal origin. Fermentation products.	Originally APF. Some intestinal synthesis. Cobalt essential for B$_{12}$ synthesis. Can be deficient for swine and early weaned calves.
Inositol $C_6H_{12}O_6$	Water	Yes	Yes	Not too well understood.	None demonstrated in livestock.	Widely distributed in farm feeds.	Not ordinarily deficient for livestock.

NAME(S) AND MOLECULAR FORMULA(S)	FAT OR WATER SOLUBLE	SYNTHE-SIZED IN RUMEN?	PRODUCED COMMER-CIALLY?	BASIC FUNCTIONS	DEFICIENCY SYMPTOMS	GOOD FARM SOURCES	ADDITIONAL INFORMATION
Para-Aminobenzoic acid $C_7H_7O_2N$	Water	Yes	Yes	Essential for growth of certain micro-organisms.	None demon-strated in livestock.	Not too well known.	Not ordinarily deficient for livestock.
Vitamin C (Ascorbic acid) (Antiscorbu-tic factor) $C_6H_8O_6$	Water	No	Yes	Catalyzes re-actions re-lated to the formation and maintenance of collagenous in-tercellular material.	None demon-strated in livestock.	Not required.	Livestock synthesize vitamin C in body tissues. Never deficient with livestock under normal conditions.

18 Physiological Phases of Livestock Production

I. **Maintenance.** By maintenance is meant *the maintaining of an animal in a state of well being or good health from day to day. A maintenance ration is the feed required to adequately support an animal doing no nonvital work, making no growth, developing no fetus, storing no fat, or yielding no product.* The nutritive requirements for maintenance are the first to be met. The nutritive needs of an animal for other purposes are for the most part over and above those for maintenance. As much as 100% of an animal's ration may go for maintenance. On the other hand, with full fed animals as little as 1/3 or even less of an animal's nutrient intake may be required for maintenance purposes. On the average about ½ of all feed fed to livestock goes for maintenance. While it is sometimes profitable to hold animals on a maintenance ration from a period of low prices to a period of high prices, generally speaking, only those nutrients fed over and above the maintenance requirements are available for economic production. The requirements for maintenance are as follows:

A. **Energy for the vital functions.**

    1.   This includes energy for the work of the heart, the work of breathing, and the work of other vital functions.

    2.   This energy must be in the form of net energy.

    3.   Energy used for the vital functions is ultimately liberated as heat.

    4.   The heat resulting from the work of the vital organs may go toward maintaining the body temperature.

    5.   The work of the vital functions is referred to as *basal metabolism.*

    6.   Basal metabolism is the heat production of an animal while at rest and digesting no food.

    7.   Basal metabolism may be determined directly or calculated from the O_2 and CO_2 exchange.

    8.   Basal metabolism is in proportion to the body surface of an animal, not its weight.

B. **Heat to maintain body temperature.**

    1.   All farm livestock are warm-blooded animals and normally maintain more or less constant body temperatures.

Horse	100.2 °F	Hog	102.6 °F
Cow	101.5 °F	Sheep	103.5 °F

    2.   Heat for the maintenance of body temperature comes from a variety of sources.

        a.   Heat from work of vital organs.

        b.   Heat of nutrient utilization.

        c.   Heat from work of normal activity.

        d.   Heat as a by-product of economic work.

        e.   Heat from the work of shivering.

    3.   The temperature at which body oxidations must be increased to maintain the body temperature is referred to as the *critical temperature.* It is that temperature at which shivering starts. It is quite variable depending on:

Species of animal	Activity of animal
Hair or wool coat of animal	Air movement
Fatness of animal	Humidity
Level of feed of animal	

    4.   The critical temperature is seldom reached with animals on a liberal feed allowance, except in extremely cold weather.

C.   **Protein for the repair of body tissues.**

    1.   There is a constant breakdown of body tissue protein.

    2.   The by-products of body protein breakdown are excreted largely in the urine.

    3.   Because of this constant breakdown and loss of protein from the body, protein must be replenished, and this is equivalent to the maintenance requirement.

    4.   The maintenance requirement of an animal for protein is equal to the nitrogen excreted in the urine during starvation.

    5.   The protein requirement for maintenance is in proportion to body surface area.

    6.   Protein for maintenance must be of good quality—that is, it must contain the proper proportion of essential amino acids.

D.   **Minerals to replace mineral losses.**

    1.   There is a constant loss of all of the essential minerals from the body—even calcium and phosphorus from the bones.

    2.   Maintenance rations must contain sufficient minerals to replace these losses.

    3.   Most farm feeds contain adequate minerals for maintenance, except for salt.

E.   **Some of all the vitamins are essential for maintenance.**

    1.   All of the vitamins are necessary for life, even if only maintenance is involved.

    2.   There is a constant destruction and/or loss from the body of all of the different vitamins.

    3.   Feeding programs must take into account the fact that animals must have a constant source of vitamins even just for maintenance.

F.   **Water.**

    1.   An animal will die more quickly from a lack of water than from a lack of any other nutritive factor.

    2.   Water is required for essentially all body functions.

    3.   There is a constant loss of water from the body through urine excretion, feces excretion, perspiration, and respiration.

G. **Certain fatty acids.** While an animal can obtain its energy requirements from any of several different sources, certain fatty acids seem to be essential for maintenance of normal health. Involved here are the unsaturated fatty acids, linoleic, linolenic, and arachidonic.

II. Growth.

A. **Growth** is largely an increase in muscle, bone, organs, and connective tissue. Since meat is basically muscle, then growth is basic for meat production. Also, it is only through growth that an animal is able to attain a mature status. It should be recognized in this connection that the nutritive requirements for growth as outlined below are in addition to those listed above for maintenance. The primary nutritive requirements for growth are as follows:

1. **Protein.**

   a. The dry matter of muscle and connective tissue, and to a considerable degree also that of bone, is primarily protein. Hence, protein is one of the major nutritive requirements of growth.

   b. Protein for growth must be of good quality—that is, it must contain the proper proportions and amounts of essential amino acids at the tissue level.

2. **Energy.**

   a. Animal tissue produced as the result of normal growth ordinarily contains a limited amount of ether extract. Energy in the form of net energy must be provided to meet this need in addition to that in the protein of tissue.

3. **Minerals.**

   a. Since bone formation is a primary activity of growth and since bone is high in calcium and phosphorus content, these two minerals are especially essential for growth.

   b. Other minerals are involved in the digestion and utilization of other nutrients needed for growth.

4. **Vitamins.**

   a. Vitamin D is essential for bone formation.

   b. Certain other vitamins function in various metabolic processes related to nutrient utilization for growth.

5. **Water.**

   a. Fat free muscle tissue is about 75-80% water. Hence, water is a major requirement for growth. Supplying adequate water for livestock growth, however, is not a major consideration since the amount needed for growth as compared to maintenance is rather inconsequential. However, it behooves the livestock producer to take advantage insofar as possible of the fact that growth, especially muscle growth, is largely water and water is cheap.

B.   **Rate of growth.**

    1.   Varies in amount per head daily among the different species— more or less in accordance with the mature size of the species.

    2.   Varies in amount per 100 lb liveweight daily among the different species, with the larger species in general having the lower rate.

    3.   Varies among different breeds largely in accordance with the size of mature animals of the respective breeds.

    4.   Daily growth rate per animal increases until puberty and then decreases until maturity.

    5.   Daily growth rate per 100 lb liveweight decreases from birth to maturity.

III.   **Fattening.**

    A.   **What is fattening?** Fattening in an animal is simply the deposition of unused energy in the form of fat within the body tissues. Fattening is of two general types (Figure 29):

        1.   **Abdominal, intermuscular, and subcutaneous deposition.** This

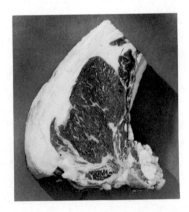

FIGURE 29.   A prime rib roast showing an abundance of marbling which unfortunately is usually accompanied by an excess of outside fat. *(Courtesy of the Greenwood Packing Co., Greenwood, S.C.)*

is for the most part undesirable but unavoidable if marbling is to be realized.

2. **Intramuscular deposition.** Commonly referred to as *marbling.* This is what is wanted. Difficult to obtain without excessive abdominal, intermuscular, and subcutaneous deposition.

B. **Object of fattening.** The object of fattening is to make the meat tender, juicy, and of good flavor. While most people do not like fat meat, they like the lean of meat only if it contains a certain amount of fat as marbling. Fat represents the most costly form of gain in livestock. Consequently, livestock are ordinarily fattened only to a point that they will be sufficiently marbled to make their meat acceptable to the consumer.

C. **Requirements for fattening.**

1. The primary requirement for fattening is energy. This energy must be in the form of net energy. Net energy for fattening may come from any of several forms of feed energy such as:

Starch   Protein
Sugars   Fat
Cellulose

2. Fattening increases an animals's need for protein over and above that required for maintenance and growth only to the extent that additional protein may be necessary to promote good digestion. Fattening increases an animal's metabolic requirements for protein little, if any.

3. Fattening may increase the need for certain of those vitamins related to energy metabolism.

4. Fattening animals are usually full fed since only that net energy over and above that required for maintenance and growth is available for fattening. There may be some loss in feed digestibility with full feeding, but this is more than offset by an overall increase in the efficiency of feed utilization for growth and fattening.

D. **Fattening vs. growth in producing weight gains.**

1. Weight gains in animals are derived from:
   a. Growth.
   b. Fattening.
   c. Fill or increase in content of feed and water.

2. Growth is a much cheaper form of gain than is fattening.

a. Gain from growth is primarily in the form of protein tissue and bone. Gain from fattening is largely in the form of fat.

b. Protein tissue is about 25% protein and 75% water. While protein is one of the more costly nutrients, water is essentially free. Hence, protein tissue is a cheap form of gain.

c. Bone formation in growth requires considerable calcium and phosphorus. Hence, growth utilizes the calcium and phosphorus of feed more completely for weight gain than does fattening. Even if the requirements for growth make it necessary to add calcium and/or phosphorus supplements, these are usually a cheap source of nutrients for gain.

d. Gain from fat is generally relatively expensive. Little, if any, water and minerals are laid down in fattening. In fact, during fattening, fat may actually replace water in the tissues, which makes for a very uneconomical exchange. Also, about 2.25 times as much net energy is required to form a kg of body fat as is required to form a kg of body protein. While protein is usually a costly source of net energy, it is seldom 2.25 times as expensive as other sources.

e. Since gain from growth is usually more efficient and cheaper than gain from fattening and since animals do most of their growing while young, young animals make more efficient and cheaper gains than do older animals.

f. Older animals are more easily fattened than are younger animals because a larger percentage of their energy consumption is available for fattening.

IV. Milk production.

A. We usually are inclined to associate milk production with the dairy cow. However, all animals produce milk upon giving birth to young (Figure 30). Milk production simply happens to be the dairy cow's specialty. She produces more milk over longer periods than do other animals. However, she certainly has no monopoly on it. In fact, certain other animals may be as efficient as the dairy cow at milk production per kg of feed consumed or per 100 kg of body weight.

B. Composition. On a dry basis the milk of all animals is fairly similar in composition, being high in protein, fat, NFE, and minerals. The milk of the mare is somewhat of an exception in that it is lower in protein and fat and much higher in NFE than the milk of other farm animals, as shown below:

FIGURE 30. While milk production is the dairy cow's specialty, all farm animals produce milk upon giving birth to young. *(Courtesy of Ralston Purina Co., St. Louis, Mo.)*

	% PROTEIN	% FAT	% NFE	% ASH
Mare's milk	21	11	63	4
Milk of other farm animals	27-36	29-36	25-38	5-6

On a fresh basis the milk of different farm animals varies considerably in percentage of water and dry matter content as shown below:

	% WATER	% DM
Cow's milk	87.2%	12.8%
Ewe's milk	80.8	19.2
Goat's milk	86.8	13.2
Mare's milk	90.6	9.4
Sow's milk	79.9	20.1

C.   **Milk secretion.** Milk is produced and secreted by the mammary glands. Nutrients for milk production are carried by the blood to the mammary glands. The nutrients are removed from the blood by the mammary glands, converted into milk, and secreted into the udder. Milk is secreted into the udder more or less throughout the day. Nutrients for milk production must come from the feed, either directly or indirectly via body reserves of nutrients, which came originally from the animal's feed.

D.   Nutritive requirements for milk production.

1.   Are in proportion to the amount of milk produced.

2.   Are over and above those for other physiological phases of production such as maintenance, growth, fattening, fetal development, etc.

3.   The major nutritive requirements are:

a.   **Protein.**

- Must be of good quality at the glandular level.
- Animals will not produce milk low in protein.
- If ration is deficient in protein, tissue reserves of protein may be used for milk production.
- Prolonged shortage of protein will limit milk production.

b.   **Energy.** Energy over and above that for milk protein is required for the formation of milk fat and milk sugar.

- Must be in the form of net energy.
- May come from ration carbohydrates, ration fat, or even excess ration protein.
- An animal will not produce milk extremely low in energy.
- If ration is low in energy, body reserves of energy may be used for milk production.
- While ration fat is not essential for milk fat, a small amount of ration fat helps milk production.
- Prolonged shortage of energy will limit milk production.

c.   **Calcium and phosphorus.**

- May come from the feed or from supplemental sources.
- Animals will not produce milk low in these minerals.
- If ration is deficient in calcium and/or phosphorus, the animal will draw upon its bones for calcium and phosphorus with which to produce milk. This will weaken bones and may result in broken bones.
- Prolonged shortage of calcium and/or phosphorus will limit milk production.

d.   **Vitamin A and/or carotene.**

- May not be essential specifically for milk production.

- Animal may produce milk low in these factors.
- Milk rich in these factors is to be preferred.
- Carotene and/or vitamin A for milk production may come from the ration or body reserves.
- Vitamin A and carotene content of milk can be increased by increasing ration vitamin A and carotene.

e. **Vitamin D.**
  - Some evidence that vitamin D is essential for assimilation of calcium and phosphorus for milk production.
  - Vitamin D content of milk can be stepped up through vitamin D feeding.

f. **Sodium and chlorine (salt).**
  - Essential for digestion of nutrients for milk production.
  - Milk normally contains some salt.

g. **Other minerals and vitamins.**
  - Not too well defined.
  - Several involved in nutrient utilization.
  - Most essential minerals and vitamins normally in milk; must come from feed, from water, or from rumen or body synthesis.

V. **Fetal development.** Two major considerations in feeding during gestation are:

A. To provide nutrients for development of fetus and fetal membranes.

1. Fetal development is basically just prenatal growth. Nutritive requirements for fetal development are qualitatively similar to those for growth—that is, protein, calcium, phosphorus, and vitamin D in particular and others indirectly.

2. Quantitatively the nutrient requirement for fetal development is not great. A newborn calf weighs on the average about 37 kg (83 lbs) and is about 25% DM, which makes its dry matter content just under 10 kg. This is the nutritive equivalent of about 75 kg of milk or about 3-4 days production for a good cow.

3. If a ration is deficient in calcium, phosphorus, protein, and/or energy, an animal may draw upon its own body for nutrients to develop the fetus.

4. If an unbred animal is severely deficient in nutrients, she probably will not come into heat.

      5.   If an animal is severely deficient in nutrients during early gestation, she may cease fetal development through resorption or abortion.

B.   A second consideration of feeding during gestation is to build up the nutrient reserve of the animal's body with which to meet the requirements for milk production following parturition.

      1.   Few animals can consume, digest, and metabolize nutrients sufficient to meet their needs during heavy lactation. Most must call on body reserves of fat, protein, calcium, phosphorus, vitamin A, and possibly other nutrients for this purpose. Hence, an animal should be fed during late gestation to encourage some buildup of reserves of these nutrients.

      2.   Termination of lactation during late gestation will promote this nutrient buildup in the body.

      3.   Too liberal feeding, however, during late gestation may result in an oversized fetus and difficult parturition.

VI.   **Wool production.**

    A.   Wool consists of two fractions.

      1.   **Wool fiber.**

         a.   Is practically pure protein.

         b.   Is sulfur-containing protein.

         c.   Makes up from 20% to 75% of a fleece (unwashed wool of a sheep).

         d.   Chemical composition:

            C–50%   N–18%   S–3%
            H– 7%   O–22%

      2.   **Yolk or grease** consists of:

         a.   **Suint.** Compounds of potassium with organic acids. Is water soluble. Makes up from 15% to 50% of unwashed wool.

         b.   **Wool fat.** Commonly known as *lanolin.* Actually a wax. Not water soluble, but is removed by scouring process. Makes up 8-30% of unwashed wool. Contains carbon, hydrogen, and oxygen.

    B.   **Nutritive requirements.**

      1.   Nutritive requirements for wool production are over and above those for other physiological phases of production.

2. With some sheep producing 10 kg (22 lbs) or more per year (average of all sheep is around 8 lbs.) the nutritive requirements for wool production can be considerable.

3. The primary nutritive requirements for wool production are:

    a. **Protein.** Must be sulfur-containing as fed or as synthesized in the rumen.

    b. **Energy.** Sheep must be provided energy over and above that in the required protein with which to produce the yolk in wool. This must be in the form of net energy. Can come from any feed energy source.

    c. **Potassium.** This mineral is an essential component of the suint in wool. Is more than adequate in most ordinary rations.

    d. **Other minerals and vitamins.** While certain other minerals and vitamins are essential either directly or indirectly for wool production, their roles are not too well defined in this connection.

VII. Work.

A. **What is work?** Work is the movement of matter through space.

B. **Types of work.**

1. **Involuntary.** Such as that of the heart and other vital organs. Essential for life. An essential part of maintenance.

2. **Voluntary.** A certain amount of voluntary work is an essential part of practical maintenance. That part of voluntary work over and above that involved in maintenance is the present consideration. It is voluntary work for recreation or economic production.

C. The primary nutritive requirement for work is energy.

1. It must be in the form of net energy.

2. Energy for work must be over and above energy for other needs.

3. Energy for work may come from carbohydrates, fats, and/or excess protein.

4. If ration energy is not adequate to meet the needs for work, fat stores of the body may be called on for this purpose.

D. The protein, mineral, and vitamin requirements of an animal are increased little, if any, by work.

1. As work may involve increased energy consumption and utilization, the need for those nutrients involved in energy digestion and metabolism may be increased.

2. As work may involve profuse perspiration, there may be a significant loss of certain water soluble nutrients and therefore an increased requirement of such nutrients due to work.

Table 14
SUMMARY OF THE PHYSIOLOGICAL PHASES OF LIVESTOCK PRODUCTION AND THEIR NUTRITIVE REQUIREMENTS

	ENERGY	PROTEIN	CALCIUM	PHOSPHORUS	VITAMIN A	VITAMIN D	OTHER MINERALS	OTHER VITAMINS
Maintenance	Yes	Yes	Yes	Yes	Yes	Yes	Yes	Yes
Growth	Mainly as protein	Yes	Yes	Yes	Yes	Yes	Yes	Yes
Fattening	Yes	Only to facilitate digestion	No	No	Possibly	No	Probably	Probably
Milk production	Yes	Yes	Yes	Yes	Not required, but desired	Probably—at least desired	Probably—at least desired	Probably—at least desired
Fetal development	Mainly as protein	Yes	Yes	Yes	Yes	Yes	Probably	Probably
Wool production	Yes	Yes	No	No	No	No	Yes—especially K and P	Probably
Work	Yes	Little, if any	No	No	Possibly	No	Yes	Yes

19 Study Questions and Problems

I. *Select by number from items at the bottom correct answers for each of the following. Use items at bottom more than once as necessary.*

1. Catalyzes iron utilization in the body.
2. Mineral element found in hemoglobin.
3. Most abundant mineral element in wool.
4. Means the opposite of macro.
5. One of the essential amino acids.
6. Prevents white muscle disease in cattle.
7. A deficiency causes goose-stepping in pigs.
8. Thiamine hydrochloride.
9. Same as cobalamin.
10. Crude protein.

11.   Second most abundant mineral element in the body.

12.   Prevents parakeratosis in swine.

13.   A good source of calcium for livestock.

14.   A condition sometimes caused by a deficiency of iodine.

15.   Converted into vitamin D by sunlight.

16.   Hydrolyzes to amino acids.

17.   A good source of zinc for livestock.

18.   The vitamin that functions in bone formation.

19.   Produced in the body by sunlight.

20.   Vitamin E.

21.   A good source of both calcium and phosphorus for livestock.

22.   Same as APF.

23.   Mineral compound with which all livestock should be supplemented.

24.   Contains 46.67% nitrogen.

25.   Functions in vitamin E absorption.

26.   A condition in older animals caused by a deficiency of calcium.

27.   Necessary to prevent losses in salt-hungry cattle given loose salt.

28.   Same as riboflavin.

29.   A good source of fluorine for livestock.

30.   An excess of this will cause defective teeth formation.

31.   Functions in blood clotting.

32.   The amino group.

33.   Most abundant mineral element in the body.

34.   Mineral element found in thyroxin.

35.   Mineral element associated with grass tetany in cattle.

36.   A condition in young pigs caused by a deficiency of iron.

37.   Prevents nightblindness in calves.

38.   The percentage of a protein usable as protein in the body.

1.	Pure urea.	*11.*	Oystershell flour.
2.	Defluorinated phosphate.	*12.*	Vitamin B_{12}.
3.	Tocopherols	*13.*	Arginine.
4.	Selenium.	*14.*	Iodine.
5.	Protein.	*15.*	Magnesium.
6.	Trace.	*16.*	Vitamin B_2.
7.	Phosphorus.	*17.*	Water.
8.	Trace-mineralized salt.	*18.*	Copper.
9.	Vitamin D.	*19.*	Fluorine.
10.	Sulfur.	*20.*	Zinc.

21.	Vitamin A.	28.	Vitamin K.
22.	N × 6.25.	29.	Anemia.
23.	NH_2.	30.	Salt.
24.	Biological value.	31.	Goiter.
25.	Pantothenic acid.	32.	Vitamin B_1.
26.	Osteoporosis.	33.	Ergosterol.
27.	Iron.	34.	Calcium.

II. Select the correct answer for each of the following:

1. Ratio of calcium to phosphorus in bone. (*1:1, 1:2, 2:1, 2:3*)

2. B.V. of a protein void of 2 essential amino acids. *(0, 20, 50, 90)*

3. Percentage of phosphorus in defluorinated phosphate. (*17.0, 33.0, 38.5, 46.7*)

4. Number of vitamins included in the B complex. *(7, 9, 11, 13)*

5. Maximum percentage of protein normally found in soybean oil meal. (*36, 41, 44, 50*)

6. A protein with a biological value of 60% and a protein with a biological value of 100% are mixed in equal amounts. What is the lowest probable biological value of the mixed protein? (*0, 60, 80, 100*)

7. What is the highest possible biological value of the mixed protein in No. 6? (*0, 60, 80, 100*)

8. Percentage of calcium in defluorinated rock phosphate. (*14.5, 30.0, 38.5, 46.7*)

9. Number of IU of vitamin A in one microgram. (*0.3, 1.0, 3.3, 6.7*)

10. Percentage of calcium in alfalfa hay, as fed. (*1.29, 2.29, 4.52, 6.73*)

11. Daily salt consumption of a cow in ounces. *(1, 10, 28, 44)*

12. Highest possible biological value of a protein. *(50, 80, 100, 291.6)*

13. Percentage of salt usually included in a complete swine ration. *(0.5, 1.5, 2.5, 3.5)*

14. Number of fat soluble vitamins that function in animal nutrition. (*4, 11, 16, 24*)

15. Number of water soluble vitamins. (*4, 12, 16, 24*)

16. Total number of vitamins that function in animal nutrition. (*4, 11, 16, 24*)

17. Percentage of nitrogen in crude protein. *(6.25, 16.0, 46.6, 291.6)*

18. Percentage of phosphorus in ground limestone. *(0. 14.5, 30.0, 35.9)*

19. Number of mineral elements that function in animal nutrition. (*16, 20, 24, 33*)

20. Number of chemical elements found in vitamin D. (*3, 4, 5, 6*)

21. Number of essential amino acids. *(5, 10, 20, 24)*

22. Maximum percentage of protein normally found in cottonseed meal. (*36, 41, 44, 50*)

23. Percentage of calcium in pure calcium carbonate. (*14.5, 30.0, 38.5, 40.0*)

24. Total number of naturally occurring amino acids. (*5, 8-10, 20-22, 24+*)

25. Percentage of salt usually included in a mineral mixture for beef cattle on pasture. (*3-5, 5-10, 25-33,67-75*)

26. Number of trace-mineral elements usually included in trace-mineralized salt. (*6, 9, 16, 20*)

27. Minimum number of chemical elements found in a typical B vitamin. (*3, 4, 5, 6*)

28. Maximum percentage of ground limestone to be included in a mineral mixture for cattle on pasture. (*0, 25, 33, 50, 75*)

29. Percentage of salt usually included in trace-mineralized salt. (*20-30, 33-50, 70-80, 95+*)

30. Number of vitamins containing cobalt. (*0, 1, 3, 5*)

31. Number of macro elements that function in animal nutrition. (*6, 11, 16, 20*)

32. Number of vitamins with which ordinary swine rations are usually supplemented. (*1, 8, 12, 16*)

33. Percentage of crude protein in a steer's ration. (*7, 12, 16, 20*)

34. Number of amino acids found in animal tissue. (*8-10, 20-22, 24-26, 30-33*)

35. The crude protein equivalent of pure urea. (*16.0, 42.5, 46.7, 291.6*)

III. *In using the tables of feed composition, for what does each of the following abbreviations stand?*

1. blm	*5.* gr	*9.* mx	*13.* sol	*17.* wt	*21.* g
2. Ca	*6.* IU	*10.* pt	*14.* solv-extd	*18.* DE	*22.* comm
3. dehy	*7.* kcal	*11.* res	*15.* w	*19.* ME	*23.* hydro
4. fbr	*8.* mn	*12.* s-c	*16.* wo	*20.* mg/kg	*24.* cond

IV. *Indicate by number the relative crude protein content of each of the following* (as fed basis):

       1.   Over 22%.

       2.   10.5-22%.

       3.   6-10.5%.

       4.   Under 6%.

1. Dried whole milk.	*5.*	Dried skimmed milk.
2. Steamed bone meal.	*6.*	Tankage with bone.
3. Feather meal.	*7.*	Meat and bone meal.
4. Peanut kernels.	*8.*	Peanut oil meal.

9.	Brewers dried grains.	41.	Peanut hay.
10.	Cottonseed meal.	42.	Soybean seed.
11.	Oat hulls.	43.	Citrus molasses.
12.	Alfalfa leaf meal.	44.	Ground oats.
13.	Alfalfa meal.	45.	Cane molasses.
14.	Cottonseed.	46.	Oat straw.
15.	Wheat straw.	47.	Shelled corn.
16.	Peanut hulls.	48.	Salt.
17.	Fish meal.	49.	Hegari grain.
18.	Soybean oil meal.	50.	Cottonseed oil.
19.	Dried beet pulp.	51.	Wheat grain.
20.	Linseed oil meal.	52.	Corn husks.
21.	Alfalfa stem meal.	53.	Cane sugar.
22.	Cottonseed hulls.	54.	Ground limestone.
23.	Fresh cow's milk.	55.	Millet hay.
24.	Alfalfa hay.	56.	Starch.
25.	Crimson clover pasture.	57.	Cellulose.
26.	Bromegrass hay.	58.	Lignin.
27.	White potatoes.	59.	Petroleum oil.
28.	Coastal bermudagrass hay.	60.	Timothy hay.
29.	Alfalfa pasture.	61.	Ground coal.
30.	Cowpea hay.	62.	Ground barley grain.
31.	Coastal bermudagrass pasture.	63.	Glucose.
32.	Ground corn cob.	64.	Oat hay.
33.	Sorgo silage.	65.	Alsike clover hay.
34.	Johnsongrass hay.	66.	Stearin.
35.	Wheat middlings.	67.	Sericea hay.
36.	Bromegrass pasture.	68.	Peanut oil.
37.	Wheat bran.	69.	Ground snapped corn.
38.	Corn silage.	70.	Lactose.
39.	Lespedeza hay.	71.	Riboflavin.
40.	Sweet potatoes.	72.	Tallow.

V. *Indicate by number which of the following are:*

    1. Water soluble.

    2. Synthesized in the rumen.

    3. Included in swine supplements.

1.	Vitamin A.	7.	Pantothenic acid.	13.	Niacin.
2.	Vitamin B_1.	8.	Vitamin C.	14.	Vitamin K.
3.	Folic acid.	9.	Vitamin B_2.	15.	Biotin.
4.	Vitamin B_6.	10.	Vitamin E.	16.	Inositol.
5.	Vitamin D.	11.	Carotene.	17.	Para-aminobenzoic
6.	Choline.	12.	Vitamin B_{12}.		acid.

VI. *Indicate which within each of the following groups of feeds is highest in crude protein and which is lowest, as fed basis.*

1. (1) Shelled corn, (2) alfalfa meal, (3) fresh cow's milk, (4) cottonseed meal, (5) soybean oil meal.

2. (1) Ground snapped corn, (2) cane molasses, (3) dried skimmed milk, (4) wheat bran, (5) bermudagrass hay.

3. (1) Ground oats, (2) corn silage, (3) Hegari grain, (4) peanut oil meal, (5) brewers dried grains.

4. (1) Lespedeza hay, (2) Johnsongrass hay, (3) cottonseed hulls, (4) shelled corn, (5) redtop hay.

5. (1) Soybean seed, (2) wheat middlings, (3) soybean oil meal, (4) cotton-seed, (5) dried beet pulp.

6. (1) Fish meal, (2) sweet potatoes, (3) barley grain, (4) cowpea hay, (5) alfalfa stem meal.

7. (1) Tankage with bone, (2) meat scrap, (3) oat hay, (4) wheat grain, (5) rice hulls.

8. (1) Fresh cow's milk, (2) feather meal, (3) linseed oil meal, (4) grain sorghum grain, (5) sericea hay.

9. (1) Citrus molasses, (2) wheat grain, (3) wheat bran, (4) shelled corn, (5) millet hay.

10. (1) Dried citrus pulp, (2) sorgo silage, (3) alfalfa leaf meal, (4) barley grain, (5) peanut hay.

11. (1) Bahiagrass pasture, (2) fescue hay, (3) alfalfa stem meal, (4) ground snapped corn, (5) peanut hulls.

12. (1) Shelled corn, (2) wheat bran, (3) ground oats, (4) barley grain, (5) milo grain.

13. (1) Cottonseed meal, (2) soybean oil meal, (3) linseed oil meal, (4) fish meal, (5) peanut oil meal.

14. (1) Soybean seed, (2) dried skimmed milk, (3) soybean oil meal, (4) dried whole milk, (5) cottonseed meal.

15. (1) Bermudagrass hay, (2) crimson clover pasture, (3) lespedeza hay, (4) oat hay, (5) timothy hay.

16. (1) Fresh cow's milk, (2) tankage, (3) fresh skimmed milk, (4) alfalfa hay, (5) alfalfa pasture.

17. (1) Alfalfa meal, (2) alfalfa leaf meal, (3) alfalfa stem meal, (4) alfalfa hay, (5) alfalfa pasture.

18. (1) Shelled corn, (2) ground snapped corn, (3) corn cobs, (4) corn silage, (5) corn gluten meal.

19. (1) Cottonseed, (2) cottonseed hulls, (3) cottonseed oil, (4) cottonseed meal, (5) dehulled cottonseed.

VII. *Indicate by number whether each of the following is:*

1. A micro mineral.
2. A macro mineral.
3. An essential amino acid.
4. A nonessential amino acid.
5. Other.

1. Riboflavin.	15. Magnesium.	28. Fluorine.	
2. Thyroxin.	16. Folacin.	29. Trypsin.	
3. Iodine.	17. Creatinine.	30. Biotin.	
4. Leucine.	18. Arginine.	31. Choline.	
5. Insulin.	19. Selenium.	32. Manganese.	
6. Methionine.	20. Para-amino-	33. Proline.	
7. Niacin.	benzoic acid.	34. Threonine.	
8. Ascorbic acid.	21. Inositol.	35. Olein.	
9. Cobalt.	22. Tocopherol.	36. Pepsin.	
10. Cyanocobalamin.	23. Pyridoxine.	37. Molybdenum.	
11. Lysine.	24. Cystine.	38. Potassium.	
12. Inulin.	25. Stearin.	39. Lipase.	
13. Histidine.	26. Glutamic acid:	40. Thiamine	
14. Urea.	27. Phenylalanine.	hydrochloride.	

VIII. *Using numbers match the following:*

1. Same as tocopherol.	1. Vitamin A.
2. Thiamine hydrochloride.	2. Vitamin B_1.
3. Used to spare methionine.	3. Folic acid.
4. Prevents nightblindness.	4. Vitamin B_6.
5. Same as folacin.	5. Vitamin D.
6. Functions in blood clotting.	6. Choline.
7. Precursor of vitamin A.	7. Pantothenic acid.
8. Resembles a carbohydrate in composition.	8. Vitamin C.
9. Prevents goose-stepping in pigs.	9. Vitamin B_2.
10. Same as pyridoxine.	10. Vitamin E.
11. Originally known as APF.	11. Carotene.
12. Rendered unavailable by raw egg white.	12. Vitamin B_{12}.
13. $C_7H_7O_2N$.	13. Niacin.
14. Same as riboflavin.	14. Vitamin K.
15. Same as nicotinic acid.	15. Biotin.
16. Same as ascorbic acid.	16. Inositol.
17. Anti-rachitic factor.	17. Para-aminobenzoic acid.

IX. *Indicate by number which of the following materials on an as fed basis would be among the:*

    1. 16 highest in % protein.

    2. 16 highest in % phosphorus.

    3. 12 highest in % calcium.

    4. 12 highest in carotene content/kg.

1.	Ground limestone.	*19.*	Cottonseed.
2.	Cottonseed hulls.	*20.*	Cane sugar.
3.	Orchardgrass pasture.	*21.*	Fresh cow's milk.
4.	Tankage with bone.	*22.*	Hegari grain.
5.	Ground corn cob.	*23.*	Crimson clover pasture.
6.	Peanut kernels.	*24.*	Lespedeza hay.
7.	Cottonseed oil.	*25.*	Defluorinated phosphate.
8.	White potatoes.	*26.*	Feather meal.
9.	Wheat bran.	*27.*	Sorgo silage.
10.	Oat straw.	*28.*	Steamed bone meal.
11.	Corn starch.	*29.*	Dried skimmed milk.
12.	Wheat grain.	*30.*	Corn silage.
13.	Soybean seed.	*31.*	Fish meal.
14.	Dried citrus pulp.	*32.*	Barley grain.
15.	Cane molasses.	*33.*	Red clover hay.
16.	Bermudagrass hay.	*34.*	Alfalfa meal.
17.	Alfalfa hay.	*35.*	Sweet potatoes.
18.	Soybean oil meal.	*36.*	Yellow shelled corn.

X. *Select from the figures below the proper answer for each of the following:*

1. Average percentage of nitrogen in protein.

2. What times nitrogen equals crude protein?

3. Percentage of nitrogen in pure urea.

4. Crude protein equivalent of pure urea.

5. Crude protein equivalent of urea containing 45% nitrogen.

6. Approximate number of essential amino acids.

7. Total number of naturally occurring animo acids.

8. Percentage of protein usually included in livestock rations.

9. Kg of corn required per kg of urea for protein synthesis in the rumen.

10. Maximum biological value of a mixture of equal amounts of two proteins—one with a B.V. of 60% and the other with a B.V. of 100%.

Answer possibilities: 44   10   16   80   60   6.25   46.67

6.5   291.6   25+   281.+   8-18   262   8   30-40

XI.   *Correlate by number the terms at the bottom with their respective*
          definitions. (Based on the glossary of terms. These are only a few of the
possibilities.)

1.   Near the kidney.

2.   Low count of red blood corpuscles.

3.   Another name for vitamin C.

4.   To waste or cause to waste away.

5.   Pertaining to the heart and blood vessels.

6.   Areas of tooth decay.

7.   Inflammation of the true skin.

8.   Difficult parturition.

9.   To disperse small drops of one liquid in another liquid.

10. Originating within or inside the cells or tissue.

11. Causes of a disease or a disorder.

12. The undesirable effects produced by taking an excess of a vitamin.

13. A prefix denoting less than the normal amount.

14. Easily destroyed.

15. In nutrition, any chemical substance found in feed.

16. In a dying state, near death.

17. The membrane that lines the passages and cavities of the body.

18. Refers to nerves.

19. The process of forming bone.

20. A compound that can be used by the body to form another compound.

21. Feverish condition.

22. The medical treatment of disease.

23. Movement of matter through space.

24. A yellow organic compound that is precursor of vitamin A.

25. A technique for separating complex mixtures of chemical substances.

26. Any unhealthy change in the structure of a part of the body.

27. Excessive overweight due to the presence of a surplus of body fat.

28. The colorless fluid portion of the blood in which the cells are suspended.

29. The energy produced by an individual during physical, digestive, and
     emotional rest.

30. The efficiency with which a protein furnishes the proper proportions and
     amounts of the amino acids needed.

31. The coming together of chemical building units to form new materials in
     the living plant or animal.

32. A chemical compound containing nitrogen, carbon, hydrogen, and oxygen
     which is present in the urine and results from the metabolism of proteins.

33. Swelling of a part or the entire body due to the presence of an excess of water.

34. A substance belonging to the class of sterols that on exposure to ultraviolet light is converted to vitamin D_2.

35. A protein in the blood that contains iron and carries oxygen from the lungs to the tissues.

1. Hypervitaminosis.	18. Protein.	35. Pyrexia.
2. Work.	19. Putrefaction.	36. Etiology.
3. Ergosterol.	20. Chromatography.	37. Therapy.
4. Adrenal.	21. Atrophy.	38. Obese.
5. Dermatitis.	22. Carcinogen.	39. Antibiotic.
6. Plasma.	23. Caries.	40. Asphyxia.
7. Basal metabolism.	24. Hyper.	41. Buffer.
8. Biological value.	25. Cystitis.	42. Cardiovascular.
9. Biosynthesis.	26. Factor.	43. Chlorophyll.
10. Creatinine.	27. Lesion.	44. Cholesterol.
11. Edema.	28. Aerobic.	45. Labile.
12. Hemoglobin.	29. Mucosa.	46. Fortify.
13. Hydrolysis.	30. Abscess.	47. Hormone.
14. Dystocia.	31. Ossification.	48. Moribund.
15. Anemia.	32. Emulsify.	49. Hypo.
16. Ascorbic acid.	33. Endogenous.	50. Neuritic.
17. Carotene.	34. Precursor.	

XII. *Indicate whether each statement is true or false.*

1. Young animals make more efficient gains than do older animals.

2. Salt is a mineral which is deficient in most farm-produced feeds.

3. Salt should never be mixed with feed for livestock.

4. A deficiency of iron will produce a condition in young pigs known as anemia.

5. A calcium-phosphorus ratio of 1:2 is about right.

6. Salt should never be provided for livestock in the loose form.

7. Copper, cobalt, maganese, and zinc are all required by livestock in micro amounts.

8. A mineral mixture consisting of 2 parts limestone flour and 1 part trace-mineralized salt would be satisfactory for cattle on pasture.

9. Vitamin A is required by all classes of farm livestock.

10. A mineral mixture of 2 parts defluorinated phosphate, 2 parts limestone, and 1 part TM salt would be satisfactory for steers on a fattening ration.

11. Carotene, vitamin A, vitamin D, vitamin K, and vitamin E are all fat soluble.

12. The ultraviolet rays of sunlight convert carotene into vitamin D.

13. Livestock on good pasture should never suffer from a vitamin D deficiency.

14. Skim milk is a very poor source of vitamin A.

15. Alfalfa hay is richer in vitamin D than alfalfa pasture.

16. A mature mule that is just being maintained requires no protein in the ration.

17. The feed cost of a lb of growth is greater than the feed cost of a lb of fattening.

18. Alfalfa hay is higher in percentage of calcium than shelled corn.

19. Bermudagrass hay is higher in percentage of carotene than dried skim milk.

20. Bone meal is higher in percentage of calcium than limestone.

21. All of the water-soluble vitamins are referred to as the vitamin B complex.

22. Calcium and phosphorus are not required for maintenance.

23. Sunlight will convert the carotene of plants into vitamin A during the hay making operation.

24. Livestock not on pasture frequently suffer from a deficiency of vitamin C.

25. Potassium is a mineral element which is frequently deficient in livestock rations.

26. A ration of cottonseed meal and cottonseed hulls would be deficient in vitamin A.

27. Green pasture is rich in carotene and vitamin D.

28. Defluorinated phosphate is high in both calcium and phosphorus.

29. Nonlegume hays in general are rich in phosphorus.

30. A cow on a ration low in carotene may produce milk which is low in this substance.

31. A cow on a ration low in calcium will produce milk which is low in this mineral.

32. Livestock can produce protein tissue in their body from fat in the ration.

33. Livestock can produce body fat from starch in their ration.

34. Livestock can produce body fat from protein in their ration.

35. The requirements for fattening are similar to those for work.

36. Energy is the principal requirement for fattening.

37. The nutrients required for growth are qualitatively quite similar to those required for fetal development.

38. Working animals require much more protein than do idle animals for repair of their muscle tissues.

39. Heat from the heat increment is valuable for keeping the body warm during very cold weather.

40. Cottonseed meal usually contains over 36% crude protein.

41. Trace-mineralized salt is a good source of cobalt and copper.

42. Livestock can store considerable quantities of vitamin A in their livers.

43. Most of the B vitamins are synthesized in the rumen of ruminants.

44. Livestock are usually fattened before being sent to market because this is usually a profitable practice.

45. The nutrient requirements of work are similar to those of fattening.

46. The principal nutrient requirement of work is plenty of net energy.

47. Approximately 1/2 of all feed fed to livestock on the average goes for maintenance.

48. The grains in general are good sources of vitamin D.

49. Green pasture is high in vitamin A value.

50. Yellow butter always contains more vitamin A than white butter.

51. Deficiency of vitamin A in cattle may affect rate of gain.

52. A fluorine deficiency in cattle may be avoided by feeding ordinary trace-mineralized salt.

53. The hays in general usually contain more calcium than phosphorus.

54. A deficiency of calcium in the ration may result in the disease known as osteomalacia.

55. Along with calcium and phosphorus, vitamin D is an important constituent of bone.

56. Skim milk is a very good source of vitamin D.

57. Copper, iodine, fluorine, and zinc are all required by livestock in very small amounts.

58. A calcium-phosphorus ratio of 1:2 may result in a calcium deficiency.

59. A cow on a ration low in quality of protein will produce milk which is low in protein quality.

60. The nutrients required for growth are qualitatively similar to those required for maintenance.

61. The percentage of crude protein in a feed may exceed 100% under certain conditions.

62. Mixing two proteins together will always improve their overall biological value.

63. Livestock have no metabolic requirement for nonessential amino acids.

64. A protein is said to be of good quality if it is high in digestibility.

65. Urea should not be fed to swine because it is very toxic to this class of livestock.

66. Cottonseed meal will kill cattle if it is fed to them in large amounts.

67. What used to be referred to as the "animal protein factor" is now known to be vitamin B_{12}.

68. Phosphorus makes up approximately one-fourth of the mineral content of an animal's body.

69. Almost all farm-produced feeds should be supplemented with salt for livestock.

70. Caution should be used in feeding "hot meal" to pregnant cows since it will sometimes cause abortion.

71. Steamed bone meal is a good source of protein for livestock.

72. Ground limestone and oystershell flour are about equal in feeding value.

73. Steamed bone meal, defluroinated phosphate, and oystershell flour are all good sources of phosphorus for livestock.

74. A mixture of equal parts of trace-mineralized salt, steamed bone meal, and ground limestone would make a good mineral mix for cows on pasture.

75. Legume roughages are in general good sources of calcium.

76. Grains are usually higher in phosphorus than in calcium.

77. Hays are usually higher in calcium than in phosphorus.

XIII. *One hundred pounds of urea is the crude protein equivalent of how many pounds of 44% soybean oil meal?*

SOLUTION:

Crude protein content of 100 lb of 44% SOM   = 45.8 lb (from Table 37)

Crude protein equivalent of 100 lb of urea (45% N) = 45 × 6.25 = 281.25 lb

$$\frac{281.25}{45.8} \times 100 = 614.08 \text{ lb}$$

XIV. *A feed company has on hand 30 tons of urea (45% N). This would be the crude protein equivalent of how many tons of 41% cottonseed meal?*

Crude protein content of one ton of 41% CSM $= \frac{41.4}{100} \times 2000$

$$= 828.0 \text{ lb}$$

Crude protein equivalent of 30 tons urea (45% N) $= \frac{281.25}{100} \times 2000 \times 30$

$$= 168,750 \text{ lb}$$

$$\frac{168,750 \text{ lb}}{828 \text{ lb.}} = 203.8 \text{ tons}$$

XV.   *If urea is $180.00 per ton and shelled corn is $2.24 per bushel, at what price per ton for 44% soybean oil meal would it become economical to use urea in the ration for finishing steers?*

SOLUTION:

1 lb urea + 6.5 lb corn $\geq$ 7.5 lb SOM

@ $180.00 per ton, urea is 9¢ per lb

@ $2.24 per bu, corn is 4¢ per lb

$$\frac{(1 \times 9) + (6.5 \times 4)}{7.5} = 4.667¢+ \text{ per lb} \qquad \text{or} \qquad \$93.33+ \text{ per ton } (answer)$$

XVI.   *If 44% soybean oil meal is $119.00 per ton, urea (45%) is $200.00 per ton, and shelled corn is $3.00 per bushel, would it be economical to use urea in a steer-feeding program under these circumstances?*

ANSWER:

Since 7.5 lb of SOM can be bought more cheaply than 1 lb of urea plus 6.5 lb of corn, it would not be economical to use urea in the ration.

XVII. *Calculate the percentage of phosphorus in tri-calcium phosphate.*

SOLUTION:

Chemical formula of tri-calcium phosphate
$$Ca_3(PO_4)_2$$

Atomic weights
  Calcium   40
  Phosphorus   31
  Oxygen   16

Molecular weight of $Ca_3(PO_4)_2$ = $(3 \times 40) + (2 \times 31) + (2 \times 4 \times 16)$

$$= 120 + 62 + 128 = 310$$

% phosphorus in $CA_3(PO_4)_2$ $= \dfrac{62}{310} \times 100 = 20.0\% \ (answer)$

XVIII. *What is the ratio of calcium to phosphorus in tri-calcium phosphate?*

ANSWER:

From the solution to Problem XIX, it is apparent that the calcium-phosphorus ratio in tri-calcium phosphate is

  120:62 or 2:1.03

XIX. *Calculate the percentage of calcium in pure calcium carbonate ($CaCO_3$).*

ANSWER: 40%

XX. *The molecular formula for the amino acid lysine is $C_6H_{14}N_2O_2$. What is the percentage of nitrogen in this amino acid?*

Atomic weights

C	12
H	1
N	14
O	16

Molecular weight of lysine = $(6 \times 12) + (1 \times 14) + (2 \times 14) + (2 \times 16)$

         = 72 + 14 + 28 + 32

         = 146

% N in lysine = $\dfrac{28}{146} \times 100 = 19.18\%$ *(answer)*

XXI. *Calculate the percentage of nitrogen in the amino acid cystine which has the following molecular formula: $C_6H_{12}N_2O_4S_2$.*

ANSWER: 11.67%

20 Feeds and Feed Groups— General

Hundreds of different products have been used from time to time over the years for feeding purposes. However, a relatively limited number of these products makes up a bulk of the nation's feed supply. This fact is borne out by Table 15 which presents annual usage figures for the country's more important feeds.

I.   It will be noted that the feed grains make up around 80% of the feed concentrates. Of these, corn is by far the most important with nearly six times as much corn being used as grain sorghum, which is in second place. While oats and barley are both important feed grains, both make a rather minor contribution to the overall feed supply. Very little wheat grain is being used as feed under present-day circumstances, but a considerable amount of wheat mill by-products is used for feeding purposes. The molasseses also make an important contribution to our overall supply of energy feeds.

Table 15
ANNUAL USE OF THE MORE IMPORTANT FEEDS IN THE U.S.

1000 TONS

*Feed grains**	
Corn	119,420
Grain sorghum	20,104
Oats	10,672
Barley	5,616
*Oil seed meals**	
Soybean oil meal	13,450
Cottonseed meal	2,200
Linseed oil meal	150
Peanut oil meal	150
Copra meal	100
*Animal protein feeds**	
Tankage and meat meal	1,775
Fish meal and solubles	450
Dried milk products	315
*Grain protein feeds**	
Gluten feed and meal	1,775
Brewers dried grains	375
Distillers dried grains	460
*Other processed feeds**	
Wheat millfeeds	4,350
Rice millfeeds	450
Dried and molasses beet pulp	1,375
Alfalfa meal	1,775
Fats and oils	475
Molasses	3,760
Miscellaneous	1,100
*Hays***	
Alfalfa and alfalfa mixtures	78,343
All others (principally clover and/or timothy hay, wild hay, grain hay, and coastal bermudagrass)	56,265
*Silages***	
Corn	109,848
Sorghum	9,557

*May 1974 feed use forecast for 1973-74 feed year. Economic Research Service, USDA (*Feed Situation*, May 1974).

**Production estimates for 1973. Crop Reporting Board, SRS, USDA (*Annual Crop Summary*, January 1974).

II.   Soybean oil meal is by far the most important high protein feed with over
       six times as much of it being used as cottonseed meal, which is second
among the high protein feeds in tonnage fed. However, tankage and meat meals,
along with the gluten feeds and meals, make important contributions to the high
protein feed supply. The fish meals and the dried milk products along with
linseed oil meal, peanut oil meal, and copra meal also make significant
contributions.

III.   Alfalfa and alfalfa mixtures make up approximately 58% of the nation's
       hay production, with mixtures of clover and timothy making up about
17%. Of the many different types of hay that go to make up the other 25% of
the country's overall hay supply, grain hay, and coastal bermudagrass are
probably the more important from a tonnage standpoint.

IV.   On an air-dry tonnage basis about three times as much hay is produced as
       silage. Most of the silage is either corn or sorghum, with over 11 times as
much corn as sorghum silage being made. While the figures in Table 15 for the
hays and silages are for production rather than use, essentially all hay and silage
produced is ultimately fed to livestock, and very little is ever exported.
Consequently then, on the average, annual production of hay and silage is
essentially equal to annual use of these products for feeding purposes.

On the following pages are presented some of the more important facts about
most of the more important feed materials. In the presentation of these facts the
different feeds have been considered by groups, based on certain similarities in
composition and feeding value. Within each group the various feeds are taken up
in no particular order except that in general the more important feeds within
each group have been given first consideration.

21 Air-Dry Energy Feeds

I.   **Shelled corn.**

    A.   Corn is the most extensively produced feed grain.

    B.   It is the most widely grown feed grain crop.

    C.   It is the most widely used energy feed.

    D.   It is unexcelled as an energy feed.

    E.   Of the common feed grains, it excels in pounds of TDN produced per acre.

    F.   Corn is extremely low in calcium but fair in phosphorus content.

    G.   It is quite deficient in vitamin B_{12} and low in riboflavin and pantothenic acid.

FIGURE 31. Most corn grown for grain is harvested with combines today. Note harvested shelled corn in the grain tank; also that the cobs and shucks, as well as the stalks and leaves, remain in the field. *(Courtesy of Ford Tractor Operations, Ford Motor Co., Troy, Mich.)*

H. It must be supplemented with protein for most classes of livestock.

I. It is especially low in the essential amino acids methionine, lysine, and tryptophan, as may be noted from Table 9.

II. **Sorghum grain,** including hegari grain, kafir grain, milo grain, and most other types and varieties of grain sorghum.

A. About 15-20% as much grain sorghum produced in the U.S. as corn.

B. Most grain sorghum production located in the semi-arid regions of the West where corn does not do too well.

C. Most grain sorghum is very similar to shelled corn in chemical composition except that most grain sorghum is slightly higher in

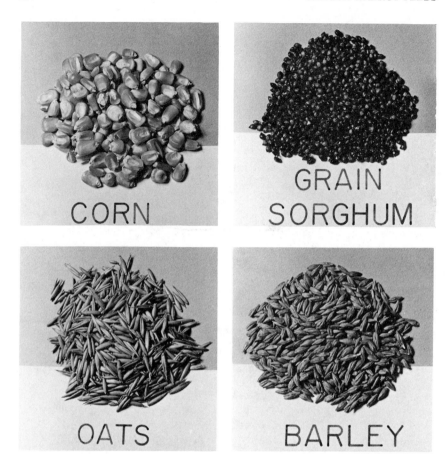

FIGURE 32. The four major feed grains. *(Courtesy of the University of Georgia College of Agriculture Experiment Stations)*

protein and contains little if any carotene.

D. Grain sorghum may be used to replace as much as 50% of the corn in the ration for most livestock without affecting animal performance.

E. If grain sorghum is used to replace all the corn in the ration, gains may be reduced by as much as 10% or possibly more.

F. Grain sorghum is usually rolled or ground for most classes of livestock.

G. Grain sorghum is quite drouth resistant.

H. It usually presents a harvesting problem. If allowed to dry in the head, much is lost to birds. If harvested early to prevent bird losses, it must be artificially dried for storing. Certain bird-resistant varieties of grain sorghum have been developed, but these have proved to be of low feeding value because of their low palatability and low digestibility.

III. **Oats grain.**

A. About 10-12% as much oats fed to livestock in U.S. as corn.

B. Oats are widely grown, but a large portion of oats grain is produced in the Midwestern and North Central states.

C. Oats are higher than corn in crude fiber (10.6% vs. 2.0%, as fed) and accordingly are lower in TDN (66.3% vs. 78%, for cattle, as fed).

D. Oats are also somewhat higher than corn in crude protein (11.7% vs. 8.8%) and a little higher in Ca and P.

E. Oats contain little, if any, carotene.

F. Oats are used extensively in rations for horses, young growing stock, show stock, and breeding animals.

G. Oats are not a good fattening feed and ordinarily are used only to a limited extent, if at all, in fattening rations.

H. Oats are usually rolled, crimped, or ground for feeding.

IV. **Barley grain.**

A. About 5-6% as much barley fed to livestock in U.S. as corn.

B. Most of the barley is produced in the North Central and Far West states.

C. Barley is similar to oats in protein content and intermediate to oats and corn in content of fiber and TDN and calcium and phosphorus.

D. Barley is used extensively along with or in the place of oats in rations for horses, young growing stock, show stock, and breeding animals.

E. Barley may be used to replace up to one-half of the corn in rations for fattening animals without materially affecting their performance.

F. When barley is used to replace all the corn in fattening rations, gains may be reduced by as much as 10%.

G. Barley is usually steam rolled (flaked), crimped, or coarsely ground for feeding.

H. Barley is sometimes cooked for beef show animals—supposedly to improve its acceptability.

V.   Wheat grain.

A.   Wheat, like corn, is widely grown in U.S.

B.   Wheat ranks second only to corn in acres produced (54 million vs. 47 million acres).

C.   Over the years, wheat has been too expensive in most instances to be fed to livestock—too much in demand as a human food.

D.   At times, however, for various reasons wheat becomes available at prices which make it competitive with corn and other feed grains as an economical source of energy for livestock.

E.   Wheat, except for being considerably higher in protein (12.8% vs. 8.8%) and having little, if any, carotene, is quite similar to corn in composition and feeding value.

F.   For best results wheat is best mixed at relative low levels with other grains, especially for feeding cattle and horses.

G.   It is usually coarsely ground or cracked for feeding.

VI.   Ground ear corn (corn and cob meal).

A.   Ground ear corn consists of whole ears of corn (grain and cob) ground to varying degrees of fineness.

B.   Ear corn has traditionally been considered to be about 80% grain and 20% cob, although modern-day corn will sometimes run somewhat higher than this in percentage of grain and lower in cob.

C.   Composition of ground ear corn approximates a weighted average of the composition of shelled corn and the composition of corn cobs.

D.   Ground ear corn is somewhat higher in fiber and lower in TDN than ground shelled corn.

E.   It is an excellent energy feed for ruminants and for horses and mules, provided other roughages are reduced sufficiently to allow for its content of cob.

F.   It should not be fed to growing-fattening pigs since this class of animal can make little, if any, use of the cob, and the cob may irritate the digestive tract and open the way for intestinal infections.

G.   It may be used in rations for mature swine.

H.   Most corn grain is harvested with combines today, leaving the cob in the field; hence, not too much ear corn is available for feeding purposes at the present time.

VII. Ground snapped corn.

    A. Ground snapped corn consists of whole ears of snapped corn (grain, cob, and shucks).

    B. Snapped corn is found mostly in the southern states and not too much found even in the South today.

    C. Handling snapped corn involves too much labor—most corn for grain is combined today, leaving cob and shuck in the field.

    D. Snapped corn has traditionally been considered to be about 70% grain, 17.5% cob, and 12.5% shuck. However, modern-day corn will sometimes vary significantly from these proportions, having slightly less grain, considerably less cob, and much more shuck.

    E. Composition of ground snapped corn approximates a weighted average of the composition of shelled corn, corn cobs, and corn shucks.

    F. Ground snapped corn is considerably higher in fiber and lower in TDN and protein than ground shelled corn; it is comparable to oats as an energy feed.

    G. It is a satisfactory energy feed for ruminants and for horses and mules, provided other roughages are reduced sufficiently to allow for its content of cob and shuck.

    H. It should not be fed to growing-fattening pigs since this class of animal can make little, if any, use of the cob or shuck—also, both may irritate the digestive tract and open the way for intestinal infections.

    I. It may be fed to mature swine, but it is not too satisfactory for this class of animals.

VIII. Wheat bran.

    A. Wheat bran consists primarily of the seed coat of wheat which is removed in the manufacture of wheat flour.

    B. It is used in livestock feeding:

        1. Primarily—

            a. As a source of bulk.

            b. As a mild laxative.

            c. As a source of phosphorus.

        2. Secondarily—

            a. As a source of energy.

            b. As a source of protein.

FIGURE 33. While most wheat is used for the manufacture of flour for human consumption, some of its by-products—such as wheat bran, as well as wheat shorts and wheat middlings—are widely used for feeding purposes. *(Courtesy of the University of Georgia College of Agriculture Experiment Stations)*

C. It is used primarily in rations for horses, dairy cows, brood sows, and beef show animals.

D. It is seldom used in rations for feedlot steers or growing-fattening pigs.

E. Its use in livestock feeding is usually limited to about 10% of the ration in view of its bulkiness, its laxative nature, and its usual relatively high price.

IX. **Wheat middlings and shorts.**

A. There are several types of middlings, shorts, and similar products resulting from the manufacture of wheat flour from various types of wheat by different processes and/or processors.

B. All consist of mixtures of fine particles of bran and germ, the aleurone layer, and coarse flour in varying proportions.

C. Such products are ordinarily lower in fiber and higher in TDN than bran and are used primarily as a source of energy in the ration.

D. The fiber content of such feeds will vary from about 2% to 9% with their TDN values varying accordingly.

E. Most such products are used in rations for swine since their content of flour causes them to become gummy upon being eaten, making them unrelished by other classes of livestock.

F.   As a component of swine rations, such feeds are amply high in total protein content but require some supplementation from an essential amino acid standpoint.

X.   **Dried citrus pulp.**

A.   Citrus pulp is a by-product of the citrus processing industry consisting usually of the remains of the fruit after the juice has been removed and sometimes cull fruit.

B.   While sometimes fed in the wet form in the vicinity of the processing plant, it is usually dried in order to facilitate its use at distant points and during off-season periods.

C.   Dried citrus pulp has over the years been fed mainly to dairy cattle but can also be fed to beef cattle. It is not usually fed to other classes of livestock.

D.   While it is relatively high in fiber, it is ordinarily regarded primarily as an energy feed.

E.   It is usually used to make up not more than about 20-25% of the ration.

F.   Fed as outlined above, it has a feeding value, comparable to that of dried beet pulp, corn and cob meal, rolled barley, and similar type feeds.

FIGURE 34.   Two important by-products widely used for feeding purposes—especially in dairy cow rations. *(Courtesy of the University of Georgia College of Agriculture Experiment Stations)*

XI. Dried beet pulp.

    A.   Most of the beet pulp produced in this country is dried to facilitate its handling, shipment, and use.

    B.   Most of the beet pulp available in this country is fed to dairy cattle but is sometimes fed also to horses and to fattening cattle and sheep.

    C.   It is used:

        1.   Primarily—

            a.   As a bulk factor.

            b.   As an appetizer.

            c.   As a mild laxative.

        2.   Secondarily—

            a.   As a source of energy.

    D.   It is usually used to replace not more than about 20% of the grain in the ration.

    E.   It is frequently reconstituted with water before being fed.

XII. Dried sweet potatoes or sweet potato meal.

    A.   This has been largely an experimental product to date—it is not generally available.

    B.   It has attracted considerable attention over the years from an experimental standpoint because of the high yielding ability of sweet potatoes for producing digestible carbohydrates.

    C.   Sweet potatoes will outproduce most other crops (even corn) in yield of digestible carbohydrates per acre—however, they are relatively costly to produce.

    D.   Dried sweet potatoes are quite high in content of digestible starch and also carotene.

    E.   It is low in most all other nutritive fractions.

    F.   When used to replace not more than 50% of the corn in rations for cattle and sheep, dried sweet potatoes have been found to have a feeding value approaching that of corn.

    G.   At higher levels dried sweet potatoes have a feeding value somewhat below that of corn.

    H.   Swine do not seem to relish and do not perform particularly well on this product.

XIII. Dried bakery product.

A.   Dried bakery product consists primarily of stale bakery products and certain other bakery wastes which have been blended together, dried, and ground into a meal.

B.   There are companies which make a business of assembling such materials from over extensive areas and converting them into the feed which is officially recognized as *dried bakery product.*

C.   The production of some companies of this type of feed will amount to thousands of tons annually.

D.   Where careful control is exercised over the blending operation, a product of considerable uniformity can be produced from such materials.

E.   Dried bakery product is similar to corn in composition except that it is usually much higher in fat and may contain a considerable amount of salt.

F.   Dried bakery product is an effective substitute for corn in rations for cattle and swine, but in view of its relatively high salt content, its use should be limited to not over about 20% of the total ration.

XIV. Hominy feed.

A.   Hominy feed is a by-product of the manufacture of hominy, hominy grits, and corn meal for human consumption.

B.   It consists of a mixture of the corn bran, the corn germ (with or without some of the fat removed), and varying amounts of the finer siftings of the starchy portion of the corn grain.

C.   It is similar to corn in composition, although it usually contains slightly more protein and somewhat more fat, and it is higher in fiber.

D.   Hominy feed is about equal to corn in feeding value for the various classes of livestock and can be substituted for corn on about a lb-for-lb basis in most livestock rations.

E.   Hominy feed which contains the usual amount of fat will tend to produce soft pork when used as a major component of swine rations.

XV. Oat groats.

A.   Oat groats are oats grain from which the hull has been removed—in other words, the oat kernel.

B.   Since most of the feeding value of oats grain is found in the kernel, oat groats are very high in feeding value.

    C.   They are usually too expensive for general livestock feeding.

    D.   Their use is usually restricted to special diets such as early weaning rations for pigs.

XVI. Potato meal.

    A.   Cull and/or surplus potatoes are sometimes dried and ground to produce a feed known as potato meal.

    B.   Potato meal is satisfactory as a substitute for a part of the grain in beef and dairy rations.

    C.   Potato meal produced from potatoes which were thoroughly cooked prior to or during the drying process may be used as a partial substitute for the grain in swine rations.

XVII. Rye grain.

    A.   Less than 0.5% as much rye as corn is produced in the U.S.

    B.   A large portion of the rye produced is used for bread making—hence, not much is available for feeding purposes.

    C.   Rye grain is similar to corn, wheat, and grain sorghum in composition but is generally somewhat less valuable as a livestock feed.

    D.   Rye grain is less palatable than other grains and is sometimes contaminated with the fungus ergot which tends to accentuate this characteristic.

    E.   Rye contaminated with ergot is sometimes toxic to livestock—especially swine—if fed at high levels.

    F.   High levels of rye in dairy rations tend to produce a hard, unsatisfactory butter.

    G.   All things considered, rye has its greatest feeding value when it does not make up over about one-third of the ration.

    H.   Up to this level, it will usually approach the other low fiber grains in feeding value.

    I.   Rye grain should be coarsely ground or rolled when fed to livestock.

XVIII. Animal fat, feed grade.

    A.   Meat slaughtering, poultry dressing, and rendering plants are frequently burdened with surpluses of animal fat that can be bought

FIGURE 35.   Feed grade animal fat is frequently used at low levels in mixed rations. When used, it is normally added in the melted form. *(Courtesy of the University of Georgia College of Agricultural Experiment Stations)*

at prices which will permit its economical use in livestock feeds.

B.   Since such products are essentially pure fat, they have nutritional value primarily as a source of energy.

C.   Animal fat is used extensively in the manufacture of commercially mixed feeds at the rate of 1% to about 7%, depending on the type of feed, as a source of energy and also for the following reasons:

    1.   To reduce dustiness.

    2.   To improve color.

    3.   To improve texture.

    4.   To improve palatability.

    5.   To reduce machinery wear.

    6.   To increase pelleting rate.

D.   Beef cattle can make effective use of animal fat up to about 5% of the ration.

E.   Swine can make effective use of animal fat up to about 10% of the ration.

F.   Animal fat to be used for feeding purposes should be treated with an anti-oxidant to prevent it from becoming rancid.

XIX. Dried whey.

A.   Whey is that portion of milk which remains after most of the casein and fat have been removed for the manufacture of cheese.

B.   Dried whey is very high in milk sugar and so must be regarded primarily as an energy feed.

C.   It contains a fair amount of protein which is of excellent quality.

D.   It is also relatively high in calcium and phosphorus, as well as several of the B vitamins.

E.   Dried whey is used primarily in poultry rations and in early weaning diets for other classes of livestock.

XX. Corn starch and corn oil.

A.   These are relatively pure forms of starch and oil, respectively, obtained from the corn grain.

B.   They are not generally used as livestock feeds.

C.   They are included in the feed composition table (Table 37) to illustrate certain points for instructional purposes only.

D.   Corn starch and corn oil are sometimes used as ingredients in purified diets for experimental animals.

XXI. Rice bran.

A.   Rice bran consists primarily of the seed coat and germ which are removed from rice grain in the manufacture of polished rice for human consumption.

B.   Rice bran, while not as high in protein, is otherwise comparable to wheat bran in feeding value.

C.   Rice bran is fed primarily to dairy cows but may be fed to other classes of livestock with satisfactory results.

D.   In view of its relatively high content of fat and fiber and its frequent lack of palatability, the use of rice bran is usually limited to not over approximately one-third of the ration concentrates.

22 High Protein Feeds

I. Those of animal origin.

   A. Dried skimmed milk.

      1. Dried skimmed milk is defatted, dehydrated cow's milk.

      2. It usually contains around 34% protein.

      3. The protein of skimmed milk is of good quality.

      4. Dried skimmed milk is usually too high priced for general livestock feeding.

      5. It is sometimes used in early weaning diets for calves and pigs.

   B. Digester tankage. Also called *wet rendered tankage* or just *tankage.*

      1. Tankage is primarily a by-product of the meat packing industry—also available from dead animal rendering plants.

    2.   It consists of otherwise unusable animal tissue, including bones, which has been cooked under steam pressure, partially defatted, dried, and ground.

    3.   It usually contains from 55 to 60% protein.

    4.   The protein quality varies according to the processing, but it is usually good.

    5.   Because of its bone content, it is also a good source of calcium and phosphorus.

    6.   Digester tankage is used primarily in feeds for swine.

C.   **Tankage with bone.**

    1.   This product is similar to digester tankage except that it contains a greater proportion of bone, and consequently is higher in calcium and phosphorus and lower in protein.

    2.   It is used primarily in rations for swine.

D.   **Meat scrap or meat meal.**

    1.   This product is similar to digester tankage in origin except that it is cooked in steam jacketed kettles in its own fat—not under steam pressure.

    2.   It is similar to digester tankage in overall composition and general feeding value.

    3.   No dried blood is normally added to meat scrap, making it a more acceptable product than tankage.

    4.   The trend is toward the production of more meat scrap and less tankage.

    5.   Meat scrap is used primarily in rations for swine and poultry.

E.   **Meat and bone scrap or meal.**

    1.   This product is similar to meat scrap except it contains more bone, and consequently is higher in calcium and phosphorus and somewhat lower in protein.

    2.   It is used primarily in rations for swine and poultry.

F.   **Fish meal.**

    1.   Fish meal consists of fish or fish by-products which have been dried and ground into a meal.

    2.   There are several types, depending on the type of fish used.

    3.   The protein content of fish meal is usually around 60%.

    4.   Fish meal protein is usually of good quality.

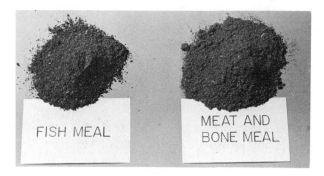

FIGURE 36. Two of the more commonly used high protein feeds of animal origin. *(Courtesy of the University of Georgia College of Agriculture Experiment Stations)*

    5.   It is also normally high in calcium and phosphorus.

    6.   It is used primarily in rations for swine and poultry.

G.   **Feathermeal.**

    1.   Feathermeal consists of poultry feathers which have been cooked under steam pressure, dried, and ground into a meal.

    2.   This product is extremely high in protein, usually containing well over 80%.

    3.   Feathermeal protein is of fair quality if it has been properly processed.

    4.   This product is used primarily in rations for swine and poultry.

H.   **Poultry by-product meal.**

    1.   This is a by-product of the poultry dressing industry.

    2.   It is produced from otherwise unusable portions of poultry carcasses which are cooked, dried, and ground into a meal.

    3.   Poultry by-product meal is similar to meat scrap in appearance, composition, and feeding value.

    4.   It is used primarily in rations for swine and poultry.

I.   **Dried whole milk.**

    1.   Dried whole milk is fresh whole milk which has been dried to a powder.

    2.   It is a good source of excellent-quality protein (25%+) and at the same time is very high in digestible energy.

    3.   Dried whole milk is so much in demand as a human food that it is usually too expensive to be fed to livestock.

4.   It is excellent as a component of milk replacers and early weaning diets for calves and pigs if and when its price will permit its use for these purposes.

J.   **Blood meal.**

1.   Blood meal is coagulated packing house blood which has been dried and ground into a meal.

2.   It is extremely high in protein (80%+), but the protein is lower in digestibility and quality than most other animal protein feeds.

3.   Also, blood meal is not very palatable to most livestock.

4.   For the above reasons blood meal is not very popular as a protein supplement, and its use as a ration component is usually restricted to a relatively low level.

II.   **Protein feeds of plant origin.**

A.   **Soybean oil meal (SOM) or soybean meal (SBM).**

1.   Soybean oil meal consists of fat-extracted soybeans which have been ground to a meal and sometimes pelleted.

2.   Most soybean oil meal is solvent-extracted today.

3.   It is of two grades: 44% and 49% protein.

4.   44% SOM is produced from whole soybeans; 49% SOM from soybeans that have had the seed coat removed.

5.   49% SOM is produced primarily for the broiler industry, secondarily for swine. It is satisfactory for cattle if the price is right.

6.   SOM is an excellent source of protein for all livestock classes.

7.   It is the most widely used high protein feed, making up approximately two-thirds of the country's high protein feed supply.

B.   **Cottonseed meal (CSM).**

1.   Cottonseed meal consists of dehulled, fat-extracted cottonseed ground to a meal with a certain amount of ground cottonseed hulls added.

2.   It is usually of either 41% or 36% crude protein grade, depending on the amount of hull added.

3.   41% protein is a practical maximum—36% protein is the legal minimum in most states.

FIGURE 37. Two of the more commonly used high protein feeds of plant origin. *(Courtesy of the University of Georgia College of Agriculture Experiment Stations)*

    4. Cottonseed meal is an excellent high protein feed for ruminants.

    5. However, it may kill growing swine if included in the ration at levels over 9% because of a toxic factor known as *gossypol.*

    6. It is the second most available high protein feed.

C. **Peanut oil meal (POM) or peanut meal (PNM).**

    1. Peanut oil meal consists of fat-extracted peanut kernels ground to a meal with a certain amount of ground peanut hulls added.

    2. Its composition and feeding value vary considerably, depending on the quality of the nuts, the method of fat extraction used, and the amount of hull included.

    3. Good quality peanut oil meal is comparable to soybean oil meal in feeding value.

    4. It will usually run around 45% protein.

    5. It is satisfactory as a source of protein for all livestock classes.

    6. Not too much available.

D. **Linseed (oil) meal, (LSM, LSOM, or LOM).**

    1. Linseed meal consists of fat-extracted flax seed.

    2. It is produced mostly in the Dakotas, Minnesota, and Texas.

    3. It is a satisfactory source of protein for almost all livestock classes when available, containing around 35% protein.

    4. Linseed meal is excellent for putting bloom on animals being fitted for show.

5.   Very little linseed meal is used outside of the flax producing
sections except in fitting beef cattle for show.

FIGURE 38.   Three high protein feeds frequently used—
especially in rations for dairy cows. *(Courtesy of the University of
Georgia College of Agriculture Experiment Stations)*

E.   Corn gluten meal.

1.   Corn gluten meal is a by-product of corn starch and corn oil
manufacture.

2.   It is usually around 40%+ in protein content.

3.   It is satisfactory as a source of supplemental protein for
ruminants.

4. However, it has a low essential amino acid supplemental value with most grains—hence, it is very poor as a protein supplement for nonruminants.

F. **Brewers dried grains.**

1. Brewers grains is the residue which remains after most of the starches and sugars have been removed from the barley and possibly other grains in carrying out the brewing process.

2. Some may be fed in the wet form near the brewery; however, most brewers grains is dried to facilitate handling, shipping, and storage.

3. Brewers dried grains is fed mainly to dairy cows, primarily as a source of protein and secondarily as a source of energy, up to about one-third of the concentrate mix.

4. It sometimes is used to replace a part of the grain in rations for horses, but it is too high in protein to be used very extensively for this class of livestock.

5. In view of its high fiber content, brewers dried grains is seldom fed to swine and poultry.

G. **Distillers dried grains.**

1. Distillers grains consists of the residue which remains in the use of grain for the production of alcohol—after the alcohol has been distilled off and other liquid materials have been removed.

2. It is sometimes fed in wet form near the distillery; however, it is usually dried to facilitate handling, shipping, and storage.

3. Distillers dried grains is fed mainly to dairy cattle, primarily as a source of protein and secondarily as a source of energy.

4. It is also satisfactory as a source of protein for beef cattle and sheep.

5. In view of its fiber content, distillers dried grains is seldom fed to swine and poultry.

H. **Copra meal.**

1. Copra meal is what remains after the dried meats of cocoanuts have been subjected to fat extraction and ground.

2. While it is regarded as a protein supplement, it is rather low in protein (22%+) compared with other high protein feeds.

3. The quality of its protein is not as good as that of soybean oil meal.

4.   The relatively low quality of its protein and its relative high content of fiber restrict its use in rations for swine and poultry.

5.   Copra meal makes an excellent dairy feed and is largely fed to this class of livestock.

I.   **Safflower meal without hulls.**

1.   This is what remains after most of the hull and the oil has been removed from safflower seed.

2.   It is fairly high in protein (40%+), but the quality of its protein is not as good as that of soybean oil meal.

3.   The relatively low quality of its protein and its relative high fiber content restrict the use of this product for swine and poultry.

4.   It is a satisfactory source of supplemental protein for other classes of livestock.

J.   **Sesame oil meal.**

1.   Sesame oil meal is produced from what remains following the production of oil from sesame seed.

2.   It is comparable to soybean oil meal in protein content, but the protein is not as good as soybean oil meal for supplementing farm grains for nonruminants.

3.   When available, it makes a very satisfactory source of supplemental protein for ruminants.

K.   **Sunflower meal or sunflower oil meal.**

1.   Sunflower meal is produced from what remains following the production of oil from dehulled sunflower seed.

2.   It is comparable to soybean oil meal in protein content, but it is not as good as soybean oil meal for supplementing farm grains for nonruminants.

3.   When available, it makes a very satisfactory source of supplemental protein for ruminants.

L.   **Soybeans, cottonseed, and peanuts.**

1.   Soybeans, cottonseed, and peanuts are usually subjected to oil extraction before being used for feeding purposes.

2.   Sometimes under circumstances of low vegetable oil prices and high protein meal prices, it becomes more economical to use

FIGURE 39. Mature soybeans being combined at harvest time. *(Courtesy of Sperry-New Holland, New Holland, Pa.)*

the unextracted beans, seed, or nuts as a source of supplemental protein than it is to use the oil meal.

3. While not as high as the respective oil meals in crude protein, the oil seeds are still fairly potent sources of protein, and with a higher fat content they contribute more energy to the ration on a pound-for-pound basis than do the meals.

4. Whole soybeans (37.9% crude protein) have been found to be a very satisfactory source of supplemental protein for beef cattle, horses, and sheep and also for dairy cattle except that large amounts of soybeans will tend to produce a soft butter.

5. Whole soybeans are not a satisfactory source of supplemental protein for swine unless the beans are thoroughly cooked, which significantly improves the quality and digestibility of the protein.

6. Whole soybeans, even though they are cooked, will tend to produce soft carcasses in growing-fattening pigs if they make up over about 10% of the ration.

7. Whole cottonseed (23.1% crude protein) have been used with satisfactory results as a source of supplemental protein for beef cattle and presumably could be used in a similar manner for dairy cattle, horses, and sheep.

8. Whole cottonseed are not a satisfactory feed for swine because of their high fiber content (16.9%) and because of their tendency to produce gossypol poisoning.

9. Harvested peanuts (28.4% crude protein) are usually too valuable as a human food to be fed to livestock, but there is no

apparent reason why they could not be used as a source of supplemental protein for all classes of livestock, should the price situation seem to warrant.

10. If used at levels above about 5% of the ration, peanuts are likely to cause dairy cows to produce soft butter and growing-fattening pigs to produce soft carcasses.

11. Peanuts are sometimes used as a swine feed through a hogging-off program, in which case they serve very effectively as the animal's primary source of both protein and energy except that soft carcasses will almost invariably result.

SOYBEANS   COTTONSEED   PEANUTS

FIGURE 40.   The three major oil seeds. *(Courtesy of the University of Georgia College of Agriculture Experiment Stations)*

III. **Commercial protein supplements.**

A. These are specially formulated protein supplements prepared by commercial companies.

B. Each is usually for a particular class of livestock.

C. They are usually blends of animal and vegetable high protein feeds, and probably include urea, if for ruminants.

D. They may also include minerals, vitamins, and/or antibiotics.

E. Those for swine will usually contain 30-40% crude protein while those for ruminants may run 2-3 times this high in crude protein as the result of a relatively high content of urea.

IV. **Hot meals.**

A. These are usually mixtures of a high protein feed and plain salt, with the latter added at levels ranging from 10 to 50%, but usually from 20 to 30%.

B.   They are used primarily for beef brood cows on low protein pasture.

C.   The salt serves as a consumption regulator, thus permitting free choice feeding.

D.   The level of salt is adjusted to bring about the desired consumption.

E.   With such a feeding program a cow will consume up to 5, and possibly more, times its normal salt consumption.

F.   The extra salt will cause no harmful effects as long as plenty of water is available.

G.   The cost of the extra salt is justified by the feeding convenience provided. Only a short trough is required with this method of feeding; trough space per animal is much less. All animals share in the feed on an equal basis.

H.   Some users of hot meal report an increase in appetite for low grade roughage from its use.

I.   This method of supplying supplement does not facilitate checking the herd at feeding time.

V.   **Liquid supplements.**

A.   This is simply an alternative form in which to provide supplemental nutrients.

B.   Nutrients are supplied either in water solution or in suspension.

C.   Minerals and vitamins as well as protein may be included.

D.   Blackstrap molasses and urea are usually the primary components of liquid cattle supplements, which usually carry a crude protein equivalent content of about 30-35%.

E.   Such supplements are used most extensively for beef brood cows on low protein forage.

F.   They are usually provided to such animals in a manner whereby they are consumed by licking—either from a wheel turning in a tank containing the supplement or from a rough-bottomed trough covered with a thin layer of the supplement.

G.   Their main advantage for brood cows is convenience of feeding while their main disadvantage is overconsumption, frequently making the cost prohibitive; also, since there is no definite feeding time, they do not facilitate the checking of the herd at the time of feeding as is possible with something like range cubes.

H.   Liquid supplements for reasons of economy and convenience are
also frequently used for adding supplemental nutrients to rations for
steers in the feedlot.

FIGURE 41.   Range cubes being fed as a supplement to beef
brood heifers being wintered on low quality pasture. *(Courtesy of
Ralston Purina Co., St. Louis, Mo.)*

VI.   Range cubes.

A.   Available on the market today are numerous products designed to be
used as supplements for beef brood cows and stocker animals
running on low-quality grazing such as dead growth of summer pasture or
on low grade hay.

B.   Such products are frequently provided in the form of so-called range
cubes which are simply large pellets, about 3/4"-1" in diameter and
1½"-2½" in length.

C.   They are usually fed on the ground in clean areas, thus making
troughs unnecessary and facilitating sufficient scattering to permit
all animals to get their proportionate share.

D.   Since some brands of range cubes are designed to provide supplemental energy as well as protein, they sometimes run as low as 20%
crude protein.

E.   They are usually fed at a level of 1-3 lb per head daily, depending on the quality and availability of the forage material.

F.   Range cubes have some definite advantages over liquid supplements and hot meals.

   1.   No facilities are required for feeding range cubes.

   2.   Level of consumption of range cubes can be precisely controlled.

   3.   The use of range cubes facilitates bringing the herd together at regular intervals for easy checking.

   4.   Range cubes may be used to facilitate moving the herd from one pasture to another.

23 Air-Dry Roughages

I. Legume hays and meals.

A. Alfalfa hay.

1. Alfalfa is the most widely produced hay crop.

2. It is a perennial which under favorable conditions will maintain a good stand for several years.

3. It will provide three or more cuttings per year.

4. It is one of the highest yielding hay crops.

5. It is the most nutritious of all hays.

6. It is relished by all hay-eating animals.

7. It is very drouth resistant.

FIGURE 42. Alfalfa being mowed, conditioned, and windrowed in one operation, preparatory to being baled for hay. *(Courtesy of Sperry-New Holland, New Holland, Pa.)*

8. Alfalfa is grown extensively throughout the Midwestern and Western states.

9. Not much alfalfa is grown in the Southern Coastal states.

   a. It requires fertile, well-drained soil and so is not adapted to the Coastal Plain.

   b. It is severely attacked by insects and diseases in other sections of the South.

   c. Coastal bermudagrass is much better adapted to southern conditions.

B. **Red clover hay.**

1. Red clover is one of the more important hay crops throughout much of the Corn Belt and the northeastern U.S.

2. It does not require as fertile soil as does alfalfa, it is less drouth resistant than alfalfa, and accordingly it is not ordinarily as high yielding as alfalfa.

3. Red clover is usually grown in combination with timothy and/or other nonlegume crops.

4.   Red clover-timothy mixtures will vary from almost all red clover the first year to almost all timothy after the first couple of years.

5.   Hay which is predominantly red clover will approach alfalfa in feeding value if harvested under favorable conditions.

C.   **Alsike clover hay.**

1.   Alsike clover in many respects is similar to red clover as a hay crop.

2.   It somewhat resembles red clover in appearance but is less rank growing and has a lighter colored blossom.

3.   It is grown in much the same area of the country as red clover but on a much less extensive scale.

4.   It too is usually seeded with timothy and other grasses for hay production.

5.   It will thrive on soils which are too wet and/or too acid for red clover.

6.   While not normally as high yielding as red clover, alsike clover hay compares favorably with red clover hay in feeding value.

D.   **Lespedeza hay.**

1.   Lespedeza hay approaches alfalfa hay in feeding value if of good quality.

2.   At one time lespedeza was widely grown as a hay crop throughout the southern states but not much of this crop is grown today.

  a.   Lespedeza is usually good for only one cutting per season.

  b.   Lespedeza is relatively low yielding.

  c.   Lespedeza hay often is not of good quality as the result of shattering and a high content of weeds.

  d.   Other more desirable hay crops have been developed—especially coastal bermudagrass.

E.   **Sericea hay.**

1.   Sericea is a perennial lespedeza grown in the southern states.

2.   Top quality sericea hay approaches alfalfa hay in feeding value.

3.   To make top quality sericea hay, it must be mowed when very immature and harvested in a manner to avoid shattering of the leaves.

    4.   While there is presently a considerable acreage of sericea to be found through the southern states, mainly on land formerly in the government soil bank, very little of it is used for hay-making purposes.

    5.   Sericea hay which is overmature and badly shattered is very low in feeding value and not readily eaten by livestock because of its stemmy nature and high content of tannin.

F.   **Peanut hay.**

    1.   Peanut hay usually consists of what remains of the peanut plant after the nuts, which grow underground, have been harvested.

    2.   Its quality varies greatly, depending on the harvesting procedure and conditions.

    3.   Top quality peanut hay will approach alfalfa hay in feeding value.

    4.   Most peanut hay is medium to low in quality and much inferior to alfalfa hay.

    5.   While it was once an important hay crop in the peanut sections, not much peanut hay is made today.

        a.   New harvesting procedures do not lend themselves to making peanut hay.

        b.   Most people rely on coastal bermudagrass and bahiagrass hay in those sections where peanut hay was formerly used.

G.   **Sweet clover hay.**

    1.   Sweet clover is not ordinarily seeded as a hay crop but is sometimes harvested for hay.

    2.   Sweet clover generally is too coarse and stemmy, and shatters its leaves too readily to make a quality hay.

    3.   Also, sweet clover tends to harbor a mold which causes a condition known as *sweet clover poisoning* in which an animal will bleed to death as the result of consuming excessive levels of dicoumerol present in some sweet clover hay.

H.   **Cow pea and soybean hays** (discussed together since they are similar in several respects).

    1.   Both crops are annuals and must be freshly seeded for each cutting.

    2.   Neither crop is used very extensively as a hay crop.

3.   Both are frequently used as emergency hay crops.

4.   The feeding value of both crops varies greatly, depending on harvesting conditions.

5.   Both are inclined to lose their leaves badly during harvest.

6.   Both crops normally contain considerable amounts of coarse stems which livestock tend to refuse.

7.   The overall consumption of both types of hay is greatly improved by fine chopping or coarse grinding.

FIGURE 43.   Dehydrated alfalfa meal. *(Courtesy of the University of Georgia College of Agriculture Experiment Stations)*

I.   **Dehydrated alfalfa meal.**

1.   Dehydrated alfalfa meal is alfalfa which has been harvested at an early stage of maturity, artificially dried, and ground into a meal.

2.   It is produced mostly in the western states.

3.   It is used for the most part in diets for poultry and swine, in early weaning diets, and in horse rations.

4.   It is used especially as a source of various minerals and vitamins and certain unidentified nutritive factors—also as a source of yellow pigments (carotene and xanthophyll) in layer diets.

5.   It is usually used at levels not to exceed about 5% of the overall diet.

6.   The most common grade of dehydrated alfalfa meal is that containing not less than 17% crude protein.

J. **Alfalfa leaf meal and alfalfa stem meal.**

1. Sometimes dehydrated alfalfa is processed whereby the leaves are separated from the stems to produce alfalfa leaf meal.

2. The remaining stems are then ground to produce alfalfa stem meal.

3. Alfalfa leaf meal is much lower than alfalfa meal in fiber content and is used primarily in certain special diets in which the nutritive content of alfalfa is desired but a low fiber level is required.

4. Alfalfa stem meal is produced strictly as a by-product of alfalfa leaf meal production.

5. While alfalfa stem meal consists almost entirely of alfalfa stems, this product has fairly good nutritive value as dry roughages go.

FIGURE 44. Large round bales of coastal bermudagrass hay remain in the field amid regrowth of grass without significant damage from the weather. *(Courtesy of E.R. Beaty of the University of Georgia College of Agriculture Experiment Stations)*

II. **Nonlegume hays.**

A. **Coastal bermudagrass hay.**

1. Coastal bermudagrass is the most extensively produced hay crop over much of the Deep South.

2. It is a medium-height perennial which, when properly fertilized, maintains its stand indefinitely.

3. Under favorable conditions it will provide three or more cuttings per year and is one of the highest yielding hay crops.

4. When adequately fertilized and cut at the proper stage, coastal bermudagrass will make a top quality nonlegume hay.

5. Coastal bermudagrass is a hybrid and does not produce viable

seed—hence, it must be reproduced vegatatively from root stolons.

6. It will sometimes suffer from winter kill in the colder sections of the South.

7. When allowed to become overmature before being cut, it loses much of its feeding value.

8. It is recommended that for quality hay production, coastal bermudagrass be cut about every 4-6 weeks.

B. **Timothy hay.**

1. Timothy is a medium-tall perennial grass.

2. It is one of the more widely grown grasses through the Corn Belt and the northeastern states.

3. It is usually seeded in combination with red clover but normally predominates the meadow after the second year.

4. Timothy is especially popular as a hay for horses, but a combination of timothy with red clover or alfalfa is generally preferred for other classes of livestock.

5. Like most hay crops, it should be cut not later than the early bloom stage for maximum yield of digestible nutrients.

C. **Redtop hay.**

1. Redtop is a fairly short-growing perennial grass which is found over much of the Southern Corn Belt and most of the northeastern states.

2. It is used extensively and primarily in pasture mixtures for these sections.

3. It will tolerate a wide range of soil and moisture conditions and is sometimes grown for hay where conditions are not favorable for more desirable hay crops.

4. While redtop cut at the proper stage and harvested under favorable conditions makes a fairly satisfactory nonlegume hay, the hay is not very palatable and is usually produced only as a last resort.

5. Redtop gets its name from the fact that the mature heads lend a reddish tinge to a meadow made up of this crop.

D. **Bromegrass hay.**

1. Bromegrass is a medium-tall perennial which is grown extensively throughout most of the North Central states but usually in combination with alfalfa and/or other legumes.

2.   It is especially valuable for preventing bloat in cattle and sheep grazing on such mixtures.

3.   When such mixtures are harvested for hay, bromegrass makes up an important part of the nutritive value of the overall forage combination.

4.   While bromegrass is not grown extensively in pure stand for hay production, when so grown and cut at the proper stage it makes a very satisfactory nonlegume hay crop.

E.   **Orchardgrass.**

1.   Orchardgrass is a medium-tall perennial which is widely grown in many sections.

2.   It is grown mostly in that area lying between the northern and southern border states.

3.   Orchardgrass is used primarily as a component of pasture mixtures but makes a very satisfactory hay if harvested no later than the early bloom stage.

4.   As with most forage crops, overmature orchardgrass makes a low-quality hay.

F.   **Reed canarygrass hay.**

1.   Reed canarygrass is a tall, rank-growing coarse-stemmed perennial.

2.   It is grown primarily in the northern states.

3.   It thrives on land too wet for most other hay crops.

4.   If harvested just prior to heading, it makes a fairly satisfactory hay.

5.   It is not one of the more extensively grown hay crops.

G.   **Bermudagrass (common) hay.**

1.   Bermudagrass is a low-growing perennial found extensively throughout the South.

2.   While it is widely used for pasture, it is used only to a limited extent for hay.

3.   When it has been well fertilized, bermudagrass hay approaches coastal bermudagrass hay in feeding value.

4.   Bermudagrass is not as drouth resistant or as high yielding as coastal bermudagrass.

5.   Much so-called bermudagrass hay is surplus pasture growth

which has been cut for hay; such hay is usually overmature and low in digestibility and feeding value.

H.   **Bahiagrass hay.**

1.   Bahiagrass is a medium-tall perennial which is widely grown over much of the southern Coastal Plain.

2.   It is grown primarily for pasture, secondarily for hay.

3.   Good quality bahiagrass hay approaches good quality coastal bermudagrass hay in feeding value.

4.   Since most bahiagrass hay is made from surplus pasture growth, it is usually overmature and low in quality.

5.   Bahiagrass hay is used primarily as a wintering feed for the beef breeding herd.

6.   Bahiagrass has the advantage over coastal bermudagrass of producing viable seed and so being reproducible from seed.

7.   It will not live through the winter above the southern Coastal Plain and so is found mostly in the Coastal Plain area.

I.   **Fescue hay.**

1.   Extensive areas of alta and Kentucky 31 fescue are found in various sections of the U.S.—especially in the southern states above the Coastal Plain.

2.   These two strains of fescue are very similar and are used almost interchangeably with each other.

3.   They are both medium-tall perennials which remain green throughout the winter in many sections.

4.   They are used primarily for pasture—especially for winter grazing.

5.   Other varieties of tall fescue are found in some sections, and while similar in composition to alta and Kentucky 31, they are as a rule less palatable.

6.   Excess growth of fescue pasture is sometimes harvested for hay in the late spring.

7.   The quality of fescue hay varies greatly, depending on the stage of maturity at harvest and harvesting conditions.

8.   Most fescue hay is of low quality, being high in fiber and low in protein and TDN.

9.   That fescue hay which is made is usually used as a wintering feed for the beef breeding herd.

J.   **Dallisgrass hay.**

   1.   Dallisgrass is a perennial bunch grass grown primarily as a pasture crop throughout much of the South.

   2.   Surplus growth of Dallisgrass pasture is sometimes harvested for hay.

   3.   Such growth is usually rather mature when cut for hay—hence, Dallisgrass hay is usually of only fair quality.

K.   **Johnsongrass hay.**

   1.   Johnsongrass, a tall, rank-growing southern perennial, over the years has been regarded as a weed and extensive acreages throughout the South have been infested with this plant.

   2.   While seldom, if ever, planted as a hay crop, it has been used for such frequently when available and needed.

   3.   Johnsongrass has coarse, sappy stems which make it difficult to cure into hay.

   4.   Livestock tend to refuse a large portion of the stems of Johnsongrass hay.

   5.   Johnsongrass is about as low in protein as any nonlegume crop.

   6.   Johnsongrass is higher in calcium than most nonlegumes.

   7.   Johnsongrass is usually found only on fertile soil.

   8.   Johnsongrass can be eliminated from a field by continuous grazing over a period of 2-3 years.

   9.   Very little Johnsongrass hay is made today.

L.   **Sudangrass hay.**

   1.   Sudangrass is a tall-growing annual which is frequently used as an emergency hay and/or pasture crop.

   2.   As to hay crop, sudangrass in many respects is similar to Johnsongrass.

       a.   Both are members of the sorghum family.

       b.   They resemble each other in appearance.

       c.   Both have coarse, sappy stems which present curing problems.

       d.   Livestock tend to refuse the coarse stems of both.

       e.   The two hays are similar in feeding value.

   3.   Sudangrass hay is normally available in very limited quantities

and is usually fed on the farm where it is produced.

M.   Millet hay.

1.   The millets are rapid-growing summer annuals of several types and varieties.

2.   Millet of one type or another is grown in most areas as a temporary pasture and/or hay crop.

3.   Millet cut at an early stage of maturity makes a satisfactory dry roughage for cattle but is not too satisfactory for lambs.

4.   Millet hay is not recommended for horses since it sometimes causes swollen joints and lameness in this class of livestock.

5.   Millet is strictly an annual but may produce more than one cutting per season.

6.   Not very much millet is grown for hay.

N.   Oat hay.

1.   Oat hay is oats which have been cut while still green, usually in the dough stage, and made into hay.

2.   The grain remains as a part of the hay.

3.   If cut at a fairly early stage of maturity, oats make a very satisfactory hay crop.

4.   More often than not, it is harvested as an emergency hay crop when it is apparent that other hay sources are in short supply.

5.   Oat hay should not be confused with the oat straw which remains after the grain has been harvested from the mature oat plant and which is much lower than oat hay in feeding value.

6.   Only a limited amount of oat hay is normally made in most areas.

O.   Prairie and meadow hays.

1.   Extensive acreages of native grasses are harvested for hay in many sections of the western states.

2.   Such forages are referred to as prairie hay or meadow hay, depending on the section.

3.   When cut at the proper stage and harvested under favorable conditions, such forages can be used to make hay of excellent quality, suitable for feeding all classes of hay-consuming livestock.

4.   Some such forages, however, are allowed to become overripe before being cut and, as a result, produce hay which is low in feeding value.

P.   Kentucky bluegrass hay.

1.   While Kentucky bluegrass is grown extensively throughout the North Central and the Northeastern states as a pasture and lawn grass, it is quite low yielding as a hay crop and is used very little for this purpose.

2.   Most Kentucky bluegrass that is harvested for hay is usually surplus pasture growth which is overmature and consequently low in feeding value.

III.   Miscellaneous air-dry roughages.

A.   Corn cobs, husks, and stover.

1.   Recent data by Lowry of the Georgia Station indicate that the dry matter of the present-day mature corn plant divides itself approximately as follows:

Grain	38%
Cobs	7%
Husks	12%
Leaves	13%
Stalks	30%

2.   With an annual production of corn grain of around 125 million tons, there then theoretically could be available in this country annually for feeding purposes as much as:

Cobs	23 million tons
Husks	39 million tons
Leaves	48 million tons
Stalks	99 million tons
	204 million tons

3.   While corn cobs, husks, and stover are at best all very low in feeding value, they do contain considerable amounts of digestible fiber and NFE, and, if properly supplemented with protein, minerals, and vitamins, can serve as a useful roughage in rations for at least certain classes of ruminant animals.

4.   The extent to which cobs and shucks have been used in the past for feeding purposes has been for the most part as components of ground snapped corn and ground ear corn or corn and cob meal.

5.   Very little corn is harvested as snapped corn or ear corn today;

most corn is presently combined, leaving the cob and husks in the field where they are for the most part unused.

6. The use of corn stover to date has been limited for the most part to grazing stalk fields with stocker animals.

7. For the most effective use of corn cobs, husks, and stalks, all should be finely chopped or coarsely ground to facilitate their consumption.

8. The use of such products may be facilitated also by ensiling.

B. **Soybean hulls and soybean mill feed.**

1. In the manufacture of 49% soybean oil meal and in the manufacture of soybean flour or grits, the outer coat of the soybean seed is removed and made available as soybean hulls for feeding purposes.

2. In some instances soybean hulls are steam rolled into a flake and marketed as "soybran flakes."

3. In other instances varying amounts of other by-products of soybean manufacture are mixed with the soybean hulls and the mixture marketed as "soybean mill feed."

4. Soybean hulls have an average as fed content of:

    Crude protein   11.3%
    Crude fiber   33.1%
    TDN   58.8%

5. As may be noted from the above, soybean hulls are fairly digestible considering their high fiber content.

6. The composition of soybean mill feed will vary, depending on the amount of soybean by-products other than hulls present, but it is normally somewhat lower in fiber and higher in protein and TDN than soybean hulls.

7. Since soybean mill feed usually consists chiefly of soybean hulls, the two products are ordinarily not too different in composition and feeding value.

8. The principal use of both products is in rations for dairy cows where they serve primarily as a source of bulk and secondarily as sources of protein and energy.

C. **Oat, barley, and wheat straws.**

1. These are what remains after the grain has been harvested from

the mature oat, barley, and wheat plants, respectively.

2.   While there are tremendous tonnages of all three straws available in the U.S. each year, they are used more extensively for bedding than for feeding purposes.

3.   Oat, barley, and wheat straws are sometimes used in maintenance rations for cattle and horses, but all three are very low in feeding value and must be supplemented with protein and minerals, and possibly vitamin A, for satisfactory results.

4.   Oat, barley, or wheat straw may be used as a satisfactory source of long roughage for fattening steers otherwise receiving a low roughage ration.

FIGURE 45.   Two commonly used low quality roughages. *(Courtesy of the University of Georgia College of Agriculture Experiment Stations)*

D.   **Cottonseed hulls, peanut hulls, and oat hulls.**

    1.   There is a considerable tonnage of all three of these products available in the U.S. each year.

       a.   Most cottonseed are subjected to oil extraction with cottonseed hulls as a by-product.

       b.   Peanut hulls is a by-product of the manufacture of peanut butter, peanut oil, and shelled peanuts.

       c.   Oat hulls results as a by-product of the manufacture of oat groats and rolled oats from oats grain.

    2.   All are usually mixed with more palatable feeds to induce their consumption.

    3.   All three are very low in nutritive value and are used, when fed, primarily as a bulk factor in the ration.

    4.   They may provide limited nutritive value to the animal if properly supplemented.

5. Of these three feeds, cottonseeds hulls is the most palatable and has the greatest feeding value.

E. **Rice hulls.**

1. Rice hulls is one of the by-products from the manufacture of polished rice from rice grain.

2. This material is one of the lower quality feeds fed to livestock.

3. Rice hulls are usually ground and mixed with more palatable feeds to induce their consumption.

4. About the only feeding value of rice hulls is as a bulk factor or filler in the ration.

24 The Molasseses

There are several types of molasses. All are concentrated water solutions of sugars, hemicelluloses, and minerals obtained usually as by-products of various manufacturing operations involving the processing of large amounts of the juices or extracts of plant materials.

I.   The various types of molasses are:

    A.   **Cane or blackstrap molasses.** This product is obtained as a by-product of the manufacture of cane sugar from sugar cane. It is what remains after as much sugar has been removed from the sugar cane juice as is possible following standard procedures. Considerable amounts of cane molasses are produced in southeastern U.S.; it is also imported from Hawaii and the Philippines.

    B.   **Beet molasses.** Beet molasses is obtained as a by-product of the manufacture of beet sugar from sugar beets. It is produced primarily in the sugar beet sections of the western states, and its use is largely restricted to that area of the country.

C.   **Citrus molasses.** Citrus molasses is produced from the juice of citrus wastes. At one time is was available in significant quantities in the citrus-producing sections. Citrus molasses is not as available as it once was since increasing amounts of it are being used in the production of citrus pulp with molasses, dried.

D.   **Wood molasses.** In the manufacture of paper, fiber board, and pure cellulose from wood, there results an extract which contains the more soluble carbohydrates and minerals of the wood material. In some instances this extract is processed into a molasses suitable for livestock feeding purposes.

II.   **Use of molasses in livestock feeding.**

   A.   The different types of molasses are similar in feeding value, pound for pound, of dry matter.

   B.   Cane molasses is used by far the most extensively.

   C.   Molasses is usually used in rations for cattle, sheep, and horses:

      1.   To improve ration acceptability.

      2.   To improve rumen microbial activity.

      3.   To reduce dustiness of ration.

      4.   As a binder for pelleting.

      5.   As a source of energy.

      6.   As a source of unidentified factors.

   D.   Molasses is usually used in amounts not to exceed about 10-15% of the ration.

      1.   It has its greatest feeding value at or below these levels.

      2.   More than 15% molasses will cause a ration to become sticky and difficult to handle.

      3.   High levels of molasses will tend to disrupt rumen microbial activity.

      4.   Used at levels not to exceed 10-15% of the ration, molasses approaches corn in feeding value; at higher levels its feeding value decreases considerably.

   E.   The different types of molasses are sometimes available in dehydrated forms.

1. Dehydration is usually accomplished by spray drying where a mist of molasses is sprayed into a blast of hot air.

2. The use of dehydrated forms will simplify ration mixing.

3. However, most forms of dehydrated molasses are quite deliquescent and must be used accordingly.

4. The same limitations apply to the use of dehydrated or dried molasses as apply to the use of the liquid product.

F. Molasses is seldom used in rations for swine since it tends to cause scouring in this class of livestock.

III. Molasses Brix.

A. *Brix* is a term commonly used in referring to the composition of molasses. Brix is expressed in degrees and was originally used to indicate the percentage by weight of sugar in sucrose solutions, with each degree Brix being equal to one percent of sucrose.

B. One way to determine the Brix of a sucrose solution is to measure its specific gravity and then refer to a conversion table to arrive at the degrees Brix or the level of sucrose present. The Brix of a molasses is determined in the same way, but since in molasses there are considerable amounts of soluble materials other than sucrose present which influence its specific gravity, the Brix value is not a true measure of the sugar content of molasses. It does, however, reflect the relative level of sugar present and so has over the years been used as a convenient basis for expressing molasses quality.

Generally accepted Brix values for cane, beet, and citrus molasses, along with corresponding specific gravities and weights per gallon for these products, are as shown below:

	DEGREES BRIX	SPECIFIC GRAVITY ($20°$ C)	WT PER GALLON
Cane molasses	79.5	1.4109	11.75 lb
Beet molasses	79.5	1.4109	11.75 lb
Citrus molasses	71.0	1.3558	11.29 lb

25 Other High Moisture Feeds

I.   **High moisture grain.**

A.   Frequently it is necessary or desirable to harvest feed grains, especially corn and grain sorghum at moisture levels too high to permit safe bin storage without artificial drying. In some instances, the moisture level is so high as to make artificial drying uneconomical. In such instances, the grain is sometimes stored in an airtight silo and preserved in a semi-ensiled state. The moisture level of such grain will usually run from about 22% to 30%, and it is commonly referred to as high moisture grain—not that it is extremely high in moisture content, but it is simply somewhat higher than that normally stored under bin conditions. High moisture grain has been found to be equal to and in some instances superior to air-dry grain as a feed for most classes of livestock on a dry matter basis. Especially is this true for grain sorghum.

B.   Since the dry matter content of high moisture grains varies considerably, when using such feeds for ration formulation it is usually best to work on a dry matter basis. If composition figures for high moisture grain as such are not available, the dry basis composition figures for the air-dry grain can be used in balancing the ration. The dry weights must then be converted to the moisture level of the product as fed before carrying out the actual feeding operation. In order to be able to do this, it is necessary, of course, to determine the dry matter content of the particular high moisture grain on hand. However, by working on a dry matter basis, only the dry matter content and not the complete analysis of the high moisture grain being used needs to be known.

II.   Haylage.

A.   Frequently weather conditions are not favorable for harvesting green forages for hay. On the other hand, such crops often are too high in water content to make good silage. A practice which is sometimes followed in such instances is to make the crop into what has become known as haylage. Haylage is the feed produced by storing in an airtight silo a forage crop which has been dried to a moisture level of about 45-55%. Such forages undergo no seepage in storage and usually produce a silagelike product of excellent quality. On a dry matter basis haylages have been found to be equal, and sometimes superior, to hay made from the same crop.

B.   Since the moisture content of haylages varies considerably between batches, when such feeds are used in ration formulation, as with high moisture grains, it is usually best to proceed on a dry matter basis. If composition figures for the particular haylage under consideration are not available, they can usually be fairly closely approximated on a dry basis from the dry basis composition of hay from the same forage.

C.   In any event, the dry weights, obtained from the ration-balancing process must be converted to an as fed basis in carrying out the actual feeding operation. In order for this to be done, however, it is necessary that the moisture level of the haylage be determined.

III.   Silages. On many livestock farms, forages are harvested in a high moisture state and stored under anaerobic conditions to produce a feed commonly known as silage. Most any crop may be made into silage, although some are superior to others for this purpose. Conventional silage is a roughage feed and must be used accordingly in the feeding program. Since silage normally contains only about 25% to 35% dry matter, about 3 lb of silage must be used to replace a lb of air-dry hay in the feeding program. While this sometimes requires an animal to consume a considerable poundage of feed per head daily, a large part of this feed is moisture, the intake of which the animal compensates for by

drinking less water. For further details about silos, silage making, and silage utilization, refer to section 28, "Silage—Its Production and Use".

IV. **Fresh forages.**

A.   Green forages are utilized for feeding in the fresh form primarily as pasture. Harvested green forages are usually made into hay, haylage, or silage for use in the feeding program. However, sometimes crops are harvested in the green state and fed in the fresh form as so-called *greenchop*. This method of forage utilization has been confined to dairy operations for the most part in the past. Up to the present it has received almost no use in beef cattle feeding programs. It could be, however, that problems of pollution, production efficiency, and/or energy conservation might necessitate a wider use of this method of forage utilization in feeding cattle of all kinds in the future.

B.   On the other hand, until such time as this method of forage feeding is more generally followed and the use of the different forage crops for this type of feeding program has been more thoroughly studied, it does not seem appropriate to attempt to discuss in detail at this point the use of various types of greenchop for feeding purposes. It seems that it should suffice for the present simply to say that the dry matter of greenchop should have a feeding value comparable to that of hay, haylage, or silage made from the same crop. It should be kept in mind, however, that individual forage crops will vary in composition from day to day, depending on weather conditions and stage of maturity. Especially variable is the percentage of dry matter in the fresh forage material, which tends to complicate the matter of knowing just how much fresh forage to feed on any certain day to provide the desired amount of forage dry matter.

V.   **Wet by-products.** Under air-dry feeds, reference was made to such products as dried citrus pulp, dried beet pulp, brewers dried grains, and distillers dried grains. These same materials are sometimes available at their point of production in the fresh or wet state and may be used in this form for feeding purposes. Their feeding value in the wet form is not materially different on a dry basis from that of the air-dry products. However, in view of their high water content it is uneconomical to transport such feeds any considerable distance. Also, they cannot be stored effectively, and consequently fresh supplies must be obtained daily. Since, as with high moisture grain and haylage, the moisture content of such products is not too consistent, it is usually best to work with such feeds on a dry basis when using them in the feeding program.

VI. **Roots and tubers**

A.   Extensive use has been made of roots and tubers for livestock feeding in northern Europe over the years. However, roots and tubers have never been used on an extensive scale for livestock feeding in the United States, and the amount of such crops used for feeding purposes in this country today is almost insignificant.

B.   When used for feeding purposes, roots and/or tubers are usually used as a substitute for silage. Sometimes such crops are used simply as a relish for livestock being fed for top performance or being fitted for show. However, barring major changes in our nation's agricultural situation, it is not conceivable that roots and tubers will ever play a significant role in this country's livestock feeding of the foreseeable future. Consequently, it does not seem a worthwhile use of space to discuss these crops individually from the standpoint of the different classes of livestock in this text.

The compositions of a representative group of such crops have been included in Table 37. This has been done not because these crops are now or are apt to become important livestock feeds but purely so that such information might be available to the teacher and student for instructional purposes.

VII. **Fresh milk.**

A.   All farm livestock normally produce significant quantities of milk upon giving birth to young. Only the milk of dairy cows, however, is ever available in quantities such as might permit its use on occasions for feeding purposes on other than a nursing basis. Even then rarely is it ever fed in any significant amount in the fresh form. Usually any fresh milk which is in surplus of that needed for current human consumption is dehydrated and held in powdered form for later use, primarily as food for humans and secondarily as a component of certain specialty diets for livestock. Consequently very little fresh milk, other than that consumed by nursing young, ever finds its way into the feeding program.

B.   Even so, however, all of our farm livestock do produce milk upon giving birth to young, and milk of one kind or another is the primary feed for up to several weeks of most newly born farm animals. Also, surplus milk is sometimes used in other ways for feeding purposes. In other words, a large part of the feed fed to livestock goes to support the production of milk—milk which in turn is used to a very significant degree as a feed for farm animals. Consequently, it behooves a student of feeds and feeding to become aware of the important role which milk plays either directly or indirectly in most livestock production programs and to prepare himself for coping with the various aspects of its production and/or utilization.

C.   Since the milks of the different farm animal species are fairly similar in composition and since quantitatively cow's milk is by far the most important, composition figures for only cow's milk in its different forms and its by-products have been included in Table 37.

26 Identifying Feeds from Their Composition

While tables of composition are usually available from which information on the composition and nutritive value of individual feeds may be obtained, students of feeds and feeding will find it helpful to be able to relate individual feeds to a certain general composition or to associate some specific composition with a particular feed group without the aid of composition tables. In Table 16 the author has attempted to summarize some of the more distinguishing composition characteristics of the various feed groups, and a knowledge of the information provided in this table will greatly facilitate one's ability to work with feeds and feed composition figures.

In the use of this information, however, it should be realized that in many instances there is considerable overlapping of different feed groups with respect to their content of the various nutritive fractions, and frequently there is no single point which will provide a clearcut line of separation of the feed groups with respect to certain nutritive components. Even so, information such as is presented in Table 16 can be most helpful to a student of feeds and feeding as he attempts to become familiar with the composition and feeding value of different feeds and relate them to each other.

An interesting and helpful learning exercise in this connection is the systematic use of such information in associating a particular composition with a specific feed group or possibly even with an individual feed.

I. **Air-dry vs. high moisture feeds.** The first step in carrying out such an exercise is to look at a feed's dry matter content to determine if it is an air-dry feed or a high moisture feed.

 A. If a feed contains over about 80% DM, it should be regarded as an air-dry feed and would probably fall into one of the following feed groups:

Mineral products	Air-dry protein feeds of plant
Legume hays	origin, defatted
Nonlegume hays	Air-dry protein feeds of plant
Low-quality air-dry roughages	origin, not defatted
Air-dry energy feeds	(oil seeds)
Air-dry protein feeds of animal	
origin	

 B. On the other hand, if a feed contains less than about 80% DM, it should be regarded as a high moisture feed and would probably fall into one of the following feed groups:

High moisture grains	Fresh forages
Molasseses	Wet by-products
Haylages	Root crops
Silages	Fresh whole or skimmed milk

 C. **Mineral products.** If a feed on the basis of its composition falls into the air-dry feed category, the next step is to decide whether it is a roughage, a concentrate, or a mineral feed. Mineral feeds or products are readily recognized as such from their composition in that they are ordinarily quite high in ash (usually over 80%) and except for steamed bone meal contain very little, if any, of the organic nutrients. They may or may not be high in calcium and/or phosphorus, depending on the product under consideration.

 D. **Air-dry roughages vs. air-dry concentrates.**

 1. An air-dry nonmineral feed may be classed as a roughage or a concentrate by referring to its crude fiber and/or its TDN content. As a general rule air-dry roughages will run over 18% in crude fiber and contain less than 60% TDN. Air-dry concentrates, on

the other hand, will usually contain less than 18% crude fiber and over 60% TDN.

2.   Air-dry roughages may be further divided on the basis of their composition into the following groups:

Legume hays   Low-quality air-dry roughages
Nonlegume hays

E.   Legume vs. nonlegume vs. low-quality air-dry roughages. The hays are characterized by having from 18% to 34% crude fiber, 40-60% TDN, and a considerable amount of carotene whereas the low-quality dry roughages are on the average higher in fiber and lower in TDN with little or no carotene and a protein content of under 6.0%, except for peanut hulls at 6.6%. The legume hays are easily distinguished from the nonlegume hays on the basis of composition in that the former ordinarily contain over 10.5% crude protein and over 0.9% calcium while the latter will usually contain from 6% to 10.5% protein and less than 0.9% calcium.

F.   Energy concentrates vs. protein concentrates. If a feed on the basis of its composition turns out to be an air-dry concentrate, it must then be decided whether it is a protein concentrate or an energy feed, since both on the basis of TDN and crude fiber contents qualify as air-dry concentrates. Air-dry concentrates with over about 18% crude protein are usually classed as protein concentrates and those with less than 18% crude protein as energy feeds. The protein concentrates may be further subdivided into those of animal origin and those of plant origin.

G.   Protein supplements of animal vs. plant origin. Those of animal origin. except for dried whole milk and dried skimmed milk, all run over 47% crude protein, whereas those of plant origin, except for 49% soybean oil meal, all run under 47% protein. Also, most of those of animal origin run over 1.0% calcium and 1.5% phosphorus and under 2.5% fiber, while those of plant origin usually contain less than 1.0% calcium and 1.5% phosphorus and over 2.5% fiber.

Those protein concentrates of plant origin may be divided into the defatted meals, on the one hand, and the oil seeds, on the other. These two groups are easily distinguished from each other in that the defatted meals will always run under 7% and usually much lower in ether extract, whereas the oil seeds all exceed 17% in this nutritive fraction.

II.   High moisture feeds. If a feed contains less than 80% DM, it will normally be classed as one of the high moisture feeds, the subgroups of which were listed previously. The separation of these different subgroups on the basis of their composition is not nearly as clearcut as is that for the air-dry feeds. There are, however, numerous composition differences among some of these various subgroups which the student will find interesting and helpful.

A.   For example, if the composition of a feed shows it to contain between 60% and 80% DM, the feed is probably either a high moisture grain or a molasses. To which of these two groups the feed belongs may be determined from its protein and/or crude fiber content. The molasseses normally contain no crude fiber, whereas high moisture grains will always show from small to significant amounts of this feed constituent. Also, the molasseses will normally run under 7% crude protein, unless some product such as urea has been added, while the high moisture grains will usually exceed this level of crude protein, even on a wet basis.

B.   Any feed containing between 45% and 60% DM will almost always be one of the haylages. The other high moisture feeds including the silages, the fresh forages, the wet by-products, the root crops, and the fresh milks all normally contain less than 45% DM. However, there is considerable overlapping of the latter groups in DM content. Silages will in general range from about 25% to 45% in DM content and, being more mature, will usually surpass fresh forage crops in this regard. Also, they will usually contain less carotene. Otherwise, there is little basis for distinguishing between these two feed groups on the basis of the composition information usually available.

C.   Wet by-product feeds might be confused with fresh forages, on the one hand, and root crops, on the other, based on their content of DM. However, the fresh forages can usually be distinguished from the wet by-products on the basis of carotene content, if this information is available, since the wet by-products will usually contain little, if any, carotene, whereas the fresh forages will normally be quite high in this regard. On the other hand, root crops can usually be distinguished from wet by-product feeds from their composition in that the latter will usually contain over 3% crude fiber on a wet basis, while the root crops will usually run below this figure in crude fiber content.

D.   While some of the root crops might be confused with the fresh milks on the basis of DM content, it is easy to distinguish between these two groups on the basis of composition in that the root crops will always contain at least a small amount of fiber whereas the fresh milk always exhibit a 0.0% fiber content.

III.   **Identifying individual feeds.** With more experience it is frequently possible to distinguish between two or more feeds within the same feed group on the basis of their composition. For example, fresh whole milk would be easy to distinguish from fresh skimmed milk on the basis of fat content. Ground snapped corn, ground ear corn, and ground shelled corn could be separated on

Table 16
SUMMARY OF SOME DISTINGUISHING COMPOSITION CHARACTERISTICS OF DIFFERENT FEED
GROUPS (all composition figures on an as fed basis)

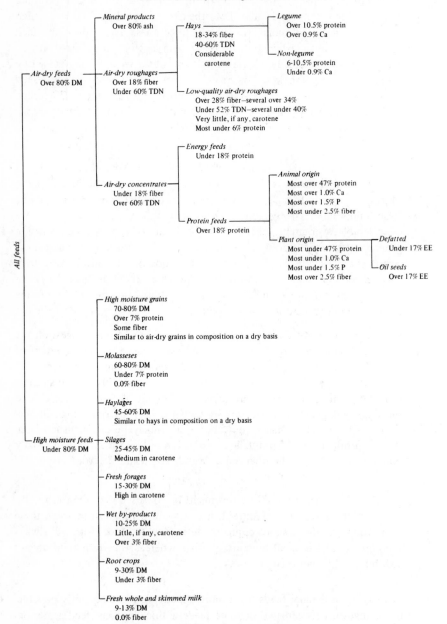

the basis of their relative contents of crude fiber. Ground snapped corn or ground ear corn could be distinguished from wheat grain or grain sorghum grain on the same basis. Yellow shelled corn might be distinguished from the other feed grains from its content of carotene. Meat scrap might be distinguished from meat and bone meal through the higher bone content and accordingly higher calcium and phosphorus contents for the latter product. Solvent-extracted meals are always lower in ether extract than are the mechanically extracted meals in view of the greater efficiency of the solvent process for fat extraction. Alfalfa meal, alfalfa leaf meal, and alfalfa stem meal are easily distinguished from each other since the leaves of the alfalfa plant are higher in protein and lower in crude fiber than are the stems, and the three meals vary in composition accordingly.

The above are only a few of the more obvious illustrations of how feeds might be distinguished from each other on the basis of logical differences in composition. Many others which might not be so clearcut could be sighted. In fact, with experience a student of feeds and feeding can develop his proficiency along this line to a point where he can usually recognize a certain composition as being that of any one of about two or three feed possibilities, and frequently he can even specify the particular feed.

27 Hay and Hay Making

I.   **Object in hay making**: To reduce the moisture content of green forage crops sufficiently to permit their safe storage without spoilage or serious loss of nutrients.

    A.   The maximum permissible water content for baling hay is around 18-22%, depending on the fineness of the hay, the tightness of the bale, the prevailing humidity, and the amount of air movement.

    B.   Freshly mown hay which is stored before it is sufficiently dry can be destroyed by fire resulting from spontaneous combustion.

    C.   Hay which is stored containing excessive moisture may tend to mold, making it unsuitable for feeding purposes.

    D.   Hay which is improperly harvested may suffer losses due to:

1. **Shattering.** This is the loss of leaves, which represent the most nutritious part of the hay plant. Legumes tend to shatter badly.

2. **Leaching.** Rain on hay during the curing period will tend to leach out and cause the loss of the more soluble nutrients.

3. **Bleaching.** While a certain amount of exposure to sunlight is essential and desirable for making good hay, excessive exposure to sunlight will cause heavy losses of certain nutrients—especially carotene.

II. **Proper hay-making procedures.**

A. A forage crop to be harvested for hay should be mowed just as soon after reaching an early bloom stage of maturity as circumstances will permit. Undue delay in harvesting hay will result in low-quality hay. However, it is sometimes necessary to delay the harvesting of a hay crop because of extended periods of rain or forecasts of rain and because of other extenuating farm circumstances.

B. Every effort should be made to select periods of rain-free weather for harvesting hay. A minimum of one and normally about two days of good drying weather are required for curing hay. Weather forecasts can frequently be of great assistance in this connection.

C. "Conditioning" will reduce the curing time required for certain crops, especially crops with coarse sappy stems, by as much as 50%.

FIGURE 46. Coastal bermudagrass being mowed, conditioned, and windrowed in one operation preparatory to being harvested for hay. *(Courtesy of E.R. Beaty of the University of Georgia College of Agriculture Experiment Stations)*

FIGURE 47.   A modern hay-making operation in progress. Note
that as the bales are formed, they are automatically ejected into the
wagon at the rear. *(Courtesy of Ford Tractor Operations, Ford
Motor Co., Troy, Mich.)*

To "condition" a hay crop, it is passed through a set of rollers to crack
open the stems and thereby facilitate the drying process.

D.   A forage crop being harvested for hay should be raked before it is
      completely dry to avoid excessive shattering and overexposure to the
sun.

E.   Turning of a windrow, when necessary to facilitate drying, should be
      done when the dew is on—especially with hays subject to serious
shattering.

F.   Baling should be carried out just as soon as the hay is sufficiently
      dry. Square bales should be stored as soon as possible and in any
event before they are rained on. Round bales, however, will shed rain and
may be left in the field for extended periods without serious damage to
the hay.

G.   The high cost of baling, storing, and feeding conventional hay bales

FIGURE 48. Feed handling is largely mechanized today. Chain-drag type elevators such as the one shown are widely used for handling almost all types of feeds, ranging from small grains to baled hay. *(Courtesy of Kewanee Machinery & Conveyor Co., Kewanee, Ill.)*

during the past few years has prompted many producers to start putting hay into large round bales or into mechanically formed stacks.

1.   Such bales and stacks will vary greatly in size, depending on the circumstances, with the bales ranging from about 500 lb up to around 3,000 lb in size and the stacks ranging up to as much as 3 tons or more.

2.   Both bales and stacks are ordinarily stored in the open and normally do not suffer excessive weather damage from such exposure.

3.   Grass hays usually suffer less damage than do legume hays from outside storage.

4.   Stacks completely eliminate the use of twine, and the large round bale reduces its use to a minimum.

5.   The harvesting and feeding of both large bales and stacks are usually completely mechanized.

6.   Where cattle are simply turned in to such bales and stacks, there is usually considerable loss of hay from trampling.

7.   Large round bales (Figures 49-52) lend themselves to being unrolled for rack or pasture feeding, and there are mechanical devices available to facilitate this practice.

8.   The equipment for making and handling large bales and stacks

FIGURE 49.   A large-round-bale baler in operation. *(Courtesy of Vermeer Manufacturing Company, Pella, Ia.)*

FIGURE 50.   A large-round-bale unroller in operation. Such a procedure is sometimes followed for feeding large round bales on the ground (or snow). *(Courtesy of Vermeer Manufacturing Co., Pella, Ia.)*

(Figures 53 and 54) is relatively expensive and can be justified only where a considerable volume of hay is involved.

FIGURE 51.   Hereford brood cows eating hay from large round bales which have been unrolled on the snow. Wastage from trampling is sometimes excessive with such a practice—especially if the hay is fed too liberally. *(Courtesy of Vermeer Manufacturing Co., Pella, Ia.)*

FIGURE 52.   An effective arrangement for feeding large round bales from the standpoint of holding wastage down. *(Courtesy of Vermeer Manufacturing Co., Pella, Ia.)*

(a)

(b)

FIGURE 53.   A hay stacker in operation. (a) Stacker being filled. (b) Stacker being unloaded. *(Courtesy of Farmhand, Inc., Hopkins, Minn.)*

III.   **Special characteristics of hay as a feed.**

    A.   Hay is classed as a dry roughage and is used primarily as a source of bulk in the ration, secondarily as a source of nutrients.

(a)   (b)

FIGURE 54.   A stack mover in operation. Such equipment may be used for moving either (a) stacks or (b) large round bales. *(Courtesy of Farmhand, Inc., Hopkins, Minn.)*

    B.   Hay of average quality will usually run from 25% to 32% crude fiber and 45-55% TDN.

    C.   Hay is primarily a cattle, horse, and sheep feed. Very little hay of any kind is ever fed to swine. Dehydrated alfalfa meal is sometimes included in swine rations up to 5-10% of the ration, but its use for this purpose is on the decline.

28 Silage— Its Production and Use

I. **What is silage?** Silage is a fermented feed resulting from the storage of high moisture crops, usually green forages, under anaerobic conditions in a structure called a silo.

II. **What is a silo?** A silo is an airtight to semi-airtight structure designed for the storage and preservation of high moisture feeds as silage.

III. Silos are of several types.

    A. **Horizontal.**

        1. **Trench.**

            a. Usually consists of a trench dug into the side of a hill—sometimes all the way through the top of a hill.

            b. They are of all different sizes but usually range from 10 ft to 20 ft in depth, 15 ft to 20 ft in width, and 100 ft to 300 ft in length.

      c.   They are usually narrower at the bottom than at the top to facilitate effective packing of silage material.

      d.   The floor and walls may vary from plain dirt to all concrete. Usually at least the floor is concreted in order to avoid the problem of mud as the silage is removed.

      e.   The floor is usually sloped toward the downhill end to permit drainage.

      f.   Such silos are usually filled from the upper end to the lower end through the use of dump wagons and/or trucks.

      g.   A tractor is usually used to pack the silage as the silo is filled.

      h.   When the silo is filled, the silage is usually covered over with a heavyweight polyethylene film which is weighted down and held in place with soil, boards, posts, old tires, or some other material. As an alternative to plastic film, a thin layer of cane molasses has been used to provide an effective seal.

      i.   Trench silos are usually emptied from the lower end to the upper end using a tractor with a front-end loader.

      j.   Trench silos are sometimes provided with movable stanchions and used on a self feeding basis.

      k.   A trench silo which extends all the way through the top of a hill would be treated much as two trench silos end to end with each other.

  2.   **Bunker.**

      a.   This type of silo is sometimes used on very flat, rocky, and/or pervious soil not well suited to a trench silo.

      b.   The side walls are usually made with posts and boards lined inside with building paper or plastic film. The floor is usually concreted, and the ends are left open.

      c.   A bunker silo is used much in the same manner as a trench silo, the main difference being that the former is above ground and the latter below ground.

B.   **Vertical or upright**—sometimes called *tower silos.*

  1.   **Conventional upright** (semi-airtight).

      a.   Most present-day conventional upright silos are constructed of reinforced poured concrete or of concrete staves.

      b.   All upright silos are circular in shape.

FIGURE 55.   Shown in the picture are two upright, concrete stave, top unloading silos along with a completely mechanized feeding setup. *(Courtesy of Badger Northland Inc., Kaukauna, Wis.)*

    c.   Conventional upright silos may or may not have a roof; they usually do, primarily to protect the unloading equipment.

    d.   A conventional upright silo is ordinarily equipped with a series of doors about 2 ft square approximately every 6 ft up one side of the silo. These are closed as the silo is filled and opened as the silo is emptied.

    e.   Conventional upright silos vary greatly in size but are usually from about 12 ft to 20 ft in diameter and from 40 ft to 80 ft in height.

    f.   This type of silo is normally unloaded from the top—usually with a mechanical unloader.

    g.   Conventional upright silos are usually emptied before being refilled, but a partially empty silo of this type may be refilled.

    h.   For effective preservation of ensilage in a conventional upright silo, the ensilage should contain from 25% to 35% dry matter.

2.   **Airtight or sealed silos.**

   a.   To date these have all been of an upright type.

   b.   They are constructed of protected metal with rubber-cemented joints.

   c.   When properly constructed, they are completely airtight.

   d.   They vary in size from about 12 ft to 24 ft in diameter and from about 40 ft to 100 ft in height.

   e.   Forages varying in dry matter content anywhere from about 25% to 75% may be effectively preserved and stored in an airtight silo.

FIGURE 56.   Oxygen-free type silos equipped with a completely mechanized feeding system. *(Courtesy of A. O. Smith Harvestore Co., Arlington Heights, Ill.)*

   f.   There are several makes of airtight silos.

      •   The A.O. Smith *Harvestore* (Figure 56) was the first and is probably the most widely used airtight silo. In many respects this silo approaches the ideal in silage preservation and storage. It holds ensiling losses to a minimum. It facilitates continuous harvesting and feeding. It has a breather bag in the top to allow for interior gas expansion and contraction. It unloads from the bottom by means of a mechanical unloader. It has two major drawbacks: (1) Its initial cost is relatively high, and, (2) The feeder is completely dependent on the mechanical unloader for silage.

● The *Herd King* manufactured by Butler is another widely used airtight silo. It differs from the Harvestore in that it unloads from the top through a hole at the center which is formed as the silo is filled by means of an elevatable core. Excessive interior gas pressures are avoided by means of a two-way valve between the inside and the outside of the silo. To a more or less degree the Herd King has the same drawbacks as the Harvestore.

IV. **Kinds of crops used for silage.**

A. Practically any crop may be made into silage, provided it contains an appropriate level of moisture, adequate amounts of readily fermentable carbohydrates, and adequate levels of other nutrients, and can be sufficiently packed.

B. The most commonly used silage crops are:

1. **Corn.**

   a. Most extensively used silage crop—about 10 times as much corn as sorghums is grown for silage.

FIGURE 57.   A forage harvester being used for chopping corn in the field for making silage. Corn is by far our most widely used silage crop. *(Courtesy of Allis-Chalmers, Milwaukee, Wisc.)*

   b. Corn silage is unexcelled in quality.

   c. Corn is not as high yielding as some of the forage sorghums.

2. **Forage sorghum.**

   a. Included here are the traditional sweet sorghums as well as some new hybrid sorghum varieties.

      b.   Most are very high yielding but do not produce the best silage; such silage is inclined to contain excess water and acid, and hence is not too palatable and nutritious.

      c.   Probably not over about 5% of the silage produced in the U.S. is forage sorghum.

  3.   **Grain sorghum.**

      a.   Probably not over about 5% of the silage produced in the U.S. is grain sorghum.

      b.   Grain sorghum silage is between corn silage and forage sorghum silage in palatability but because of a higher dry matter content is superior to both in nutritive value.

      c.   The grain sorghums are generally low yielding.

  4.   **Small grains and hay crops.**

      a.   None of these is used very much as a silage crop.

      b.   The quality of silage resulting from these materials is quite variable, depending on various factors such as stage of maturity, moisture content, and additives used, if any.

      c.   Yields are variable but are usually below corn and the sorghums.

V.   **Preparation of forage for making silage.**

  A.   Most crops to be used for silage are permitted to mature or field dry to a moisture level of around 65-75% (25-35% DM). For corn this is about the early dent stage of maturity and for grain sorghum the late dough stage at the earliest. This is when the moisture level is about right for good silage formation. Silage materials containing under 25% dry matter (over 75% moisture) will form a very sour silage and will usually lose considerable amounts of silage juices during storage, involving a considerable loss of nutrients. Silage materials with over 35% dry matter do not pack well and will frequently develop spots of mold during storage as the result of excess entrapped oxygen. This is especially true for nonairtight silos.

  B.   Silage crops are usually chopped up into fairly small pieces for making silage. The pieces will usually vary from a fraction of an inch to over an inch in length. This permits good packing and facilitates the mechanization of silage handling. Ground shelled corn, cane molasses, limestone, urea, and/or various commercial preparations are sometimes added to reduce fermentation losses, improve silage quality, and/or increase nutritive value.

FIGURE 58.   Field chopped corn fodder being moved from a field wagon through a blower into a silo in the process of making silage. *(Courtesy of Sperry-New Holland, New Holland, Pa.)*

VI.   How silage is formed.

A.   Basic for silage formation is the early establishment and maintenance of a relatively oxygen-free (anaerobic) condition.

1.   All silos are constructed to be as airtight as practicable with the type of silo being used.

2.   When practicable, silage material is packed as it goes into the silo to reduce air spaces to a minimum and eliminate O_2 in so far as possible.

3.   Most silage is made from green to semi-green forage material which uses up the oxygen present during the dying process.

B.   Following storage, the forage material normally undergoes an anaerobic fermentation to change it into silage.

1.   The fermentation normally sets in within hours after storage.

2.   The heat produced by the living cells of the forage material during the dying process tends to create a favorable environment for the anaerobic bacteria responsible for the fermentation process.

3. The fermentation process accelerates over the first 2-3 days, then levels off, and normally will for the most part terminate itself within the first 2-3 weeks.

4. Products of the fermentation are:

    a. Organic acids—primarily lactic, acetic, and butyric.

    b. Ethanol in varying amounts.

    c. Fermentation gases—primarily CO_2, CH_4, CO, NO, and NO_2.

    d. Water.

    e. Heat.

5. The fermentation is normally terminated by the accumulation of certain by-products of bacterial metabolism which tend to preserve the forage material indefinitely unless air is permitted to enter.

6. If air is able to permeate the silage, a prolonged fermentation and possibly molding and spoilage may result.

VII. Use of silage as a livestock feed.

A. Silage is primarily a beef and dairy animal feed, where it ordinarily is used as a part of or the only roughage in the ration. It is also a good feed for sheep, but it is seldom produced for this purpose alone. Horses will eat silage, but it is not too good a horse feed. Silage is seldom fed to hogs although it is sometimes fed to brood sows.

B. About 2.5-3.0 lb of silate are required to replace 1 lb of hay. This is because of the lower dry matter content of the silage.

C. From 2 inches to 5 inches of silage per day should be fed off the exposed surface of a conventional upright silo (nonairtight) in order to keep the silage fresh. The warmer the weather, the more it will be necessary to feed. With a bunker or trench silo, even more may have to be fed. With an airtight silo, the amount fed is not too important.

D. Once you start feeding from a conventional silo, it is best to continue feeding from that silo on an uninterrupted basis until it is empty; otherwise, there will be considerable spoilage of the exposed surface.

E. Silage is usually used when pastures are short.

F. In addition to serving as a roughage, silage is a good appetizer and tends to keep cattle on feed during hot weather.

FIGURE 59.   A top-unloading silo unloader in operation in an uncovered concrete-stave silo. *(Courtesy of Badger Northland Inc., Kaukauna, Wisc.)*

VIII. **Advantages of silage.**

    A.   More TDN per acre. When only the grain of a crop is harvested, up to one-half of the TDN is left in the field.

    B.   Maintains feed in a succulent form. Dry feed must become saturated with water before it can be digested. Silage never loses its wetness.

    C.   No losses from shattering, leaching, or bleaching.

    D.   No waste in feeding. Even the cobs, shucks, and coarsest stems are eaten.

    E.   Even weedy crops make good silage.

    F.   Can make silage in almost any kind of weather.

    G.   Permits early re-use of land. Facilitates double cropping.

    H.   Less fire hazard with silage than with hay and other dry crops.

    I.   Less internal parasite trouble with silage than with pasture.

    J.   A silage program facilitates complete mechanization of forage harvesting and feeding.

IX.   **Disadvantages of silage.**

    A.   Extra labor is required at silo-filling time.

B.   Considerable costly equipment is required for the harvesting, storing, and feeding of silage.

C.   Must handle considerable water.

D.   Not well suited for intermittent use.

E.   Silage in less airtight silos may undergo considerable deterioration if held for extended periods.

29 Processing Feeds

I.   Mechanical processing of grains.

   A.   Most grains are ground, rolled, or crimped before being fed to livestock.

      1.   Grinding is usually accomplished with a *hammer mill* (Figure 60). A hammer mill grinds by beating grain until it is fine enough to pass through a screen. The size of screen will determine the degree of fineness obtained.

      2.   Rolling is accomplished simply by passing grain between a closely fitted set of rollers. This leaves the grain in the form of a flake, and the process is sometimes referred to as *flaking*. Cattle seem to prefer flaked grain to ground grain and usually do somewhat better on the flaked compared to the ground product.

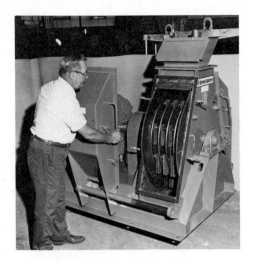

FIGURE 60. A commercial type hammer mill with interior exposed. A perforated metal screen that encloses the grinding chamber has been removed to reveal the rotor, hinge pins, and hammers. The inlet at the top includes a magnet for the removal of tramp metal. *(Courtesy of Sprout-Waldron, Muncy, Pa.)*

3. Crimping is accomplished in a manner similar to rolling except that instead of rollers with smooth surfaces being used, rollers with corrugated surfaces are used. The end result is much the same.

B. Mechanical processing of grains is especially recommended for:

1. Very young animals before their teeth are fully developed. Grain is usually ground for such animals.

2. Very old animals with badly worn teeth. For such animals grain is also usually ground.

3. All cattle over six months of age unless the roughage content of the ration is very low. Grinding, rolling, and flaking are all used extensively in processing grain for cattle, horses, and sheep.

C. The following classes of livestock usually chew grain fairly thoroughly and thereby dispense with the need for mechanical processing.

Calves 2-6 months of age	Horses and mules
Sheep and lambs	Fattening pigs

However, even with these classes, the improvement of feeding value from processing will usually offset the processing costs.

    D.   Fineness of grinding. Grinding to a medium fineness (equivalent to about a coarse meal) is usually best. Finely ground feeds (flour fine) are dusty and unpalatable to livestock. Fine grinding might be desirable for grains containing small weed seeds, the viability of which should be destroyed. Coarse grinding or cracking may be more trouble than it is worth.

    E.   Cost of grinding under present-day circumstances will usually run around 30-40¢ per 100 lb.

II.   **Grinding hays.**

    A.   It is not necessary to grind good quality hay to realize its effective use.

    B.   Grinding coarse, stemmy hays will encourage their total consumption by livestock but will not improve their digestibility.

    C.   Hay must be ground for incorporation into complete ration mixtures for livestock.

    D.   The coarser the hay is ground, the more it retains its bulk value.

    E.   Grinding their hay will cause a drop in the butterfat level of dairy cows.

III.   **Pelleting feeds.**

    A.   **Pelleting grains and other concentrates.**

        1.   Feeds to be pelleted are usually ground first—hence, they enjoy the benefits of grinding.

        2.   Pelleting returns the feed to a free-flowing form, thus facilitating the mechanization of handling it and also its use in a self-feeder.

        3.   Pelleted feeds are usually less dusty and more palatable, thus making for greater consumption.

        4.   Pelleting reduces storage space requirement of a feed.

        5.   Pelleting of ground concentrate mixtures will add $2.00-4.00 per ton to the cost of the feed under present-day circumstances.

    B.   **Pelleting hay and other roughages.**

        1.   Hays must be ground prior to pelleting—hence, pelleting embraces most of the advantages and disadvantages of grinding.

        2.   Pelleting of hays and other roughages converts them into a

FIGURE 61.   Pelleted feeds have some definite advantages as a form in which to provide feed for livestock. *(Courtesy of Ralston Purina Co., St. Louis, Mo.)*

free-flowing form which can be handled mechanically and fed in a self-feeder.

3.   Pelleting of hay and other roughages reduces the space requirement for storage by as much as 75%.

4.   Pelleting of hay and other roughages increases consumption and performance in beef cattle.

5.   Pelleting eliminates the air spaces in hays and other roughages, and increases their density, thus causing them to lose much of their roughage value.

6.   Pelleting reduces dustiness of hay and other roughages.

7.   Pelleting roughages is about twice as costly as pelleting concentrates.

C.   **Pelleting complete rations.**

1.   Same advantages as with individual feeds.

2.   Pelleting of complete rations for cattle destroys the roughage value of any hay or roughage in the ration.

    3.   Unpelleted roughage will need to be fed along with the pellets for good results with cattle and probably sheep.

    4.   Pelleting of complete rations for cattle and sheep will usually not be economical. It may be practical for horses and swine.

D.   **Cubing supplements for brood cows on low-quality pasture.**

    1.   "Cubes" are nothing more than large pellets.

    2.   Cubes may be fed on the ground in clean pastures; no troughs are needed.

    3.   The cost of the cubing is usually more than offset by the added convenience.

IV.   **Cooking feeds.**

    A.   Cooking may take on different forms.

        1.   Conventional cooking with or without steam pressure.

        2.   Short period steaming—usually in connection with rolling or flaking.

        3.   Roasting and "popping".

    B.   While cooking will frequently improve the palatability and nutritive value of feeds, it has not proved to be a profitable practice for feeds in general, except with a few large feedlots.

    C.   Cooking seems to have its greatest benefit with:

        Irish potatoes   Field beans
        Soybeans

    D.   Barley is frequently cooked for show cattle to improve ration acceptability.

    E.   Garbage is required by law to be cooked in most states. This law was provoked by an outbreak of Vesicular Exanthema (VE disease) more than 20 years ago. This is a disease of hogs which is transmitted by hogs eating uncooked flesh of animals with the disease. The disease was promptly brought under control by cooking all garbage. Cooking of garbage no doubt helps control other diseases also.

V.   **Wetting feeds.** Not usually considered to be practical except possibly with—

        Very dusty feeds   Rations for show cattle
        Oats for horses and mules

VI. **Soaking feeds.** Not usually considered to be practical except possibly with—

Very hard grains which are not mechanically processed. Tender or sore mouthed horses and mules.	Dried beet pulp, soybran flakes, and similar type feeds normally fed in the wet form.

VII. **Fermenting feeds (other than silage).** No benefit has ever been demonstrated experimentally from the fermenting of feeds. Efforts to improve ration digestibility and protein quality and to detoxify certain feeds by fermentation have been to no avail.

VIII. **Germinating grains.** Not generally considered to be practical. It may serve as a means of providing the grass juice factor for dairy cows when pasture is not available, but this has not been demonstrated experimentally to be of practical value.

IX. **Liquid feeds and slurries.**

A. Supplements are sometimes provided to brood cows on pasture through feeders in the liquid form. This is a convenient way of feeding supplements to such animals, provided the cost is in line and consumption can be adequately controlled.

B. Supplements are sometimes added to complete ration mixes for livestock in the liquid form. This is a satisfactory way of adding supplemental nutrients to the ration, provided it is more convenient and/or cheaper than the dry form and does not add excessive moisture to the mix.

C. Some swine producers follow a practice of feeding the complete ration in the form of a slurry or gruel. Some of the reasons given for following such a practice are:

1. It helps mechanize the feeding operation.

2. It eliminates dust.

3. It reduces feed wastage.

4. The animals prefer wet feed to dry feed.

5. Early weaned pigs do better on wet feed than they do on dry feed.

30 Uniform State Feed Bill

Except for certain federal regulations dealing primarily with the use of new drugs in animal feeds, the manufacture and distribution of commercial feeds are regulated for the most part in most states by the respective State Departments of Agriculture in accordance with the laws of the respective states. Fortunately, the feed laws, rules, and regulations for most of the different states are quite similar. To a very large degree this has come about as the result of the preparation and publication of a so-called "Uniform State Feed Bill" by the Association of the American Feed Control Officials. This Bill was designed as a pattern to be used by the different states in the writing of their respective feed laws, rules, and regulations. The Uniform State Feed Bill as published in the *1975 Official Publication of the AAFCO* reads as follows:*

*Reprinted with the permission of the Association of American Feed Control Officials.

UNIFORM STATE FEED BILL

Officially Adopted By
ASSOCIATION OF AMERICAN FEED CONTROL OFFICIALS
And Endorsed By
AMERICAN FEED MANUFACTURERS ASSOCIATION
NATIONAL FEED INGREDIENTS ASSOCIATION
PET FOOD INSTITUTE

Although this Bill and the Regulations have not been passed into law in all the states the subject matter covered herein does represent the official policy of this Association.

AN ACT

To regulate the manufacture and distribution of commercial feeds in the State of —————— BE IT ENACTED by the Legislature of the State of ——————.

Section 1. Title

This Act shall be known as the "—————— Commercial Feed Law of 19———."

Section 2. Enforcing Official

This Act shall be administered by the —————— of the State of ——————, hereinafter referred to as the "——————".

Section 3. Definitions of Words and Terms

When used in this Act:
(a) The term "person" includes individual, partnership, corporation, and association.
(b) The term "distribute" means to offer for sale, sell, exchange, or barter, commercial feed; or to supply, furnish, or otherwise provide commercial feed to a contract feeder.
(c) The term "distributor" means any person who distributes.
(d) The term "commercial feed" means all materials except whole seeds unmixed or physically altered entire unmixed seeds, when not adulterated within the meaning of Sec. 7(a), which are distributed for use as feed or for mixing in feed: Provided, That the ——————by regulation may exempt from this definition, or from specific provisions of this Act, commodities such as hay, straw, stover, silage, cobs, husks, hulls, and individual chemical compounds or substance when such commodities, compounds or substances are not inter-mixed or mixed with other materials, and are not adulterated within the meaning of Section 7(a), of this Act.
(e) The term "feed ingredient" means each of the constituent materials making up a commercial feed.
(f) The term "mineral feed" means a commercial feed intended to supply primarily mineral elements or inorganic nutrients.
(g) The term "drug" means any article intended for use in the diagnosis, cure, mitigation, treatment, or prevention of disease in animals other than man and articles other than feed intended to affect the structure or any function of the animal body.
(h) The term "customer-formula feed" means commercial feed which consists of a mixture of commercial feeds and/or feed ingredients each batch of which is manufactured according to the specific instructions of the final purchaser.
(i) The term "manufacture" means to grind, mix or blend, or further process a commercial feed for distribution.

(j) The term "brand name" means any word, name, symbol, or device, or any combination thereof, identifying the commercial feed of a distributor or registrant and distinguishing it from that of others.

(k) The term "product name" means the name of the commercial feed which identifies it as to kind, class, or specific use.

(l) The term "label" means a display of written, printed, or graphic matter upon or affixed to the container in which a commercial feed is distributed, or on the invoice or delivery slip with which a commercial feed is distributed.

(m) The term "labeling" means all labels and other written, printed, or graphic matter (1) upon a commercial feed or any of its containers or wrapper or (2) accompanying such commercial feed.

(n) The term "ton" means a net weight of two thousand pounds avoirdupois.

(o) The terms "per cent" or "percentages" means percentages by weights.

(p) The term "official sample" means a sample of feed taken by the _____ or his agent in accordance with the provisions of Section 11 (c), (e), or (f) of this Act.

(q) The term "contract feeder" means a person who as an independent contractor, feeds commercial feed to animals pursuant to a contract whereby such commercial feed is supplied, furnished, or otherwise provided to such person and whereby such person's remuneration is determined all or in part by feed consumption, mortality, profits, or amount or quality of product.

(r) The term "pet food" means any commercial feed prepared and distributed for consumption by pets.

(s) The term "pet" means any domesticated animal normally maintained in or near the household(s) of the owner(s) thereof.

(t) The term "specialty pet food" means any commercial feed prepared and distributed for consumption by specialty pets.

(u) The term "specialty pet" means any domesticated animal pet normally maintained in a cage or tank, such as, but not limited to, gerbils, hamsters, canaries, psittacine birds, mynahs, finches, tropical fish, goldfish, snakes and turtles.

Section 4. Registration

(a) No person shall manufacture a commercial feed in this State, unless he has filed with the _____ on forms provided by the _____, his name, place of business and location of each manufacturing facility in this State.

(b) No person shall distribute in this State a commercial feed, except a customer-formula feed, which has not been registered pursuant to the provisions of this section. The application for registration shall be submitted in the manner prescribed by the _____. Upon approval by the _____ the registration shall be issued to the applicant. All registrations expire on the 31st day of December of each year. (Option: A registration shall continue in effect unless it is cancelled by the registrant or unless it is cancelled by the _____ pursuant to Subsection (c) of this section.)

(c) The _____ is empowered to refuse registration of any commercial feed not in compliance with the provisions of this Act and to cancel any registration subsequently found not to be in compliance with any provision of this Act: Provided, That no registration shall be refused or canceled unless the registrant shall have been given an opportunity to be heard before the _____ and to amend his application in order to comply with the requirements of this Act.

Section 5. Labeling

A commercial feed shall be labeled as follows:

(a) In case of a commercial feed, except a customer-formula feed, it shall

be accompanied by a label bearing the following information:
(1) The net weight.
(2) The product name and the brand name, if any, under which the commercial feed is distributed.
(3) The guaranteed analysis stated in such terms as the _____ by regulation determines is required to advise the user of the composition of the feed or to support claims made in the labeling. In all cases the substances or elements must be determinable by laboratory methods such as the methods published by the Association of Official Analytical Chemists.
(4) The common or usual name of each ingredient used in the manufacture of the commercial feed: Provided, That the _____ by regulation may permit the use of a collective term for a group of ingredients which perform a similar function, or he may exempt such commercial feeds, or any group thereof, from this requirement of an ingredient statement if he finds that such statement is not required in the interest of consumers.
(5) The name and principal mailing address of the manufacturer or the person responsible for distributing the commercial feed.
(6) Adequate directions for use for all commercial feeds containing drugs and for such other feeds as the _____ may require by regulation as necessary for their safe and effective use.
(7) Such precautionary statements as the _____ by regulation determines are necessary for the safe and effective use of the commercial feed.
(b) In the case of a customer-formula feed, it shall be accompanied by a label, invoice, delivery slip, or other shipping document, bearing the following information:
(1) Name and address of the manufacturer.
(2) Name and address of the purchaser.
(3) Date of delivery.
(4) The product name and brand name, if any, and the net weight of each registered commercial feed used in the mixture, and the net weight of each other ingredient used.
(5) Adequate directions for use for all customer-formula feeds containing drugs and for such other feeds as the _____ may require by regulation as necessary for their safe and effective use.
(6) Such precautionary statements as the _____ by regulation determines are necessary for the safe and effective use of the customer-formula feed.

Section 6. Misbranding

A commercial feed shall be deemed to be misbranded:
(a) If its labeling is false or misleading in any particular.
(b) If it is distributed under the name of another commercial feed.
(c) If it is not labeled as required in Section 5 of this Act.
(d) If it purports to be or is represented as a commercial feed, or if it purports to contain or is represented as containing a commercial feed ingredient, unless such commercial feed or feed ingredient conforms to the definition, if any, prescribed by regulation by the _____.
(e) If any word, statement, or other information required by or under authority of this Act to appear on the label or labeling is not prominently placed thereon with such conspicuousness (as compared with other words, statements, designs, or devices in the labeling) and in such terms as to render it likely to be read and understood by the ordinary individual under customary conditions of purchase and use.

Section 7. Adulteration

A commercial feed shall be deemed to be adulterated:
(a) (1) If it bears or contains any poisonous or deleterious substance which

may render it injurious to health; but in case the substance is not an added substance, such commercial feed shall not be considered adulterated under this subsection if the quantity of such substance in such commercial feed does not ordinarily render it injurious to health; or

(2) If it bears or contains any added poisonous, added deleterious, or added nonnutritive substance which is unsafe within the meaning of Section 406 of the Federal Food, Drug, and Cosmetic Act (other than one which is (i) a pesticide chemical in or on a raw agricultural commodity; or (ii) a food additive); or

(3) If it is, or it bears or contains any food additive which is unsafe within the meaning of Section 409 of the Federal Food, Drug, and Cosmetic Act; or

(4) If it is a raw agricultural commodity and it bears or contains a pesticide chemical which is unsafe within the meaning of Section 408 (a) of the Federal Food, Drug, and Cosmetic Act: Provided, That where a pesticide chemical has been used in or on a raw agricultural commodity in conformity with an exemption granted or a tolerance prescribed under Section 408 of the Federal Food, Drug, and Cosmetic Act and such raw agricultural commodity has been subjected to processing such as canning, cooking, freezing, dehydrating, or milling, the residue of such pesticide chemical remaining in or on such processed feed shall not be deemed unsafe if such residue in or on the raw agricultural commodity has been removed to the extent possible in good manufacturing practice and the concentration of such residue in the processed feed is not greater than the tolerance prescribed for the raw agricultural commodity unless the feeding of such processed feed will result or is likely to result in a pesticide residue in the edible product of the animal, which is unsafe within the meaning of Section 408 (a), of the Federal Food, Drug, and Cosmetic Act.

(5) If it is, or it bears or contains any color additive which is unsafe within the meaning of Section 706 of the Federal Food, Drug and Cosmetic Act.

(b) If any valuable constituent has been in whole or in part omitted or abstracted therefrom or any less valuable substance substituted therefor.

(c) If its composition or quality falls below or differs from that which it is purported or is represented to possess by its labeling.

(d) If it contains a drug and the methods used in or the facilities or controls used for its manufacture, processing, or packaging do not conform to current good manufacturing practice regulations promulgated by the ——————— to assure that the drug meets the requirement of this Act as to safety and has the identity and strength and meets the quality and purity characteristics which it purports or is represented to possess. In promulgating such regulations, the ——————— shall adopt the current good manufacturing practice regulations for medicated feed premixes and for medicated feeds established under authority of the Federal Food, Drug, and Cosmetic Act, unless he determines that they are not appropriate to the conditions which exist in this State.

(e) If it contains viable weed seeds in amounts exceeding the limits which the ——————— shall establish by rule or regulation.

Section 8. Prohibited Acts

The following acts and the causing thereof within the State of ——————— are hereby prohibited:

(a) The manufacture or distribution of any commercial feed that is adulterated or misbranded.

(b) The adulteration or misbranding of any commercial feed.

(c) The distribution of agricultural commodities such as whole seed, hay, straw, stover, silage, cobs, husks, and hulls, which are adulterated within the meaning of Section 7(a), of this Act.

(d) The removal or disposal of a commercial feed in violation of an order under Section 12 of this Act .

(e) The failure or refusal to register in accordance with Section 4 of this Act.

(f) The violation of Section 13(f) of this Act.

(g) Failure to pay inspection fees and file reports as required by Section 9 of this Act.

Section 9. Inspection Fees and Reports

(a) An inspection fee at the rate of _____ cents per ton shall be paid on commercial feeds distributed in this State by the person who distributes the commercial feed to the consumer, subject to the following:

    (1) No fee shall be paid on a commercial feed if the payment has been made by a previous distributor.

    (2) No fee shall be paid on customer-formula feeds if the inspection fee is paid on the commercial feeds which are used as ingredients therein.

    (3) No fee shall be paid on commercial feeds which are used as ingredients for the manufacture of commercial feeds which are registered. If the fee has already been paid, credit shall be given for such payment.

    (4) In the case of a commercial feed which is distributed in the State only in packages of ten pounds or less, an annual fee of _____ shall be paid in lieu of the inspection fee specified above.

    (5) The minimum inspection fee shall be _____ per quarter.

    (6) In the case of specialty pet food which is distributed in the state in packages of one pound or less, an annual fee of _____ shall be paid in lieu of an inspection fee.

(b) Each person who is liable for the payment of such fee shall:

    (1) File, not later than the last day of January, April, July, and October of each year, a quarterly statement, setting forth the number of net tons of commercial feeds distributed in this State during the preceding calendar quarter; and upon filing such statement shall pay the inspection fee at the rate stated in paragraph (a) of this Section. Inspection fees which are due and owing and have not been remitted to the _____ within 15 days following the date due shall have a penalty fee of _____ per cent (minimum _____) added to the amount due when payment is finally made. The assessment of this penalty fee shall not prevent the _____ from taking other actions as provided in this chapter.

    (2) Keep such records as may be necessary or required by the _____ to indicate accurately the tonnage of commercial feed distributed in this State, and the _____ shall have the right to examine such records to verify statements of tonnage.

Failure to make an accurate statement of tonnage or to pay the inspection fee or comply as provided herein shall constitute sufficient cause for the cancellation of all registrations on file for the distributor.

(c) Fees collected shall constitute a fund for the payment of the costs of inspection, sampling, and analysis, and other expenses necessary for the administration of this Act.

Section 10. Rules and Regulations

(a) The _____ is authorized to promulgate such rules and regulations for commercial feeds and pet foods as are specifically authorized in this Act and such other reasonable rules and regulations as may be necessary for the efficient enforcement of this Act. In the interest of uniformity the _____ shall by regulation adopt, unless he determines that they are inconsistent with the provisions of this Act or are not appropriate to conditions which exist in this state, the following:

    (1) The Official Definitions of Feed Ingredients and Official Feed Terms adopted by the Association of American Feed Control Officials and published in the Official Publication of that organization, and

    (2) Any regulation promulgated pursuant to the authority of the Federal

Food, Drug, and Cosmetic Act (U.S.C. Sec. 301, **et seq.**): Provided, That the _____ would have the authority under this Act to promulgate such regulations.

(b) Before the issuance, amendment, or repeal of any rule or regulation authorized by this Act, the _____ shall publish the proposed regulation, amendment, or notice to repeal an existing regulation in a manner reasonably calculated to give interested parties, including all current registrants, adequate notice and shall afford all interested persons an opportunity to present their views thereon, orally or in writing, within a reasonable period of time. After consideration of all views presented by interested persons, the _____ shall take appropriate action to issue the proposed rule or regulation or to amend or repeal an existing rule or regulation. The provisions of this paragraph not withstanding, if the _____, pursuant to the authority of this Act, adopts the Official Definitions of Feed Ingredients or Official Feed Terms as adopted by the Association of American Feed Control Officials, or regulations promulgated pursuant to the authority of the Federal Food, Drug, and Cosmetic Act, any amendment or modification adopted by said Association or by the Secretary of Health, Education and Welfare in the case of regulations promulgated pursuant to the Federal Food, Drug and Cosmetic Act, shall be adopted automatically under this Act without regard to the publication of the notice required by this paragraph (b), unless the _____, by order specifically determines that said amendment or modification shall not be adopted.

Section 11. Inspection, Sampling, and Analysis

(a) For the purpose of enforcement of this act, and in order to determine whether its provisions have been complied with, including whether or not any operations may be subject to such provisions, officers or employees duly designated by the _____, upon presenting appropriate credentials, and a written notice to the owner, operator, or agent in charge, are authorized (1) to enter, during normal business hours, any factory, warehouse, or establishment within the State in which commercial feeds are manufactured, processed, packed, or held for distribution, or to enter any vehicle being used to transport or hold such feeds; and (2) to inspect at reasonable times and within reasonable limits and in a reasonable manner, such factory, warehouse, establishment or vehicle and all pertinent equipment, finished and unfinished materials, containers, and labeling therein. The inspection may include the verification of only such records, and production and control procedures as may be necessary to determine compliance with the Good Manufacturing Practice Regulations established under Section 7(d) of this Act.

(b) A separate notice shall be given for each such inspection, but a notice shall not be required for each entry made during the period covered by the inspection. Each such inspection shall be commenced and completed with reasonable promptness. Upon completion of the inspection, the person in charge of the facility or vehicle shall be so notified.

(c) If the officer or employee making such inspection of a factory, warehouse, or other establishment has obtained a sample in the course of the inspection, upon completion of the inspection and prior to leaving the premises he shall give to the owner, operator, or agent in charge a receipt describing the samples obtained.

(d) If the owner of any factory, warehouse, or establishment described in paragraph (a), or his agent, refuses to admit the _____ or his agent to inspect in accordance with paragraphs (a) and (b), the _____ is authorized to obtain from any State Court a warrant directing such owner or his agent to submit the premises described in such warrant to inspection.

(e) For the purpose of the enforcement of this Act, the _____ or his duly designated agent is authorized to enter upon any public or private premises including any vehicle of transport during regular business hours to have access to, and to obtain samples, and to examine records relating to distribution of commercial feeds.

(f) Sampling and analysis shall be conducted in accordance with methods published by the Association of Official Analytical Chemists, or in accordance with other generally recognized methods.

(g) The results of all analyses of official samples shall be forwarded by the _____ to the person named on the label and to the purchaser. When the inspection and analysis of an official sample indicates a commercial feed has been adulterated or misbranded and upon request within 30 days following receipt of the analysis the _____ shall furnish to the registrant a portion of the sample concerned.

(h) The _____, in determining for administrative purposes whether a commercial feed is deficient in any component, shall be guided by the official sample as defined in paragraph (p) of Section 3 and obtained and analyzed as provided for in paragraphs (c), (e), and (f) of Section 11 of this Act.

Section 12. Detained Commercial Feeds

(a) "Withdrawal from distribution" orders: When the _____ or his authorized agent has reasonable cause to believe any lot of commercial feed is being distributed in violation of any of the provisions of this Act or of any of the prescribed regulations under this Act, he may issue and enforce a written or printed "withdrawal from distribution" order, warning the distributor not to dispose of the lot of commercial feed in any manner until written permission is given by the _____ or the Court. The _____ shall release the lot of commercial feed so withdrawn when said provisions and regulations have been complied with. If compliance is not obtained within 30 days, the _____ may begin, or upon request of the distributor or registrant shall begin, proceedings for condemnation.

(b) "Condemnation and Confiscation": Any lot of commercial feed not in compliance with said provisions and regulations shall be subject to seizure on complaint of the _____ to a court of competent jurisdiction in the area in which said commercial feed is located. In the event the court finds the said commercial feed to be in violation of this Act and orders the condemnation of said commercial feed, it shall be disposed of in any manner consistent with the quality of the commercial feed and the laws of the State: Provided, That in no instance shall the disposition of said commercial feed be ordered by the court without first giving the claimant an opportunity to apply to the court for release of said commercial feed or for permission to process or re-label said commercial feed to bring it into compliance with this Act.

Section 13. Penalties

(a) Any person convicted of violating any of the provisions of this Act or who shall impede, hinder, or otherwise prevent, or attempt to prevent, said _____ or his duly authorized agent in performance of his duty in connection with the provisions of this Act, shall be adjudged guilty of a misdemeanor and shall be fined not less than _____ or more than _____ for the first violation, and not less than _____ or more than _____ for a susequent violation.

(b) Nothing in this act shall be construed as requiring the _____ or his representative to : (1) report for prosecution, or (2) institute seizure proceedings, or (3) issue a withdrawal from distribution order, as a result of minor violations of the Act, or when he believes the public interest will best be served by suitable notice of warning in writing.

(c) It shall be the duty of each _____ attorney to whom any violation is reported to cause appropriate proceedings to be instituted and prosecuted in a court of competent jurisdiction without delay. Before the

_____ reports a violation for such prosecution, an opportunity shall be given the distributor to present his view to the _____.

(d) The _____ is hereby authorized to apply for and the court to grant a temporary or permanent injunction restraining any person from violating or continuing to violate any of the provisions of this Act or any rule or regulation promulgated under the Act notwithstanding the existence of other remedies at law. Said injunction to be issued without bond.

(e) Any person adversely affected by an act, order, or ruling made pursuant to the provisions of this Act may within 45 days thereafter bring action in the (here name the particular Court in the county where the enforcement official has his office) for judicial review of such actions. The form of the proceeding shall be any which may be provided by statutes of this state to review decisions of administrative agencies, or in the absence or inadequacy thereof, any applicable form of legal action, including actions for declaratory judgments or writs of prohibitory or mandatory injunctions.

(f) Any person who uses to his own advantage, or reveals to other than the _____, or officers of the _____ (appropriate departments of this State), or to the courts when relevant in any judicial proceeding, any information acquired under the authority of this Act, concerning any method, records, formulations, or processes which as a trade secret is entitled to protection, is guilty of a misdemeanor and shall on conviction thereof be fined not less than $_____ or imprisoned for not less than _____ year(s) or both: Provided, That this prohibition shall not be deemed as prohibiting the _____, or his duly authorized agent, from exchanging information of a regulatory nature with duly appointed officials of the United States Government, or of other States, who are similarly prohibited by law from revealing this information.

Section 14. Cooperation with other entities

The _____ may cooperate with and enter into agreements with governmental agencies of this State, other States, agencies of the Federal Government, and private associations in order to carry out the purpose and provisions of this Act.

Section 15. Publication

The _____ shall publish at least annually, in such forms as he may deem proper, information concerning the sales of commercial feeds, together with such data on their production and use as he may consider advisable, and a report of the results of the analyses of official samples of commercial feeds sold within the State as compared with the analyses guaranteed in the registration and on the label: Provided, That the information concerning production and use of commercial feed shall not disclose the operations of any person.

Section 16. Constitutionality

If any clause, sentence, paragraph, or part of this Act shall for any reason be judged invalid by any court of competent jurisdiction, such judgment shall not affect, impair, or invalidate the remainder thereof but shall be confined in its operation to the clause, sentence, paragraph, or part thereof directly involved in the controversy in which such judgment shall have been rendered.

Section 17. Repeal

All laws and parts of laws in conflict with or inconsistent with the provisions of this Act are hereby repealed. (The specific statute and specific code sections to be repealed may have to be stated.)

Section 18. Effective Date

This Act shall take effect and be in force from and after the first day of _____.

PROPOSED RULES AND REGULATIONS

under the

UNIFORM STATE FEED BILL

Pursuant to due publication and public hearing required by the provisions of Chapter ——— of the Laws of this State, the ————— has adopted the following Rules and Regulations.

Regulation 1. Definitions and Terms

(a) The names and definitions for commercial feeds shall be the Official Definition of Feed Ingredients adopted by the Association of American Feed Control Officials, except as the ————— designates otherwise in specific cases.

(b) The terms used in reference to commercial feeds shall be the Official Feed Terms adoted by the AAFCO, except as the ————— designates otherwise in specific cases.

(c) The following commodities are hereby declared exempt from the definition of commercial feed, under the provisions of Section 3(d) of the Act: raw meat; and hay, straw, stover, silages, cobs, husks, and hulls when unground and when not mixed or intermixed with other materials: Provided that these commodities are not adulterated within the meaning of Section 7(a), of the Act.

Regulation 2. Label Format

Commercial feeds shall be labeled with the information prescribed in this regulation on the principal display panel of the product and in the following general format:

(a) Net Weight

(b) Product name and brand name if any

(c) If drugs are used:
  (1) The word "medicated" shall appear directly following and below the product name in type size no smaller than one half the type size of the product name
  (2) The purpose of medication (claim statement).
  (3) The required direction for use and precautionary statements or reference to their location if the detailed feeding directions and precautionary statements required by Regulations 6 and 7 appear elsewhere on the label.
  (4) An active drug ingredient statement listing the active drug ingredients by their established name and the amounts in accordance with Regulation 4(d).

(d) The guaranteed analysis of the feed as required under the provisions of Section 5(a)(3) of the Act include the following items, unless exempted in (8) of this subsection, and in the order listed:
  (1) Minimum percentage of crude protein.
  (2) Maximum or minimum percentage of equivalent protein from non-protein nitrogen as required in Regulation 4(e).
  (3) Minimum percentage of crude fat.
  (4) Maximum percentage of crude fiber.
  (5) Minerals, to include, in the following order: (a) minimum and maximum percentages of calcium (Ca), (b) minimum percentages of phosphorus (P), (c) minimum and maximum percentages of salt (NaCl), and (d) other minerals.
  (6) Vitamins in such terms as specified in Regulation 4(c).

(7) Total Sugars as Invert on dried molasses products or products being sold primarily for their molasses content.
(8) Exemptions.
    (i) Guarantees for minerals are not required when there are no specific label claims and when the commercial feed contains less than 6½ % of mineral elements.
    (ii) Guarantees for vitamins are not required when the commercial feed is neither formulated for nor represented in any manner as a vitamin supplement.
    (iii) Guarantees for crude protein, crude fat, and crude fiber are not required when the commercial feed is intended for purposes other than to furnish these substances or they are of minor significance relating to the primary purpose of the product, such as drug premixes, mineral or vitamin supplements, and molasses.

(e) Feed ingredients, collective terms for the grouping of feed ingredients, or appropriate statements as provided under the provisions of Section 5(a)(4) of the Act.
(1) The name of each ingredient as defined in the Official Definitions of Feed Ingredients published in the Official Publication of the Association of American Feed Control Officials, common or usual name, or one approved by the ——————————.
(2) Collective terms for the grouping of feed ingredients as defined in the Official Definitions of Feed Ingredients published in the Official Publication of the Association of American Feed Control Officials in lieu of the individual ingredients; Provided that:
    (i) When a collective term for a group of ingredients is used on the label, individual ingredients within that group shall not be listed on the label.
    (ii) The manufacturer shall provide the feed control official, upon request, with a listing of individual ingredients, within a defined group, that are or have been used at manufacturing facilities distributing in or into the state.
(3) The registrant may affix the statement, "Ingredients as registered with the State" in lieu of the ingredient list on the label. The list of ingredients must be on file with the ——————————. This list shall be made available to the feed purchaser upon request.

(f) Name and principal mailing address of the manufacturer or person responsible for distributing the feed. The principal mailing address shall include the street address, city, state and zip code; however, the street address may be omitted if it is shown in the current city directory or telephone directory.

(g) The information required in Section 5(a)(1)-(5) of the Act must appear in its entirety on one side of the label or on one side of the container. The information required by Section 5(a)(6)-(7) of the Act shall be displayed in a prominent place on the label or container but not necessarily on the same side as the above information. When the information required by Section 5(a)(6)-(7) is placed on a different side of the label or container, it must be referenced on the front side with a statement such as "see back of label for directions for use." None of the information required by Section 5 of the Act shall be subordinated or obscured by other statements or designs.

Regulation 3. Brand and Product Names

(a) The brand or product name must be appropriate for the intended use of the feed and must not be misleading. If the name indicates the feed is made for a specific use, the character of the feed must conform therewith. A mixture labeled "Dairy Feed," for example, must be suitable for that purpose.

(b) Commercial, registered brand or trade names are not permitted in guarantees or ingredient listings.

(c) The name of a commercial feed shall not be derived from one or more ingredients of a mixture to the exclusion of other ingredients and shall not be one representing any components of a mixture unless all components are included in the name: Provided, That if any ingredient or combination of ingredients is intended to impart a distinctive characteristic to the product which is of significance to the purchaser, the name of that ingredient or combination of ingredients may be used as a part of the brand name or product name if the ingredient or combination of ingredients is quantitatively guaranteed in the guaranteed analysis, and the brand or product name is not otherwise false or misleading.

(d) The word "protein" shall not be permitted in the product name of a feed that contains added non-protein nitrogen.

(e) When the name carries a percentage value, it shall be understood to signify protein and/or equivalent protein content only, even though it may not explicitly modify the percentage with the word "protein": Provided, That other percentage values may be permitted if they are followed by the proper description and conform to good labeling practice. When a figure is used in the brand name (except in mineral, vitamin, or other products where the protein guarantee is nil or unimportant), it shall be preceded by the word "number" or some other suitable designation.

(f) Single ingredient feeds shall have a product name in accordance with the designated definition of feed ingredients as recognized by the Association of American Feed Control Officials unless the —————— designates otherwise.

(g) The word "vitamin", or a contraction thereof, or any word suggesting vitamin can be used only in the name of a feed which is represented to be a vitamin supplement, and which is labeled with the minimum content of each vitamin declared, as specified in Regulation 4(c).

(h) The term "mineralized" shall not be used in the name of a feed, except for "TRACE MINERALIZED SALT". When so used, the product must contain significant amounts of trace minerals which are recognized as essential for animal nutrition.

(i) The term "meat" and "meat by-products" shall be qualified to designate the animal from which the meat and meat by-products is derived unless the meat and meat by-products are from cattle, swine, sheep and goats.

Regulation 4. Expression of Guarantees

(a) The guarantees for crude protein, equivalent protein from non-protein nitrogen, crude fat, crude fiber and mineral guarantees (when required) will be in terms of percentage by weight.

(b) Commercial feeds containing $6\frac{1}{2}\%$ or more mineral elements shall include in the guaranteed analysis the minimum and maximum percentages of calcium (Ca), the minimum percentage of phosphorus (P), and if salt is added, the minimum and maximum percentage of salt (NaCl). Minerals, except salt (NaCl), shall be guaranteed in terms of percentage of the element. When calcium and/or salt guarantees are given in the guaranteed analysis such shall be stated and conform to the following:
(1) When the minimum is 5.0% or less, the maximum shall not exceed the minimum by more than one percentage point.
(2) When the minimum is above 5.0%, the maximum shall not exceed the minimum by more than 20% and in no case shall the maximum exceed the minimum by more than 5 percentage points.

(c) Guarantees for minimum vitamin content of commercial feeds and feed supplements, when made, shall be stated on the label in milligrams per pound of feed except that:

(1) Vitamin A, other than precursors of vitamin A, shall be stated in International or USP units per pound.

(2) Vitamin D, in products offered for poultry feeding, shall be stated in International Chick Units per pound.

(3) Vitamin D for other uses shall be stated in International or USP units per pound.

(4) Vitamin E shall be stated in International or USP Units per pound.

(5) Guarantees for vitamin content on the label of a commercial feed shall state the guarantee as true vitamins, not compounds, with the exception of the compounds, Pyridoxine Hydrochloride, Choline Chloride, Thiamine, and d-Pantothenic Acid.

(6) Oils and premixes containing vitamin A or vitamin D or both may be labeled to show vitamin content in terms of units per gram.

(d) Guarantees for drugs shall be stated in terms of percent by weight, except:

(1) Antibiotics present at less than 2,000 grams per ton (total) of commercial feed shall be stated in grams per ton of commercial feed.

(2) Antibiotics present at 2,000 or more grams per ton (total) of commercial feed shall be stated in grams per pound of commercial feed.

(3) Labels for commercial feeds containing growth promotion and/or feed efficiency levels of antibiotics, which are to be fed continuously as the sole ration, are not required to make quantitative guarantees except as specifically noted in the Federal Food Additive Regulations for certain antibiotics, wherein, quantitative guarantees are required regardless of the level or purpose of the antibiotic.

(4) The term "milligrams per pound" may be used for drugs or antibiotics in those cases where a dosage is given in "milligrams" in the feeding directions.

(e) Commercial feeds containing any added non-protein nitrogen shall be labeled as follows:

(1) Complete feeds, supplements, and concentrates containing added non-protein nitrogen and containing more than 5% protein from natural sources shall be guaranteed as follows:

Crude Protein, minimum, _____%

(This includes not more than _____% equivalent protein from non-protein nitrogen).

(2) Mixed feed concentrates and supplements containing less than 5% protein from natural sources may be guaranteed as follows:

Equivalent Crude Protein from Non-Protein Nitrogen, minimum, _____%

(3) Ingredient sources of non-protein nitrogen such as Urea, Di-Ammonium Phosphate, Ammonium Polyphosphate Solution, Ammoniated Rice Hulls, or other basic non-protein nitrogen ingredients defined by the Association of American Feed Control Officials shall be guaranteed as follows:

Nitrogen, minimum, _____%

Equivalent Crude Protein from Non-Protein Nitrogen, minimum, _____%

(f) Mineral phosphatic materials for feeding purposes shall be labeled with the guarantee for minimum and maximum percentage of calcium (when present), the minimum percentage of phosphorus, and the maximum percentage of fluorine.

Regulation 5. Ingredients

(a) The name of each ingredient or collective term for the grouping of ingredients, when required to be listed, shall be the name as defined in the Official Definitions of Feed Ingredients as published in the Official Publication of American Feed Control Officials, the common or usual name, or one approved by the _____.

(b) The name of each ingredient must be shown in letters or type of the same size.

(c) No reference to quality or grade of an ingredient shall appear in the ingredient statement of a feed.

(d) The term "dehydrated" may precede the name of any product that has been artifically dried.

(e) A single ingredient product defined by the Association of American Feed Control Officials is not required to have an ingredient statement.

(f) Tentative definitions for ingredients shall not be used until adopted as official, unless no official definition exists or the ingredient has a common accepted name that requires no definition, (i.e. sugar).

(g) When the word "iodized" is used in connection with a feed ingredient, the feed ingredient shall contain not less than 0.007% iodine, uniformly distributed.

Regulation 6. Directions for Use and Precautionary Statements

(a) Directions for use and precautionary statements on the labeling of all commercial feeds and customer-formula feeds containing additives (including drugs, special purpose additives, or non-nutritive additives) shall:
  (1) Be adequate to enable safe and effective use for the intended purposes by users, with no special knowledge of the purpose and use of such articles; and,
  (2) Include, but not be limited to, all information prescribed by all applicable regulations under the Federal Food, Drug and Cosmetic Act.

(b) Adequate directions for use and precautionary statements are required for feeds containing non-protein nitrogen as specified in Regulation 7.

(c) Adequate directions for use and precautionary statements necessary for safe and effective use are required on commercial feeds distributed to supply particular dietary needs or for supplementing or fortifying the usual diet or ration with any vitamin, mineral, or other dietary nutrient or compound.

Regulation 7. Non-Protein Nitrogen

(a) Urea and other non-protein nitrogen products defined in the Official Publication of the Association of American Feed Control Officials are acceptable ingredients only in commercial feeds for ruminant animals as a source of equivalent crude protein and are not to be used in commercial feeds for other animals and birds.

(b) If the commercial feed contains more than 8.75% of equivalent crude protein from all forms of non-protein nitrogen, added as such, or the equivalent crude protein from all forms of non-protein nitrogen, added as such, exceeds one-third of the total crude protein, the label shall bear adequate directions for the safe use of feeds and a precautionary statement:

"CAUTION: USE AS DIRECTED"

The directions for use and the caution statement shall be in type of such size so placed on the label that they will be read and understood by ordinary persons under customary conditions of purchase and use.

(c) On labels such as those for medicated feeds which bear adequate feeding directions and/or warning statements, the presence of added non-protein nitrogen shall not require a duplication of the feeding directions or the precautionary statements as long as those statements include sufficient information to ensure the safe and effective use of this product due to the presence of non-protein nitrogen.

Regulation 8. Drug and Feed Additives

(a) Prior to approval of a registration application and/or approval of a label for commercial feed which contain additives (including drugs, other special purpose additives, or non-nutritive additives) the distributor may be required to submit evidence to prove the safety and efficacy of the commercial feed when used•according to the directions furnished on the label.

(b) Satisfactory evidence of safety and efficacy of a commercial feed may be:

  (i) When the commercial feed contains such additives, the use of which conforms to the requirements of the applicable regulation in the Code of Federal Regulations, Title 21, or which are "prior sanctioned" or "generally recognized as safe" for such use, or

  (ii) When the commercial feed is itself a drug as defined in Section 3(g) of the Act and is generally recognized as safe and effective for the labeled use or is marketed subject to an application approved by the Food and Drug Administration under Title 21 U.S.C. 360(b).

Regulation 9. Adulterants

(a) For the purpose of Section 7(a)(1) of the Act, the terms "poisonous or deleterious substances" include but are not limited to the following:

  (1) Fluorine   and any mineral or mineral mixture which is to be used directly for the feeding of domestic animals and in which the fluorine exceeds 0.30% for cattle; 0.35% for sheep; 0.45% for swine; and 0.60% for poultry.

  (2) Fluorine bearing ingredients when used in such amounts that they raise the fluorine content of the total ration above the following amounts: 0.009% for cattle; 0.01% for sheep; 0.014% for swine; and 0.035% for poultry.

  (3) Soybean meal, flakes or pellets or other vegetable meals, flakes or pellets which have been extracted with trichlorethylene or other chlorinated solvents.

  (4) Sulfur dioxide, Sulfurous acid, and salts of Sulfurous acid when used in or on feeds or feed ingredients which are considered or reported to be a significant source of vitamin B_1 (Thiamine).

(b) All screenings or by-products of grains and seeds containing weed seeds, when used in commercial feed or sold as such to the ultimate consumer, shall be ground fine enough or otherwise treated to destroy the viability of such weed seeds so that the finished product contains no more than —————— viable prohibited weed seeds per pound and not more than ————— viable restricted weed seeds per pound.

Regulation 10. Good Manufacturing Practices

(a) For the purposes of enforcement of Section 7(d) of the Act the ————— adopts the following as current good manufacturing practices:

  (1) The regulations prescribing good manufacturing practices for medicated feeds as published in the Code of Federal Regulations, Title 21, Part 133, Sections 133.100-133.110.

  (2) The regulations prescribing good manufacturing practices for medicated premixes as published in the Code of Federal Regulations. Title 21, Part 133, Sections 133.200-133.210.

31 Balancing Rations —General

I. **What is a balanced ration?**

A. A balanced ration is one which will supply the different nutrients in such proportions as will properly nourish a given animal when fed in proper amounts.

B. A balanced daily ration is one which will supply the different nutrients in such proportions and amounts as will properly nourish a given animal for a 24-hour period.

II. **What is needed for balancing a ration?**

A. The first requirement for balancing a ration is a feeding standard. This is a table which states the amounts of nutrients which should be provided in rations for farm animals of various classes in order to secure the desired results. These requirements may be expressed as:

1.   Amounts per animal per day, or

2.   Percentage of overall feed mixture or amount per kg of ration.

B.   In order to be of use, feeding standards must be accompanied and used with feed composition tables which provide information concerning the nutritive composition of the feeds to be used in balancing the ration.

III.   **Brief history of balanced rations.**

A.   Thaer of Germany was among the first to move toward a scientific approach to livestock feeding. In 1810 he published a table of "Hay Equivalents." These were the amounts of different feeds equivalent to 100 lb of meadow hay. Such an approach naturally prompted much difference of opinion among authorities.

B.   Grouven in 1859 proposed the first feeding standard. However, it was through necessity based on total rather than digestible nutrients or some other more precise nutrient evaluation, and consequently was not too accurate.

C.   Wolff, another German, in 1864 presented the first feeding standard based on digestible nutrients. It was naturally quite limited in scope.

D.   The Wolff standards were revised in 1896 by another German named Lehman and published as the Wolff-Lehman standards. While they were widely used at the time, they contained many inaccuracies and are now out of date.

E.   Dr. W.A. Henry of the University of Wisconsin came out with the first edition of his *Feeds and Feeding* in 1898. It embraced the Henry feeding standards. Various other standards have been introduced over the years, but none has persisted as have those he originated. Beginning with the 10th edition of *Feeds and Feeding*, which came out in 1910, Dr. Henry was assisted by Dr. F.B. Morrison, a former student. Following the death of Dr. Henry in the early 1930s, Dr. Morrison continued to keep the book revised through the 22nd edition, published in 1956. However, Dr. Morrison died in 1958, and there have been no revisions of Morrison's *Feeds and Feeding* published since the 22nd edition.

F.   Back in the 1940's the National Research Council of the National Academy of Sciences initiated a series of publications on the nutrient requirements of various species of farm animals. In 1959 this same organization published a rather comprehensive set of tables on feed composition. From time to time all of these publications have been revised, and today they are regarded as the final authority on nutritive requirements and feed composition by workers in the field.

IV.   **Factors to be considered in balancing rations.**

    A.   **Nutritional.**

        1.   **Dry matter.**

           a.   A certain minimum amount of dry matter is essential to satisfy an animal's appetite and to promote the proper functioning of its digestive tract.

           b.   Animals have certain physical and physiological limitations of dry matter consumption beyond which they cannot go.

        2.   **Protein.**

           a.   All animals require protein. The amount will depend on the physiological processes of the animal. Many feeds are deficient in protein. No other nutrient can take the place of protein. The essential amino acid requirement must be met in this connection.

           b.   Adequacy of protein may be based on:

Total protein	Digestible protein plus
Digestible protein	certain critical essential amino acids

        3.   **Energy.**

           a.   All animals require energy. The amount will depend on the physiological processes of the animal. Except for keeping the body warm, an animal needs its energy in the form of net energy. This energy may come from sugar, starch, cellulose, fat, and/or excess protein.

           b.   Adequacy of energy may be based on:

Digestible energy	Metabolizable energy
Total digestible nutrients (TDN)	Net energy

        4.   **Calcium.**

           a.   All animals require calcium.

           b.   The amount of calcium required will depend on the physiological processes of the animal.

           c.   No other nutrient can substitute for calcium.

           d.   Calcium needs and allowances are ordinarily based on total calcium.

  e. Proper Ca:P ratio is important.

  f. Excess calcium may be harmful by interfering with the availability of other nutrients.

5. **Phosphorus.**

  a. All animals require phosphorus.

  b. The amount of phosphorus required will depend on the physiological processes of the animal.

  c. No other nutrient can substitute for phosphorus.

  d. Phosphorus needs and allowances are ordinarily based on total phosphorus.

  e. Proper Ca:P ratio is important.

  f. Excess phosphorus may be harmful by rendering other nutrients unavailable.

6. **Vitamin A.**

  a. All animals require vitamin A as pre-formed vitamin A, or carotene.

  b. Amount will vary, depending on the circumstances.

  c. Vitamin A may be provided as pre-formed vitamin A or as carotene.

  d. Vitamin A needs and allowances are based on total "vitamin A value" (vitamin A and carotene combined).

  e. Carotene is sometimes very inefficiently coverted to vitamin A, resulting in a vitamin A deficiency in the presence of ample carotene.

7. **Other minerals.**

  a. Calcium and phosphorus are usually the only two macro minerals that are likely to be in critical supply. No attempt is ordinarily routinely made to balance allowances precisely against requirements of other macro minerals.

  b. Minimum needs for critical trace minerals are usually met through the use of trace-mineralized salt. No attempt is ordinarily routinely made to balance allowances precisely against requirements.

8. **Other vitamins.** The need for other vitamins that may be in short supply are ordinarily met simply by adding to the overall ration the animal's minimum daily requirement for these vitamins, letting the natural content of these vitamins in the feed serve as a margin of safety. No attempt is ordinarily routinely made to balance allowances precisely against requirements.

B. **Economic.**

    1. Feeds are not always priced in accordance with their nutritive value. Some feeds may be a cheaper source of nutrients than other feeds with any given set of prices.

    2. To compare feeds as economical sources of nutrients, it is not sufficient to compare them in terms of price per bushel or even per Cwt. Different feeds have different weights per bushel as well as different contents of nutrients per Cwt.

    3. High energy feeds are usually compared pricewise on the basis of the cost per lb of TDN or per unit of energy.

    4. High protein feeds are usually compared pricewise on the basis of the cost per lb of protein, either total or digestible.

    5. For example, shelled corn, hegari grain, and barley grain could be compared as sources of TDN as follows:

	PREVAILING PRICE	PREVAILING PRICE PER CWT	TDN PER CWT	COST. PER LB TDN
Shelled corn	$1.40 per bushel	$2.50	78.0	$.032
Hegari grain	$2.75 per Cwt	$2.75	81.0	$.034
Barley grain	$1.20 per bushel	$2.50	74.0	$.034

Since in the above calculation corn is lowest in cost per lb of TDN, it would be the cheapest source of energy at the prevailing set of prices.

    6. High protein feeds may be compared in a similar manner as sources of protein simply by inserting total protein or digestible protein for TDN in the calculation.

    7. In establishing the prevailing price, care should be exercised to make sure the price used is FOB the buyer's feedlot and that the feeds are in comparable physical form.

    8. For feeds which are intermediate in protein content between that of high protein feeds, on the one hand, and high energy feeds, on the other, the above procedure will not suffice. For such feeds the Petersen method of evaluating feeds as outlined in section 32 is recommended.

32 The Petersen Method of Evaluating Feeds

An alternative method to that outlined above for comparing different feeds with respect to their nutritive worth was devised several years ago by Petersen of the University of Minnesota. The Petersen method is superior to the one outlined above in that it is more precise and is applicable to all feeds and not just to high energy feeds, on the one hand, or high protein feeds, on the other. By using the Petersen method, appropriate weight is given to both the protein and the energy contents of a feed in establishing its feeding value, whereas the relative value of a feed as established by the preceding method is based entirely on its content of protein, or energy, but not both.

I.   **Establishing constants.**

A.   The Petersen method involves the establishment of a set of constants for each feed to be considered as a potential ration component. These constants are calculated for each feed so as to reflect the relative value of each feed's content of digestible protein, on the one hand, and usable energy expressed as TDN, on the other.

B. In most instances protein is considerably more costly per unit of TDN than are carbohydrates and fats. Consequently, high protein feeds are ordinarily more expensive than are low protein feeds of comparable energy content. It is obvious then that the actual nutritive worth of most feeds will usually depend largely on their content of DP and TDN and the prevailing costs of these two nutritive fractions.

C. The most widely used high energy feed and the most widely used high protein feed are normally used as base feeds for establishing protein and energy worth. Ground shelled corn and 44% soybean oil meal would logically qualify as base feeds under prevailing conditions. Other feeds might be used, however, as base feeds, should they become more widely used as sources of protein or energy than the above.

D. Constants of the nature referred to above have been calculated for the more important feeds and are included in Table 17. In calculating these constants, ground shelled corn and 44% soybean oil meal were used as the base high energy feed and the base high protein feed, respectively. Digestible protein was used in calculating protein worth, and TDN was used in calculating energy worth. Constants were calculated for both cattle and swine.

All of the values used for calculating the two sets of constants for the various feeds are to be found in Table 37. Consequently, they are not given again in Table 17. Also, while the procedure followed in the use of DP and TDN figures for calculating the various constants is quite unique and most interesting to the scientifically minded student, knowing the procedure is not essential for the use of the constants, and so is not included herein. Anyone who desires to become familiar with this procedure is referred to the following publication: Petersen, Wm. E. "A Formula for Evaluating Feeds on the Basis of Digestible Nutrients." *Journal of Dairy Science* (1932) 15:293.

II. **Use of constants for feed evaluation.**

A. In using the constants of Table 17 for calculating the actual worth of a feed based on its content of DP and TDN, one first determines the prevailing price per ton of the two base feeds—ground shelled corn and 44% soybean oil meal. These prices are then multiplied by the respective constants for a particular feed to be evaluated, and the two products are added together to obtain the actual worth of that feed for feeding purposes in dollars per ton. By comparing this calculated worth per ton with the prevailing price per ton for any particular feed, one can readily determine if that feed is a good buy at prevailing prices.

B.   To illustrate the evaluation of a feed by the Petersen method let us assume that ground shelled corn costs $68.00 per ton and 44% soybean oil costs $120.00 per ton. One wishes to determine the actual worth per ton of wheat middlings as a concentrate feed for a milking herd.

The calculation should proceed as follows, using constants for cattle from Table 17 for the feed in question.

$68.00 (price of corn per ton) × .887 (corn constant
    for wheat middlings)   = $60.32

$120.00 (price of SOM per ton) × .127 (SOM constant
    for wheat middlings)   = $15.24

                                                        TOTAL $75.56

From the above, it is apparent that at the prevailing prices for corn and SOM, wheat middlings has a worth as a feed for a milking herd of $75.56 per ton. If this feed can be bought at a price below its worth, then it would be considered to be a good buy compared with corn and SOM as sources of protein and energy. This would not, however, rule out the possibility of some other feeds being even a better buy as determined by a similar evaluation.

III.   **Negative constants.** In some instances, the constants are negative values.

A.   Negative corn constants are always associated with feeds extremely high in the ratio of DP to nonprotein TDN content. This is because an increase in the price of corn with the price of soybean oil meal remaining constant actually tends to lower the value of DP per lb and to increase the value per lb of nonprotein TDN. As a result, those feeds extremely high in protein and low in nonprotein TDN are actually lowered more in overall nutritive worth by the reduced value of DP than they are raised by the simultaneous increase in the value per lb of nonprotein TDN.

B.   A similar situation prevails with those feeds which have negative SOM constants except that negative SOM constants are always associated with feeds having an extremely low ratio of DP to nonprotein TDN. In any case, when negative constants are involved, they are supposed to have a negative effect on the overall worth of a feed and should be so handled in carrying out the feed evaluation calculation.

IV.   **Limitations of the Petersen method.** In the use of feed evaluations calculated by the Petersen method, appropriate consideration should be given to the fact that TDN from different sources does not necessarily have the same energy value for the different body functions. For example, the TDN of roughages approaches the TDN of concentrates in energy value for maintenance but has only a fraction of the energy value of concentrates for body gain. Other similar relationships also prevail. Consequently, in comparing feed worth figures as calculated following the Petersen procedure, such comparisons should be restricted to between roughages, on the one hand, and between concentrates, on the other. In other words, one should not attempt to substitute roughage for concentrates in the ration of animals in heavy production simply on the basis of favorable feed worth values as calculated by the Petersen method.

Table 17
CONSTANTS FOR EVALUATING FEEDS

BASED ON CONTENT OF

NAME OF FEED	DP AND TDN FOR CATTLE		DP AND TDN FOR SWINE	
	44% SOM	Shelled Corn	44% SOM	Shelled Corn
Dehydrated alfalfa meal	.281	.467	.155	.275
Alfalfa hay	.185	.459	.127	.287
Dehydrated alfalfa leaf meal	.310	.478	.254	.413
Dehydrated alfalfa stem meal	.121	.507	. . . . .	. . . . .
Fresh alfalfa forage	.091	.131	. . . . .	. . . . .
Animal fat, feed grade	−.439	2.652	. . . . .	. . . . .
Blood meal	. . . . .	. . . . .	1.641	−.848
Meat scrap	1.359	−.459	1.197	-.290
Digester tankage	1.468	−.515	. . . . .	. . . . .
Meat and bone meal	1.211	−.341	1.049	.037
Bahiagrass hay	−.101	.745	. . . . .	. . . . .
Dried bakery product	−.018	1.077	.056	1.036
Barley straw	−.084	.568	. . . . .	. . . . .
Barley grain	.076	.852	.069	.830
Beet molasses	−.039	.819	. . . . .	. . . . .
Dried beet pulp	−.046	.879	−.069	.926
Wet beet pulp	0.000	.092	. . . . .	. . . . .
Dried beet pulp/molasses	.011	.858	−.111	.987
Bermudagrass hay	−.010	.582	. . . . .	. . . . .
Dehydrated coastal bermudagrass meal	.192	.522	. . . . .	. . . . .
Coastal bermudagrass hay	.024	.607	. . . . .	. . . . .
Fresh coastal bermudagrass	.045	.207	. . . . .	. . . . .
Kentucky bluegrass hay	.013	.706	. . . . .	. . . . .
Fresh Kentucky bluegrass forage	.044	.198	. . . . .	. . . . .
Bromegrass hay	.068	.576	. . . . .	. . . . .
Fresh bromegrass forage	.085	.196	. . . . .	. . . . .
Cabbage heads	.024	.084	. . . . .	. . . . .
Reed canarygrass hay	.066	.540	. . . . .	. . . . .

| | BASED ON CONTENT OF | | | |
| | DP AND TDN FOR CATTLE | | DP AND TDN FOR SWINE | |
NAME OF FEED	44% SOM	Shelled Corn	44% SOM	Shelled Corn
Carrots	−.006	.141	−.001	.135
Dried whey	.107	.899	.198	.761
Dried whole milk	.440	1.005	.356	1.352
Fresh cow's milk	.059	.152	.058	.143
Fresh skimmed milk	.072	.042	.068	.052
Dried skimmed milk	.732	.317	.735	.378
Fresh citrus pulp	−.026	.219	. . . . .	. . . . .
Citrus pulp/molasses, dried	−.131	1.108	. . . . .	. . . . .
Dried citrus pulp	−.124	1.096	−.046	.643
Citrus molasses	−.057	.726	. . . . .	. . . . .
Alsike clover hay	.104	.576	. . . . .	. . . . .
Fresh crimson clover forage	.034	.120	. . . . .	. . . . .
Fresh ladino clover forage	.072	.107	. . . . .	. . . . .
Red clover hay	.106	.578	. . . . .	. . . . .
Fresh red clover forage	.045	.146	. . . . .	. . . . .
Fresh white clover forage	.076	.081	. . . . .	. . . . .
Copra meal	.371	.602	.288	.656
Corn stover	−.060	.714	. . . . .	. . . . .
Ground corn cob	−.107	.673	. . . . .	. . . . .
Corn fodder silage	−.007	.259	. . . . .	. . . . .
Corn stover silage	−.018	.275	. . . . .	. . . . .
Corn ear silage	−.015	.411	. . . . .	. . . . .
Corn bran	−.057	.926	. . . . .	. . . . .
Ground ear corn	−.065	.991	−.012	.885
Ground snapped corn	−.059	.910	. . . . .	. . . . .
Hominy feed	.001	1.088	. . . . .	. . . . .
Corn oil	−.433	2.615	−.434	2.599
Corn starch	−.214	1.297	−.216	1.303
Corn gluten meal	.825	.205	. . . . .	. . . . .
Yellow shelled corn	0.000	1.000	0.000	1.000
Cottonseed hulls	−.136	.721	. . . . .	. . . . .
Ground cottonseed	.224	.904	. . . . .	. . . . .
36% cottonseed meal	.691	.222	.701	.241
41% cottonseed meal	.829	.178	.850	−.021
41% solvent-extracted cottonseed meal	.854	.090	. . . . .	. . . . .
Cowpea hay	.202	.497	. . . . .	. . . . .
Dallisgrass hay	−.046	.720	. . . . .	. . . . .
Fresh Dallisgrass forage	.022	.174	. . . . .	. . . . .

| | BASED ON CONTENT OF | | | |
| NAME OF FEED | DP AND TDN FOR CATTLE | | DP AND TDN FOR SWINE | |
	44% SOM	Shelled Corn	44% SOM	Shelled Corn
Fescue hay	.006	.579	. . . . .	. . . . .
Fresh fescue forage	.041	.215	. . . . .	. . . . .
Fish meal	1.323	−.352	1.421	−.467
Linseed meal	.772	.231	.745	.175
Restaurant garbage	.031	.258	.049	.252
Brewers dried grains	.427	.379	.477	.059
Wet brewers grains	.081	.129	.068	.210
Distillers dried grains	.462	.553	.392	.751
Kudzu hay	.192	.476	. . . . .	. . . . .
Lespedeza hay	.115	.574	. . . . .	. . . . .
Sericea Hay	.146	.384	. . . . .	. . . . .
Millet hay	.019	.611	. . . . .	. . . . .
Meadow hay	−.023	.741	. . . . .	. . . . .
Prairie hay	−.073	.684	. . . . .	. . . . .
Oat hay	−.025	.721	. . . . .	. . . . .
Oat hulls	−.045	.454	. . . . .	. . . . .
Oat straw	−.081	.671	. . . . .	. . . . .
Oat grain	.096	.761	.115	.675
Oat groats	.120	1.050	.182	.887
Orchardgrass hay	.042	.639	. . . . .	. . . . .
Fresh orchardgrass forage	.040	.188	. . . . .	. . . . .
Peanut hay	.047	.704	. . . . .	. . . . .
Peanut hulls	.049	.169	. . . . .	. . . . .
Peanut oil meal	1.116	−.129	1.109	−.237
Peanut kernels	.384	1.326	. . . . .	. . . . .
Potato meal	−.020	.938	.044	.924
White potatoes	−.010	.247	−.029	.273
Feather meal	. . . . .	. . . . .	1.594	−.699
Poultry by-product meal	. . . . .	. . . . .	1.144	−.086
Redtop hay	−.035	.763	. . . . .	. . . . .
Rice hulls	−.034	.160	. . . . .	. . . . .
Rice bran	.106	.657	.113	.727
Fresh rye forage	.066	.129	. . . . .	. . . . .
Rye grain	.026	.892	.063	.927
Fresh ryegrass forage	.047	.146	. . . . .	. . . . .
Safflower meal without hulls	.855	.102	. . . . .	. . . . .
Sesame oil meal	.889	.110	.980	.090
Grain sorghum silage	−.005	.230	. . . . .	. . . . .

BASED ON CONTENT OF

NAME OF FEED	DP AND TDN FOR CATTLE		DP AND TDN FOR SWINE	
	44% SOM	Shelled Corn	44% SOM	Shelled Corn
Grain sorghum grain	−.026	.938	−.019	1.013
Hegari grain	−.015	.937	−.013	1.048
Johnsongrass hay	−.039	.726	. . . . .	. . . . .
Kafir grain	.042	.795	.020	.993
Milo grain	.008	.907	.025	.952
Sorgo silage	−.026	.230	. . . . .	. . . . .
Sudangrass hay	−.019	.698	. . . . .	. . . . .
Fresh sudangrass forage	.028	.158	. . . . .	. . . . .
Soybean hay	.130	.467	. . . . .	. . . . .
Soybean hulls	.064	.695	. . . . .	. . . . .
Soybean seed	.823	.305	.656	.534
44% soybean oil meal	1.000	0.000	1.000	0.000
49% soybean oil meal	1.219	−.147	1.164	−.203
Cane molasses	−.108	1.047	. . . . .	. . . . .
Sunflower meal	1.111	−.248	1.028	−.112
Sweetclover hay	.174	.525	. . . . .	. . . . .
Sweet potato meal	−.156	1.076	. . . . .	. . . . .
Sweet potatoes	−.055	.376	. . . . .	. . . . .
Timothy hay	−.044	.667	. . . . .	. . . . .
Turnips	−.003	.109	−.007	.099
Wheat straw	−.096	.637	. . . . .	. . . . .
Wheat bran	.214	.606	.198	.532
Wheat shorts	.150	.794	.257	.595
Wheat middlings	.127	.887	.210	.722
Wheat grain	.084	.926	.103	.887
Wood molasses	−.189	.859	. . . . .	. . . . .

33 Calculating a Balanced Daily Ration— General

I.   The first step in calculating a balanced daily ration for any given animal is to determine the amount of each critical nutrient required by the animal by referring to the appropriate table of nutrient requirements in section 56 of the text. Those nutrients which are usually regarded as critical and to which consideration is always given, either directly or indirectly, in any ration balancing process are:

Dry matter	Calcium
Protein	Phosphorus
Energy	Carotene (vitamin A)

With swine, consideration must also be given to the matter of essential amino acid balance as well as to the adequacy of several of the other vitamins. In addition to vitamin A, the following vitamins are frequently deficient in

unfortified swine feeds under present-day conditions: niacin, riboflavin, pantothenic aicd, and vitamins B_{12}, D, and E.

II.   The second step in calculating a balanced daily ration for any given animal
is to formulate a suitable combination of feeds such as will provide the
critical nutrients in the amounts needed to satisfy the requirements.

    A.   Various approaches and techniques may be used in arriving at which
feeds and the quantity of each to use for balancing the ration.

        1.   The cheapest source of each critical nutrient should be
determined as discussed previously if cost of production is to
be given consideration.

        2.   If a calculator is available and preciseness is desired in the
ration-balancing process, the use of certain mathematical
formulas and equations as discussed subsequently is recommended.

        3.   If a calculator is not available and/or if considerable precise-
ness is not required, the use of feeding guides in conjunction
with a trial-and-error approach may serve the purpose.

    B.   The amount of each of the various critical nutrients supplied by each
of the feeds used is calculated by using appropriate figures from the
feed composition table (Table 37), in section 57 of the text.

    C.   The requirements should always be met within acceptable limits by
the ration formulated.

        1.   Any nutritive allowance which is not more than about 3.0%
below the minimum requirement is usually considered to be
within acceptable limits, although it is always best to at least meet
the requirement whenever practicable to do so.

        2.   The energy allowance should not be permitted to exceed the
requirement by more than about 5.0% since an animal is
definitely limited in its energy utilization capacity.

        3.   A protein allowance in excess of the requirement by as much
as 5-10% is sometimes good insurance against feeds of below
normal protein content, especially if protein feeds are not too
expensive.

        4.   A protein allowance greatly in excess of the requirement will
normally not harm the animal but will usually cause the cost
of the ration to be higher than necessary.

        5.   Excesses of calcium and phosphorus are sometimes difficult to
avoid and are permissible, provided there has been no undue
use of mineral feeds and the Ca:P ratio is held between 1:1 and 2:1.

6.   Large excesses of carotene are often impractical to avoid, are
      not normally detrimental to the health of an animal, and are
usually acceptable as a part of a balanced ration.

7.   In the routine balancing of rations for livestock, the rations are
      not ordinarily evaluated for vitamin content, other than for
vitamin A. If other vitamins are to be added to the ration, they are
usually simply added at the minimum daily requirement level or at
some lower level such as experience has shown to give satisfactory
performance.

8.   DM requirements when listed should be interpreted in
      accordance with their intended purpose.

      a.   With full-fed animals they usually represent a maximum
            DM consumption capacity on the part of the animal, not
      to be exceeded by more than about 3%.

      b.   With other animals they usually simply indicate the
            approximate amount of feed to be fed and can be
      deviated from in either direction to a considerable degree
      should circumstances seem to warrant.

34 Balancing a Daily Ration for a Steer Using As Fed Weights and Composition Figures*

I.   Let us assume that a daily ration is to be balanced for a 300 kg finishing yearling steer using as fed feed weights and composition figures. The first step then is to look up the requirements for such an animal as given in Table 20. In this instance it is decided for no reason in particular that total protein (TP) rather than digestible protein (DP) and metabolizable energy (ME) rather than digestible energy (DE) or total digestible nutrients (TDN) will be used as the bases for evaluating protein and energy adequacy, respectively.

	DM kg	TP kg	ME Mcal	Ca kg	P kg	Caro mg
Requirements for a 300 kg finishing yearling steer	8.3	0.92	21.7	0.029	0.021	46.0

*Those persons who prefer not to work with the metric system are referred to "Balancing a Steer Ration with Minimum Use of the Metric System", section 37.

II.   The next step is to determine the amount of air-dry feed that will be required to approximate the DM requirement as listed above. This is done by dividing the DM requirement (8.3 kg) by the anticipated percentage of DM in the final ration and multiplying by 100. Ordinarily an average DM figure of 90% is used for this purpose, unless prior knowledge of the feeds to be used and their DM contents would indicate that a slightly lower or higher figure would be in order. Since at this point in this ration-balancing process no such prior knowledge is available, then the average figure of 90% will be used in estimating the amount of air-dry feed needed to meet the DM requirement, as follows:

$$\frac{8.3 \text{ (kg DM required)}}{90 \text{ (avg \% DM in air-dry feeds)}} \times 100 = 9.2 \text{ (kg of air-dry feed to be used)}$$

III.   Next, let us determine if the above requirements might be met by feeding 9.2 kg of a single feed such as coastal bermudagrass hay. To do this, one must first look up the composition of coastal bermudagrass hay as given in Table 37 and as listed below:

	DM %	TP %	ME Mcal/kg	Ca %	P %	Caro mg/kg
As fed composition of coastal bermudagrass hay	91.0	9.0	1.76	0.31	0.14	74.4

Next, calculate the total nutrients in 9.2 kg of coastal bermudagrass hay by multiplying each of the above figures by 9.2 with each (%) divided by 100.

	DM kg	TP kg	ME Mcal	Ca kg	P kg	Caro mg
Nutrients in 9.2 kg coastal bermudagrass hay	8.37	0.83	16.2	0.029	0.013	684.5

Upon comparing the latter amounts with the requirements as listed above, it becomes apparent that while such a ration is about right in DM and calcium (Ca) and more than adequate in carotene (Caro), it is somwehat low in TP and quite low in ME and phosphorus (P). These deficiencies cannot be overcome by simply feeding more hay because the animal will be unable to eat more than about 9.2 kg of air-dry feed.

IV.   The only alternative then is to substitute for at least a part of the hay feeds which will tend to correct the above deficiencies. Since with any livestock ration energy is of major concern, the addition of a feed to the ration which will correct this deficiency should be the next step.

Ground shelled corn is such a feed. How much of the hay to replace with corn then is the next consideration. Certainly all of the hay should not be replaced with corn since steers will ordinarily do best with some roughage in the ration. Also, to replace all of the hay with corn would tend to provide an excess of energy and/or a shortage of DM.

However, the precise combination of hay and corn which will exactly meet the DM and ME requirements may be calculated by the use of an appropriate algebraic equation in which:

$$9.2 \text{ kg} = \text{Approximate amount of air-dry feed to be used}$$

$$21.7 \text{ Mcal} = \text{Approximate amount of ME to be provided}$$

$$x = \text{Calculated amount of hay to use}$$

$$9.2 - x = \text{Calculated amount of corn to use}$$

Then:

$$(x \text{ kg hay} \times 1.76 \text{ Mcal ME/kg hay}) + [(9.2 \text{ kg total air-dry feed} - x \text{ kg hay})$$
$$\times 2.83 \text{ Mcal ME/kg corn}] = 21.7 \text{ Mcal ME required}$$

Solving for x:

$$1.76x + 26.04 - 2.83x = 21.7$$
$$-1.07x = -4.34$$
$$1.07x = 4.34$$
$$x = 4.06 \text{ kg of hay to be used}$$
$$9.2 - x = 5.14 \text{ kg of corn to be used}$$

The preceding calculation might also be accomplished by the use of algebraic equations where:

$$x = \text{Calculated amount of hay to use}$$

$$y = \text{Calculated amount of corn to use}$$

$$x + y = 9.2 \text{ kg air-dry feed to be fed}$$

$$1.76x + 2.83y = 21.7 \text{ Mcal ME required}$$

$$(\text{minus}) \underline{1.76x + 1.76y} = 16.2 \ (9.2 \times 1.76)$$
$$0.00x + 1.07y = 5.5$$

$$y = 5.5/1.07 = 5.14 \text{ kg of corn to be used}$$
$$x = 9.2 - 5.14 = 4.06 \text{ kg of hay to be used}$$

A second trial ration is then set up in which the hay has been reduced to 4.06 kg, and 5.14 kg of ground shelled corn has been added.

It becomes necessary then to determine the adequacy of such a ration by calculating the total amount of each of the nutrients it would provide in relation to the amount required, following the same procedure as was followed when hay alone was used above.

The compositions of coastal bermudagrass hay and shelled corn as given in Table 37 with respect to the critical nutrients on an as fed basis are as follows:

	DM %	TP %	ME Mcal/kg	Ca %	P %	Caro mg/kg
Coastal bermudagrass hay	91.0	9.0	1.76	0.31	0.14	74.4
Ground shelled corn	86.0	8.8	2.83	0.03	0.27	4.1

The total amount of each of the critical nutrients which would be provided by the tentative ration would then be as follows:

	DM kg	TP kg	ME Mcal	Ca kg	P kg	Caro mg
4.06 kg coastal bermudagrass hay	3.69	0.365	7.15	0.0126	0.0057	302.1
5.14 kg ground shelled corn	4.42	0.452	14.55	0.0015	0.0139	21.1
TOTAL	8.11	0.817	21.70	0.0141	0.0196	323.2

Upon comparing the latter totals with the requirements listed at the outset of this illustration, it becomes apparent that the ME deficiency of the original all-hay ration has been corrected, but the TP and phosphorus deficiencies still prevail, and a calcium deficiency was created by bringing corn into the ration in the place of a portion of the hay.

V.   The TP deficiency can be corrected by substituting a small amount of some high protein feed, such as 44% soybeal oil meal, for an equal amount of corn, without materially altering the ration ME and DM. The amount of soybean oil meal to use can be calculated by comparing the amount of TP in the hay plus corn trial ration with the TP requirement as listed above.

It will be noted that whereas the above requirement for TP is 0.92 kg, the latter trial ration provides only 0.817 kg, indicating the need for a minimum increase of 0.103 kg of TP to the ration. For each kg of ground shelled corn (8.8% crude protein) which is replaced by an equal weight of 44% soybean oil meal (45.8% crude protein), there is a net gain to the ration of 0.37 kg of TP (0.458 − 0.088 = 0.370). It would appear then that a minimum of 0.28 kg (0.103/0.37 = 0.28) of corn should be replaced with an equal weight of soybean oil meal.

An alternative procedure to follow in arriving at the amounts of corn and SOM to use in balancing this ration involves the setting up of an equation similar to the one used in determining the proper proportion of hay to concentrates to use where:

5.14 = Approximate amount of total concentrates to be used

0.555 = Approximate amount of TP to come from concentrates (total amount of TP minus TP from 4.06 kg hay)

x = Calculated amount of ground shelled corn to be used

5.14 − x = Calculated amount of 44% soybean oil meal to be used

Then:

(x kg corn × 0.088 kg TP/kg corn) + [(5.14 kg total air-dry concentrates − x kg corn × 0.458 kg TP/kg SOM] = 0.555 kg TP to come from total concentrates

Solving for x:

$$0.088x + 2.354 - 0.458x = 0.555$$
$$-0.370x = -1.799$$
$$0.370x = 1.799$$
$$x = 4.862 \text{ kg corn to be used}$$
$$5.14 - x = 0.278 \text{ kg SOM to be used}$$

The mathematical equation approach as previously described could also be used for making the above calculation.

A third trial ration consisting of the following combination of feeds is then tested for adequacy.

	DM kg	TP kg	ME Mcal	Ca kg	P kg	Caro mg
4.06 kg coastal bermudagrass hay	3.69	0.365	7.15	0.0126	0.0057	302.1
4.86 kg ground shelled corn	4.18	0.428	13.75	0.0015	0.0131	19.9
0.28 kg 44% SOM	0.25	0.128	0.73	0.0009	0.0019	
TOTAL	8.12	0.921	21.63	0.0150	0.0207	322.0
Percentage of requirement	97.8%	100.1%	99.7%	51.7%	98.6%	700.0%

Upon examining the above figures, it is apparent that calcium is the only nutrient which is not within the acceptable range. This can be easily corrected by simply adding a small amount of ground limestone to the ration. For each kg of ground limestone that might be added to the ration, there would be an increase of 0.3585 kg of calcium (35.85/100), or for each 0.1 kg of ground limestone added, there would be an increase of 0.0358 kg of calcium. Even the latter amount is more than is needed.

The exact amount of ground limestone which should be added can be calculated by dividing 0.0140 (kg of additional calcium needed) by 35.85 (%

calcium in ground limestone) and multiplying by 100. The amount then is 0.0390 kg, which rounds off to 0.04 kg.

VI. The final ration then, along with an evaluation of its nutritive adequacy, would be as follows:

	DM kg	TP kg	ME Mcal	Ca kg	P kg	Caro mg
4.06 kg coastal bermudagrass hay	3.69	0.365	7.15	0.0126	0.0057	302.1
4.86 ground shelled corn	4.18	0.428	13.75	0.0015	0.0131	19.9
0.28 kg 44% SOM	0.25	0.128	0.73	0.0009	0.0019	
0.04 kg ground limestone	0.04			0.0143		
TOTAL	8.16	0.921	21.63	0.0293	0.0207	322.0
Percentage of requirement	98.0%	100.1%	99.7%	101.0%	98.6%	700.0

It is apparent from the above that the final ration is adequate in all respects. The amount of DM provided approaches closely but does not exceed the DM-consuming limits of this animal. The TP supplied essentially meets the requirement. The same can be said of the ME, as well as of the calcium and phosphorus.

The fact that the carotene requirement is exceeded by several times should create no complications.

VII. Since most farm scales are still calibrated in pounds, however, it is necessary that the above final ration be converted to a pound basis before it can be used under most farm conditions. This would be done simply by multiplying the amounts of each feed used by 2.205 as follows:

FINAL RATION (AS FED BASIS)	IN KG		IN LB
Coastal bermudagrass hay	4.06 × 2.205	=	8.95
Ground shelled corn	4.86 × 2.205	=	10.72
44% soybean oil meal	0.28 × 2.205	=	0.62
Ground limestone	0.04 × 2.205	=	0.09
TOTAL	9.24		20.38

VIII. In view of the relatively high content of hay in the final ration above and in view of the relatively low digestibility of the nutrients in hay, a question might be raised regarding the adequacy of this ration in digestible protein (DP). This can be readily checked out as follows:

FEEDS USED	AMOUNT OF EACH FEED USED	% DP IN FEEDS USED	AMOUNT OF DP IN FEEDS USED
Coastal bermudagrass hay	4.06 kg ×	4.9 ÷	100 = 0.199 kg

FEEDS USED	AMOUNT OF EACH FEED USED		% DP IN FEEDS USED			AMOUNT OF DP IN FEEDS USED
Ground shelled corn	4.86 kg	X	6.5	÷	100 =	0.316 kg
44% soybean oil meal	0.28 kg	X	39.0	÷	100 =	0.109 kg
Ground limestone	0.04 kg	X	0.0	÷	100 =	0.000 kg
TOTAL						0.624 kg

Upon referring to Table 20, it will be noted that the DP requirement of a 300 kg finishing yearling steer is 0.62 kg. It would appear then that the above ration is not only adequate in protein based on TP (0.92 kg required vs. 0.921 provided) but also on the basis of DP (0.62 kg required vs. .624 kg provided).

35 Balancing a Daily Ration for a Steer Using Dry Feed Weights and Composition Figures

I. Let us assume once again that a daily ration is to be balanced for a 300 kg finishing yearling steer. However, in this instance dry feed weights and composition figures are to be used rather than the as fed figures as were used in the previous example.

Whether the ration is to be balanced on an as fed or dry basis, the requirements of the animal will be the same. For the above animal they were as follows:

	DM kg	TP kg	ME Mcal	Ca kg	P kg	Caro mg
Requirements for 300 kg finishing yearling steer	8.3	0.92	21.7	0.029	0.021	46

II. From the above requirements it is apparent that the final ration should contain approximately 8.3 kg or somewhat less of total DM.

The proportion of hay DM to corn DM making up the total (8.3 kg) is calculated using the same formula as in the above illustration, except dry weights and dry composition figures are used in the place of as fed weights and composition figures. If:

$$8.3 \text{ kg} = \text{Approximate amount of DM to be used}$$

$$21.7 \text{ Mcal} = \text{Approximate amount of ME to be provided}$$

$$x = \text{Calculated amount of hay DM to use}$$

$$8.3 - x = \text{Calculated amount of corn DM to use}$$

Then:

$$(x \text{ kg hay DM} \times 1.94 \text{ Mcal ME/kg hay DM}) + [(8.3 \text{ kg total DM} - x \text{ kg hay DM})$$
$$\times 3.29 \text{ Mcal ME/kg corn DM}] = 21.7 \text{ Mcal ME required}$$

Solving for x:

$$1.94x + 27.31 - 3.29x = 21.7$$
$$-1.35x = -5.61$$
$$1.35x = 5.61$$
$$x = 4.16 \text{ kg of hay DM to be used}$$
$$8.3 - x = 4.14 \text{ kg of corn DM to be used}$$

A trial ration consisting of 4.16 kg coastal bermudagrsss hay DM and 4.14 kg of ground shelled corn DM is then set up and checked for adequacy. The first step in doing this is to look up in Table 37 the composition of each of the above feeds on a dry basis. These compositions are found to be as follows:

	TP %	ME Mcal/kg	Ca %	P %	Caro mg/kg
Coastal bermudagrass hay	9.9	1.94	0.34	0.16	81.7
Ground shelled corn	10.2	3.29	0.03	0.31	4.8

The amount of each of the critical nutrients which would be provided by this tentative ration would then be calculated as follows:

	TP kg	ME Mcal	Ca kg	P kg	Caro mg
4.16 kg coastal bermudagrass hay DM	0.412	8.07	0.0141	0.0067	339.9
4.14 kg ground shelled corn DM	0.422	13.62	0.0012	0.0128	19.9
TOTAL	0.834	21.69	0.0153	0.0195	359.8

Upon comparing the latter totals with the requirements, it is apparent that a combination of hay and corn such as will meet the ME requirement, whether calculated on an as fed or dry basis, is somewhat deficient in TP, calcium, and phosphorus.

III. As in the as fed basis calculation, correcting the protein deficiency should be the next concern. Whereas 0.92 kg of TP is required, the tentative combination of hay and corn DM supplies only 0.834 kg. A small amount of the corn in the latter tentative ration needs to be replaced with an equal weight of soybean oil meal in order to bring up the TP 0.086 kg (0.92 − 0.834) without materially altering the ME. For each kg of corn DM (10.2% TP) which is replaced with an equal weight of 44% soybean oil meal DM (51.5% TP), an increase of 0.413 kg (0.515 − 0.102 = 0.413) of TP would result. It would appear then that 0.21 kg (0.086/0.413 = 0.208) of corn DM should be replaced with an equal weight of soybean oil meal DM to overcome the TP shortage in the above tentative ration. The ration is then so adjusted and checked again for nutritive adequacy.

	TP kg	ME Mcal	Ca kg	P kg	Caro mg
4.16 kg coastal bermudagrass hay DM	0.412	8.07	0.0141	0.0067	339.9
3.93 kg ground shelled corn DM	0.401	12.93	0.0012	0.0122	18.9
0.21 kg 44% SOM DM	0.108	0.61	0.0008	0.0016	
TOTAL	0.921	21.61	0.0161	0.0205	358.8

From the above totals it is apparent that calcium is the only nutrient which is not within the acceptable range. As in the previous illustration, this may be corrected by adding a small amount of ground limestone. The exact amount is determined by subtracting the amount of calcium in the ration (0.0161 kg) from the calcium requirement (0.0290 kg) and then dividing by the amount of calcium in 1 kg of ground limestone DM (0.3589 kg). This works out to be 0.0359 kg, which is rounded off to 0.04 kg of ground limestone DM to be added to the above tentative ration. The final ration is then checked for nutritive adequacy as follows:

FINAL RATION (DRY BASIS)	TP kg	ME Mcal	Ca kg	P kg	Caro mg
4.16 kg coastal bermudagrass hay DM	0.412	8.07	0.0141	0.0067	339.9
3.93 kg ground shelled corn DM	0.401	12.93	0.0012	0.0122	18.9
0.21 kg 44% SOM DM	0.108	0.61	0.0008	0.0016	
0.04 kg ground limestone DM			0.0144		
TOTAL	0.921	21.61	0.0305	0.0205	358.8
Percentage of requirement	100.1%	99.6%	105.2%	97.6%	780.0%

IV.   It is apparent that the above ration is adequate in all respects. However, the DM figures must be converted to as fed quantities before they can be used by the feeder. This is accomplished by dividing the amount of each feed on the dry basis by the percentage of DM in that feed on an as fed basis and multiplying by 100 as follows:

FEEDS USED	DM kg		% DM as fed				As fed kg
Coastal bermudagrass hay	4.16	÷	91.0	X	100	=	4.57
Ground shelled corn	3.93	÷	86.0	X	100	=	4.57
44% soybean oil meal	0.21	÷	89.0	X	100	=	0.24
Ground limestone	0.04	÷	99.9	X	100	=	0.04
TOTAL							9.42

V.   The as fed quantities in kg are then converted to lb by multiplying by 2.205 as follows:

	Amounts as fed kg				Amounts as fed lb
Coastal bermudagrass hay	4.57	X	2.205	=	10.08
Ground shelled corn	4.57	X	2.205	=	10.08
44% soybean oil meal	0.24	X	2.205	=	0.53
Ground limestone	0.04	X	2.205	=	0.09
TOTAL					20.78

VI.   When the final ration as calculated using as fed weights and compositions is compared with the final ration as calculated using dry weight and compositions, some small differences are noted.

	FINAL RATIONS AS CALCULATED USING	
	As fed basis	Dry basis
Coastal bermudagrass hay	8.95 lb	10.08 lb
Ground shelled corn	10.72 lb	10.08 lb
44% soybean oil meal	0.62 lb	0.53 lb
Ground limestone	0.09 lb	0.09 lb
TOTAL	20.38 lb	20.78 lb

The differences that do prevail are primarily the result of the fact that the as fed basis computations resulted in slightly less total dry matter being used than was used in the dry basis computations (8.16 kg vs. 8.34 kg). This permitted a somewhat greater proportion of hay to corn to be used in meeting the ME requirement with the dry basis computations. The slightly smaller amount of total DM involved with the as fed basis computations resulted from the fact that in making these computations, a dry matter content of 90% was assumed for the overall ration when in actuality it was somewhat lower than this (88.3%), mainly as the result of corn having a dry matter content of only 86.0%. Had all of the feeds contained exactly 90% DM, the two rations would have been identical. Even so, the two rations are very similar, and each adequately meets the animal's nutritive requirements.

36 Balancing a Daily Ration for a Steer Using a Restricted Level of Roughage

I.   Under the conditions which prevailed in this country for several years prior to 1973, roughages proved to be one of the more expensive sources of nutrients in the ration. Consequently, most feedlot operators found it desirable from a cost standpoint during this period to restrict the use of roughages to a level which would barely meet the roughage factor need of the animal. This was generally somewhere between 10% and 20% of the ration on an air-dry basis, depending on the extent of acidosis and founder and the number of condemned livers and rumens which the operator was willing to tolerate. This is considerably less roughage than was used (44.1% and 48.5%) in balancing the above rations.

II.   In order to balance a steer ration using a greatly restricted roughage level, a procedure somewhat different from that used in either of the above instances is required. A procedure which might be used under such conditions is as follows:

	DM kg	TP kg	ME Mcal	Ca kg	P kg	Caro mg
Requirements for a 300 kg finishing yearling steer	8.3	0.92	21.7	0.029	0.021	46

III.   Let us assume that only 1.5 kg of roughage in the form of coastal
bermudagrass hay is to be fed to this animal daily. This amount of hay will
provide 2.65 Mcal (1.5 × 1.76) of ME daily. Subtracting this amount of ME
from the total requirement as shown above leaves 19.06 Mcal (21.7 − 2.64) of
ME which will have to be supplied by the ration concentrates. If ground shelled
corn is to be the primary concentrate in the ration, then approximately 6.73 kg
of total concentrates (19.06 Mcal ME needed ÷ 2.83 Mcal/kg corn) will be
required.

A tentative ration of 1.5 kg of hay and 6.73 kg of corn is then checked for
nutritive adequacy.

	DM kg	TP kg	ME Mcal	Ca kg	P kg	Caro mg
1.5 kg coastal bermudagrass hay	1.36	0.135	2.64	0.0046	0.0021	111.6
6.73 kg ground shelled corn	5.79	0.592	19.05	0.0020	0.0182	27.5
TOTAL	7.15	0.727	21.69	0.0066	0.0203	139.1
To correct TP						
Subtract 0.52 kg corn	−0.45	−0.046	−1.47	−0.0002	−0.0014	− 2.1
Add 0.52 kg 44% SOM	+0.46	+0.238	+1.36	+0.0017	+0.0035	
NEW TOTAL	7.16	0.919	21.58	0.0081	0.0224	137.0
To correct calcium						
Add 0.06 kg ground limestone	0.06			0.0215		
FINAL TOTAL	7.22	0.919	21.58	0.0296	0.0224	137.0
Percentage of requirement	87.0%	99.9%	99.4%	102.1%	106.7%	297.8%

FINAL RATION   **IN LB**

1.50 kg coastal bermudagrass hay	3.31
6.21 kg ground shelled corn	13.69
0.52 kg 44% soybean oil meal	1.15
0.06 kg ground limestone	0.13
8.29 kg total feed, as fed	18.28

IV.   While the above ration comes within the acceptable limits for each of the
different nutritive considerations, it is 13% below the requirement in DM
content. However, this is of no major concern since the amount of roughage fed
is sufficient with good management to produce good gains and at the same time
hold the incidence of acidosis and founder and the number of condemned livers
and rumens to an acceptable level.

37 Balancing a Steer Ration with Minimum Use of the Metric System

I.   For those who prefer to work with lb and % rather than kg, kcal, Mcal, kcal/kg, Mcal/kg, g/kg, mg/kg, etc., the following procedure is recommended. Assuming once again a ration is to be balanced for a 300 kg (661 lb) finishing yearling steer, one must first obtain the requirements of such an animal from Table 20. This is done as in previous examples except that TDN rather than ME is used for evaluating energy adequacy. Also, the requirement values for DM, TP, TDN, calcium, and phosphorous are listed in lb rather than kg, while carotene is left in mg, as shown below:

	DM *lb*	TP *lb*	TDN *lb*	Ca *lb*	P *lb*	Caro *mg*
Requirements for a 300 kg (661 lb) finishing yearling steer	18.30	2.03	13.23	0.064	0.046	46.0

FIGURE 62.   Most cattle today are finished in large feedlots such as the one shown above. *(Courtesy of Ralston Purina Co., St. Louis, Mo.)*

(Should the requirements be listed in the table in kg but not in lb, kg may be converted to lb by multiplying kg by 2.205. Should some of the requirements be listed in grams but not in kg and/or lb, grams may be converted to kg by dividing grams by 1000—in other words, by moving the decimal point three places to the left. Kilograms may then be converted to lb as indicated above.)

II.   In order to determine what precise combination of coastal bermudagrass hay and ground shelled corn will exactly meet both the DM and the TDN requirements, the equation procedure will be followed, using as fed weights and composition figures.

The first step is to determine the approximate amount of air-dry feed which meeting the 18.3 lb DM requirement will entail by dividing the latter figure by the average DM content of air-dry feeds (90%) and multiplying by 100, as follows:

$$\frac{18.3 \text{ (lb DM required)}}{90 \text{ (avg \% DM in air-dry feeds)}} \times 100 = 20.3 \text{ (lb of air dry feeds to be used)}$$

III.   The next step then is to determine what combination of hay and concentrates will provide 13.23 lbs of TDN per each 20.3 lb of air-dry feed using the equation method, where

20.3 lb = Approximate amount of air-dry feed to be used

13.23 lb = Approximate amount of TDN to be provided

$$x = \text{Calculated amount of hay to use}$$

$$20.3 - x = \text{Calculated amount of concentrates as corn to use}$$

Then:

$$(x \text{ lb hay} \times 0.491 \text{ lb TDN/lb hay}) + [(20.3 \text{ lb total air-dry feed} - x \text{ lb hay})$$
$$\times 0.78 \text{ lb TDN/lb corn}] = 13.23 \text{ lb TDN required}$$

Solving for x:

$$0.491x + 15.83 - 0.78x = 13.23$$
$$-0.289x = -2.60$$
$$0.289x = 2.60$$
$$x = 9.0 \text{ lb hay to use}$$
$$20.3 - x = 11.3 \text{ lb concentrates to use}$$

IV.   A trial ration consisting of 9.0 lb of coastal bermudagrass hay and 11.3 lb of concentrates (all corn) is checked for nutritional adequacy as follows: The compositions of coastal bermudagrass hay and ground shelled corn with respect to the following nutritive factors as obtained from Table 37 are as follows, on an as fed basis:

	DM %	TP %	TDN %	Ca %	P %	Caro mg/lb
Coastal bermudagrass hay	91.0	9.0	49.1	0.31	0.14	33.7
Ground shelled corn	86.0	8.8	78.0	0.03	0.27	1.8

The total amount of each of the above nutrients provided by the tentative ration is calculated as follows:

	DM lb	TP lb	TDN lb	Ca lb	P lb	Caro mg
9.0 lb coastal bermudagrass hay	8.19	0.810	4.42	0.0279	0.0126	303.3
11.3 lb ground shelled corn	9.72	0.994	8.81	0.0034	0.0305	30.3
TOTAL	17.91	1.804	13.23	0.0313	0.0431	323.6

V.   Next, determine the amount of corn that must be replaced with SOM to meet the TP requirement.

$$\text{TP requirement} = 2.03 \text{ lb}$$
$$\text{TP in tentative ration} = 1.804 \text{ lb}$$
$$\text{Additional TP needed} = 0.226 \text{ lb } (2.03 - 1.804)$$

Amount TP is increased for each
lb of corn replaced by SOM = 0.37 lb (0.458 − 0.088)

Amount of corn to replace with
SOM to increase TP in ration 0.226 lb = 0.61 lb (0.226/0.37)

Revise the ration and check:

	DM lb	TP lb	TDN lb	Ca lb	P lb	Caro mg
9.0 lb coastal bermudagrass hay	8.19	0.810	4.42	0.0279	0.0126	303.3
10.69 lb ground shelled corn	9.19	0.941	8.34	0.0032	0.0289	19.2
0.61 lb 44% SOM	0.54	0.279	0.44	0.0020	0.0041	
TOTAL	17.92	2.030	13.20	0.0331	0.0456	322.5
Percentage of requirement	97.9%	100.0%	99.8%	51.7%	99.1%	701.1%

It is apparent that the above are all within the acceptable limits except for calcium which is quite deficient. Whereas the requirement for calcium is 0.064 lb, the latter ration provides only 0.0331 lb, which is 0.0309 lb short of the requirement. To increase the calcium by this amount makes it necessary to add 0.086 lb ground limestone (0.0309 ÷ 0.3585 lb Ca/lb limestone).

VI. The final ration then becomes:

	DM lb	TP lb	TDN lb	Ca lb	P lb	Caro mg
9.0 lb coastal bermudagrass hay	8.19	0.810	4.42	0.0279	0.0126	303.3
10.69 lb ground shelled corn	9.19	0.940	8.34	0.0032	0.0289	19.2
0.61 lb 44% SOM	0.54	0.279	0.44	0.0020	0.0041	
0.09 lb ground limestone	0.09			0.0323		
TOTAL	18.01	2.030	13.20	0.0654	0.0456	322.5
Percentage of requirement	98.4%	100.0%	99.8%	102.2%	99.1%	701.1%

It will be noted that the above ration is in lb ready for use by the feeder without any conversion being necessary. Also, it is essentially identical with the final ration obtained following the first ration-balancing procedure in section 34.

38 The California System for the Net Energy Evaluation of Rations for Growing-Fattening Cattle

I. The metabolizable energy (ME) of feeds is used by livestock much more efficiently for maintenance than it is for gain. Also, feeds vary greatly in the proportion of their ME content which is usable by livestock for gain. Concentrates, for example, are in general superior to roughages in the gain-producing capacity of their ME energy, and good quality roughages are superior to low quality roughages in this regard. Consequently, the ME content of a ration is sometimes not an accurate measure of its capacity to produce gain. This is especially true when a relatively high proportion of the ME comes from roughages, and particularly so when these roughages are of low quality. Roughages in general, and low quality roughages in particular, yield a high proportion of their ME in the form of heat, which is useful to the extent needed for keeping the animal warm but of no value for producing gain. In fact, excess heat may actually interfere with an animal's gain-producing capacity.

II.   In an effort to provide a more accurate basis for evaluating feeds from a gain-producing standpoint, various sets of so called NE values have been proposed over the years. However, some of the earlier NE systems have been justifiably criticized for not separating body functions in the establishment of net energy values.

Recognizing the need for such a separation, workers at the California Station have proposed a system which is designed to accomplish this goal. This system separates an animal's energy requirements into NE for maintenance (NE_m) and NE for gain (NE_{gain}) and the requirements of various classes of steers and heifers for each are given in Table 22. Also, individual feeds have been assigned dual net energy values, one for maintenance (NE_m) and another for gain (NE_{gain}), and these values have been included in Table 37.

In evaluating any particular ration then from the standpoint of its energy adequacy using the latter method, separate NE_m and NE_{gain} values are calculated for the overall ration. The NE_m value is used with as much of the overall ration as is needed to meet the animal's NE_m requirements. The remainder of the ration is then available for gain based on the calculated NE_{gain} value.

III.   To illustrate the use of the California System, let us proceed to calculate the gain-producing potential of the ration developed under Section 34. The ration in question is shown at the left below.

Column 1 FINAL RATION (AS FED BASIS)	Column 2 (FROM TABLE 37) NE_m Mcal/kg	Column 3 (COL. 1 × COL. 2) TOTAL NE_m Mcal	Column 4 (FROM TABLE 37) NE_{gain} Mcal/kg	Column 5 (COL. 1 × COL. 4) TOTAL NE_{gain} Mcal
4.06 kg coastal bermuda grass hay	1.04	4.22	0.38	1.54
4.86 kg ground shelled corn	1.96	9.53	1.27	6.17
0.28 kg 44% SOM	1.72	0.48	1.15	0.32
0.04 kg ground limestone	0.00	0.00	0.00	0.00
9.24 kg total ration	(1.54)	14.23	(0.87)	8.03

Calculated NE_m value for overall ration (14.23/9.24) = 1.54 Mcal/kg

Calculated NE_{gain} value for overall ration (8.03/9.24) = 0.87 Mcal/kg

Requirement of 300 kg finishing steer for NE_m (from Table 22) = 5.55 Mcal

Amount of overall ration needed for maintenance (5.55/1.54) = 3.60 kg

Amount of overall ration available for gain (9.24 − 3.60) = 5.64 kg

Amount of NE_{gain} in 5.64 kg of ration available for gain (5.64 × 0.87) = 4.91 Mcal

Amount of gain to be expected from 4.91 Mcal of NE_{gain} for 300 kg steer
(from Table 22) = approximately 1.12 kg

On the basis of the California system, between 1.1 and 1.2 kg of daily gain could be expected from a 300 kg steer receiving the above ration. This is a somewhat lower gain than normally would be expected on such an animal receiving a balanced ration. However, there was a relatively high level of hay used in balancing the above ration, and with this amount of hay the predicted level of gain is probably about right.

IV.   For sake of comparison, it might be of interest to use the California system to evaluate from the standpoint of its gain-producing capacity a ration which has been balanced on the basis of ME but with more of the ration's ME coming from grain than was true with the above ration. The ration developed in section 36 using a restricted level of roughage would be such a ration. This ration and its evaluation using the California system would be as follows:

Ration (As fed basis)	NE_m Mcal/kg	Total NE_m Mcal	NE_{gain} Mcal/kg	Total NE_{gain} Mcal
1.5 kg coastal bermudagrass hay	1.04	1.56	0.38	0.57
6.21 kg ground shelled corn	1.96	12.17	1.27	7.89
0.52 kg 44% SOM	1.72	0.89	1.15	0.60
0.06 kg ground limestone	0.00	0.00	0.00	0.00
8.29 kg total ration	(1.76)	14.62	(1.09)	9.06

Calculated NE_m value for overall ration   = 1.76 Mcal/kg

Calculated NE_{gain} value for overall ration   = 1.09 Mcal/kg

Requirement of 300 kg finishing steer for NE_m = 5.55 Mcal

Amount of overall ration needed for maintenance = 3.15 kg

Amount of overall ration available for gain = 5.14 kg

Amount of NE_{gain} in 5.14 kg of ration available for gain = 5.60 Mcal

Amount of gain to be expected from 5.60 Mcal of NE_{gain} for 300 kg steer = approximately 1.28 kg

It will be noted from the above that the latter ration, even though it embraces 0.95 kg less total feed, actually supplies 0.69 Mcal more NE_{gain} and should support about a 0.16 kg greater average daily gain. In other words, even though the two rations were balanced to provide essentially the same amounts of ME, the latter ration, because of its greater proportion of concentrates to hay, supplies more NE_{gain} and in turn would support a greater daily gain.

39 Balancing a Ration for a Dairy Cow

I. The balancing of a ration for a dairy cow differs in certain respects from the balancing of a ration for a beef animal.

   A. In the first place, the requirements cannot be taken directly from the table but must be calculated by adding to the requirements for maintenance (including growth and fetal development, if involved) an additional set of requirements for milk production based on the amount of milk produced and its fat content.

   B. Secondly, instead of having an overall dry matter requirement, this consideration is covered simply by following the general practice of feeding not less than 1% and not more than 2% of the cow's liveweight as air-dry roughage daily or the equivalent of silage.

   C. Also, since the energy requirements of dairy cows have been estimated and published in terms of net energy for lactating cows

($NE_{lactating\ cows}$), and since $NE_{lact\ cows}$ values have been made available
for most of the more common dairy feeds, dairy cow rations are frequently balanced for energy on the basis of $NE_{lact\ cows}$ rather than on the basis
of metabolizable energy (ME) or total digestible nutrients (TDN).

Otherwise, the balancing of a dairy cow ration is similar to the
balancing of a steer ration, and most of the same techniques can be applied.
After the requirements of a dairy cow have been established, the ration-
balancing process is very similar to that followed in balancing a ration for a steer,
using a restricted level of roughage (section 36).

II.   For the sake of illustration, let us balance a ration for a 500 kg mature
dairy cow in early gestation and producing 20 kg of 4.0% milk daily. The
daily requirements for such a cow would be calculated (using figures from Table
23) as follows:

	TP kg	$NE_{lact\ cows}$ Mcal	Ca kg	P kg	Caro mg
Requirements:					
Maintenance 500 kg cow	0.638	9.0	0.0200	0.0150	53.0
20 kg of 4% milk	1.560	14.8	0.0540	0.0400	
TOTAL	2.198	23.8	0.0740	0.0550	53.0

On the basis of the general guide above, a 500 kg cow should be fed
somewhere between 5 and 10 kg of air-dry roughage daily or the equivalent of
silage. In this instance, let us assume that a medium level of roughage (7.5 kg) is
fed in the form of 2.5 kg alfalfa hay and 15 kg of corn silage (15 kg of silage to
replace 5 kg of air-dry roughage).

A.   The first step then is to determine the amount of each nutrient
provided by the above amounts of alfalfa hay and corn silage, and
then calculate by difference the amount of each nutrient which must come
from the concentrates in the ration.

	TP kg	$NE_{lact\ cows}$ Mcal	Ca kg	P kg	Caro mg
2.5 kg alfalfa hay	0.388	2.85	0.0322	0.0048	163.0
15 kg corn silage	0.330	7.05	0.0120	0.0090	66.0
TOTAL	0.718	9.90	0.0442	0.0138	229.0
Needed from concentrates	1.480	13.90	0.0298	0.0412	

B.   The next step is to determine the approximate amount of
concentrates which will need to be fed with the above roughages to
meet the $NE_{lact\ cows}$ requirement. This is done by dividing the 13.90

Mcal $NE_{lact\ cows}$ needed from the concentrates by the approximate average $NE_{lact\ cows}$ content of the contentrates to be fed. If it is assumed that 1.90 Mcal $NE_{lact\ cows}$ kg would be about right for this figure, then 13.90/1.90 would equal the approximate amount of total concentrates to be fed (7.32 kg). A ration such as follows is then set up and tested for adequacy.

| | TP | $NE_{lact\ cows}$ | Ca | P | Caro |
	kg	Mcal	kg	kg	mg
2.5 kg alfalfa hay	0.388	2.85	0.0322	0.0048	163.0
15 kg corn silage	0.330	7.05	0.0120	0.0090	66.0
7.32 kg { 3.0 kg ground shelled corn	0.264	6.24	0.0009	0.0081	12.3
4.32 kg ground barley	0.501	8.21	0.0030	0.0173	
TOTAL	1.483	24.35	0.0481	0.0392	241.3

C.   It is apparent from the above that the latter ration is low in TP, as well as calcium and phosphorus. The next step then is to correct the shortage of TP by substituting some high protein concentrate such as soybean oil meal for a part of the barley. The amount of barley to replace with soybean oil meal is arrived at by dividing the amount of additional TP needed ($2.198 - 1.483 = 0.715$ kg) by the increase in TP brought about by each kg of barley replaced by 44% soybean oil meal ($0.458 - 0.116 = 0.342$ kg). Upon dividing 0.715 by 0.342, it is determined that 2.09 kg of barley will need to be replaced with soybean oil meal.

The ration is then revised accordingly and rechecked for adequacy as follows:

| | TP | $NE_{lact\ cows}$ | Ca | P | Caro |
	kg	Mcal	kg	kg	mg
2.5 kg alfalfa hay	0.388	2.85	0.0322	0.0048	163.0
15.0 kg corn silage	0.330	7.05	0.0120	0.0090	66.0
3.0 kg ground shelled corn	0.264	6.24	0.0009	0.0081	12.3
2.23 kg ground barley	0.259	4.24	0.0016	0.0089	
2.09 kg 44% SOM	0.957	3.85	0.0067	0.0140	
TOTAL	2.198	24.23	0.0534	0.0448	241.3
Percentage of requirement	100.0%	101.8%	72.2%	81.5%	455.3%

D.   In view of the fact that the above ration is still low in both calcium and phosphorus, it is desirable that additional amounts of these nutrients be added. Since most of the commonly used phosphorus supplements are also high in calcium, it is best to correct the phosphorus shortage next and then add some additional calcium if needed.

By subtracting the amount of phosphorus in the latter ration above from the phosphorus requirement $(0.0550 - 0.0448)$, it is determined that an additional 0.0102 kg of phosphorus needs to be added to the ration. If 0.0102 kg is divided by the amount of phosphorus in 1 kg of defluorinated phosphate (0.18 kg), the amount of defluorinated phosphate that needs to be added to correct the phosphorus shortage is determined. This figures out to be 0.057 kg. This amount of defluorinated phosphate is then added to the above ration which is rechecked for adequacy as follows:

	TP kg	NE$_{lact cows}$ Mcal	Ca kg	P kg	Caro mg
2.5 kg alfalfa hay	0.388	2.85	0.0322	0.0048	163.0
15.0 kg corn silage	0.330	7.05	0.0120	0.0090	66.0
3.0 kg ground shelled corn	0.264	6.24	0.0009	0.0081	12.3
2.23 kg ground barley	0.259	4.24	0.0016	0.0089	
2.09 kg 44% SOM	0.957	3.85	0.0067	0.0140	
0.057 kg defluorinated phosphate			0.0188	0.0103	
TOTAL	2.198	24.23	0.0722	0.0551	241.3
Percentage of requirement	100.0%	101.8%	97.6%	100.2%	455.3%

It will be noted that the defluorinated phosphate which was added not only corrected the phosphorus shortage but at the same time for the most part corrected the shortage of calcium as well. To make this ration 100% adequate in calcium, there should be added to it 0.01 kg of ground limestone.

E.   Before being usable under most present-day farm conditions, however, the above ration will need to be converted to lb by multiplying each of the individual quantities of feed by 2.205. The final ration then would be:

    5.51 lb alfalfa hay
   33.08 lb corn silage
    6.62 lb ground shelled corn
    4.92 lb ground barley
    4.61 lb 44% soybean oil meal
   0.126 lb defluorinated phosphate
   0.022 lb ground limestone

F.   In balancing the above ration, as fed weights and composition figures were used for the various feeds. Essentially the same results could have been realized using dry feed weights and composition figures. Also, DP could have been used rather than TP for evaluating protein and ME or

TDN rather than $NE_{lact\ cows}$ for evaluating energy without significantly affecting the overall results.

G.   As with the balancing of a steer ration earlier, those who prefer to avoid the use of the metric system insofar as possible may for the most part do so by using TDN as the measure for evaluating energy adequacy. However, as with the steer, the requirements for all the critical nutrients except carotene must be converted from kg to lb by multiplying by 2.205 unless this has already been done in the table.

40 Balancing Daily Rations for Horses

I.   The balancing of daily rations for horses is similar in most respects to balancing rations for cattle, and most of the same general methods and techniques may be used.

A.   As in balancing rations for other livestock classes, the first step is to determine the requirements for the particular animal to be fed as given in the appropriate NRC table. The critical nutrients to be considered are essentially the same for horses as for other animals. As with lactating dairy cows and sometimes with certain classes of other livestock, no specific DM requirement is indicated. While a "Daily Feed" intake is given in the NRC tables of requirements for horses, this is based on the assumption that the ration used will contain 2.75 Mcal of DE per kg of DM and so does not necessarily apply for rations or mixtures of other energy concentrations such as are frequently used in feeding horses.

B. Consequently, in order to be certain that a horse is provided with the proper DM allowance—that is, with a ration which contains sufficient bulk, on the one hand, but does not exceed the feed-consuming capacity of the animal, on the other—a minimum of 1.0% and a maximum of 2.0% of the animal's liveweight as air-dry roughage is ordinarily included in the ration. Grain and supplement are then added as needed to balance the ration.

C. The rate of roughage feeding should be adjusted so that the animal's total daily consumption of air-dry feed does not exceed about 2.25% to 2.50% of the animal's liveweight. This is roughly the upper limit of a horse's capacity to consume feed.

II. With the above in mind let us proceed then to balance a ration for a horse such as a 500 kg mare at the peak of lactation. The requirements as obtained from Table 26 are as follows:

	TP kg	DE Mcal	Ca kg	P kg	Caro mg
Requirements of 500 kg mare, peak of lactation	1.317	27.62	0.0470	0.0386	62.5

The development of a balanced ration in keeping with the above stipulations and requirements would then proceed as follows. In this connection it should be pointed out that in using the feed composition table for balancing horse rations, cattle values for energy (also for DP, if used) should be used.

	TP kg	DE Mcal	Ca kg	P kg	Caro mg

Step 1. Feed at least 1% of body weight as hay.

| 5 kg timothy hay | 0.315 | 10.75 | 0.0180 | 0.0075 | 236.5 |

Step 2. Feed with hay enough of some grain, such as oats, to meet the DE requirement: Amount to use = (27.62 − 10.75) ÷ 2.92 Mcal DE/kg oats = 5.78 kg.

| 5.78 kg oats grain | 0.676 | 16.88 | 0.0052 | 0.0191 | *** |
| TOTAL (oats + hay) | 0.991 | 27.63 | 0.0232 | 0.0266 | 236.5 |

Step 3. Correct protein shortage by replacing some of the oats with an equal wt of some high protein feed, such as 41% CSM: Amount of oats to replace with CSM = (1.317 − 0.991) ÷ (41.4 − 11.7) = 1.1 kg.

Deduct 1.10 kg oats	−0.129	−3.21	−0.0010	−0.0036	
Add 1.10 kg 41% CSM	+0.455	+3.53	+0.0021	+0.0120	
TOTAL (oats, hay + CSM)	1.317	27.95	0.0243	0.0350	236.5

	TP	DE	Ca	P	Caro
	kg	Mcal	kg	kg	mg

Step 4. Correct phosphorus shortage by adding defluorinated phosphate as needed to do so. [0.0386 kg P required − 0.0350 kg P provided] ÷ 0.18 kg P/kg defluorinated phosphate = 0.02 kg defluorinated phosphated needed.

	TP	DE	Ca	P	Caro
Add 0.02 kg defl. phos.	0.000	00.00	0.0066	0.0036	
TOTAL	1.317	27.95	0.0309	0.0386	236.5

Step 5. Correct calcium shortage by adding ground limestone as needed to do so: (0.0470 kg Ca required − 0.0309 kg Ca provided) ÷ 0.3585 kg Ca/kg ground limestone = 0.045 kg ground limestone.

	TP	DE	Ca	P	Caro
Add 0.045 kg ground limestone	0.000	00.00	0.0161	0.0000	000.0
TOTAL	1.317	27.95	0.0470	0.0386	236.5
Percentage of requirement	100.0	101.2	100.0	100.0	378.4

Final ration:

5.00	kg timothy hay	11.02 lb
4.68	kg oats grain	10.32 lb
1.10	kg 41% cottonseed meal	2.43 lb
0.02	kg defluorinated phosphate	0.044 lb
0.045	kg ground limestone	0.099 lb
10.845 kg	TOTAL	23.913 lb

III. There are several alternatives which might have been used in balancing the above ration, but regardless of the alternative pursued, the procedure and the results would have been essentially the same. Some of these alternatives are:

A. Digestible protein (DP) rather than total protein (TP) might have been used for checking the ration's protein adequacy.

B. Total digestible nutrients (TDN) rather than digestible energy (DE) might have been used for balancing the ration with respect to energy.

C. Pounds rather than kilograms might have been used as the unit of feed weight. However, this would have necessitated multiplying all of the requirements, except carotene, by 2.205 to change them from kg to lb. Also, this would make it necessary for energy to be expressed as lb of TDN rather than Mcal of DE.

D. Hays other than timothy, grains other than oats, and a protein supplement other than cottonseed meal could have been used with equally effective results.

E. Simultaneous equations could have been used in calculating the proportion of oats to cottonseed meal to feed.

41 Balancing Daily Rations for Sheep

I.   The balancing of daily rations for sheep is similar in most respects to balancing rations for cattle and horses, and most of the same general methods and techniques are applicable.

II.   In order to provide sheep rations with the proper bulk, on the one hand, and energy density, on the other, a minimum of about 2% and a maximum of about 3% of a sheep's liveweight as air-dry roughage (or equivalent silage) is ordinarily included in the animal's daily ration. Grain and supplement are then added as needed to meet the requirements. Also, in this connection the rate of roughage feeding should be adjusted so that the total daily consumption of air-dry feed does not exceed about 3.5% to 4% of the animal's liveweight.

III.   In balancing rations for fattening lambs, it is very important that such animals be brought on to grain feeding gradually in order to avoid excessive losses from enterotoxemia or so-called "overeating disease".

IV.   Within the limits of the above stipulations, daily rations for sheep are developed much in the same manner as outlined previously for cattle and horses.

42 Balancing Daily Rations for Swine— General

I.   The balancing of daily rations for swine is similar in basic principle to balancing rations for other classes of livestock in that it is simply a matter of meeting a given set of nutritive requirements with the feeds and supplements fed. Feeding swine, however, differs from feeding other livestock in that the vitamins and, in some respects, protein and minerals are so much more critical in swine nutrition, and commercially prepared supplements have become so much a standard part of swine feeding.

   A.   Commercially prepared swine supplements are usually of three general types. The one which is most extensively used is probably the complete protein, mineral, and vitamin supplement. Many different brands of complete supplements for swine are on the market, but all have supposedly been formulated to meet the needs of the animal for the critical minerals, vitamins, and essential amino acids when fed in the

NET WEIGHT SHOWN ON BAG

2835

**PURINA® HOG
SUPPLEMENT 40%
(G)**

Carefully Follow Feeding Directions
on Reverse Side

Manufactured By

RALSTON PURINA COMPANY
General Offices, St. Louis, Mo.

GUARANTEED ANALYSIS
Crude protein not less than......40.0%
Crude fat not less than............... 1.0%
Crude fiber not more than 10.0%
Calcium (Ca) not less than...... 3.0%
Calcium (Ca) not more than.... 4.0%
Phosphorus (P) not less than.. 1.1%
Iodine (I) not less than........ 0.0003%
Salt (NaCl) not less than........ 2.0%
Salt (NaCl) not more than...... 3.0%
INGREDIENTS
Processed grain by-products, animal
protein products, plant protein products, cane molasses, dehydrated alfalfa meal, vitamin A supplement, D
activated animal sterol, vitamin B-12
supplement, riboflavin supplement,
niacin, methionine hydroxy analogue
calcium, calcium pantothenate, defluorinated phosphate, calcium carbonate, iodized salt, iron carbonate, iron
sulfate, manganous oxide, copper
sulfate, cobalt carbonate, zinc oxide.

MTW-2835

CG-2835

HOG SUPPLEMENT 40% (G)

RALSTON PURINA CO., ST. LOUIS MO

2835 M

MEAL

DIRECTIONS

Feed Purina Hog Supplement 40% with
ground grain from 50 pounds to market.
Recommended mixing levels for pigs of different weight ranges are as follows:

Approx. Wt. of Pigs	Pounds Hog Suppl./Ton	Pounds Corn/Ton
50- 85	455	1545
85-125	325	1675
125-200	200	1800
200-Market	200	1800

Keep water available at all times.

MTW-2835

FIGURE 63.   Shown above is the information on the feed tag
for a typical complete 40% supplement for growing-finishing swine.
(Courtesy of Ralston Purina Co., St. Louis, Mo.)

proper amount along with farm grown grain as the major ration
component. The amount to be fed is usually the amount required to meet
the animal's needs for protein.

B.   Also available are so-called "mineral and vitamin" supplements.
These are usually designed so that when fed at the recommended
level along with farm grown grain and soybean oil meal, or other suitable
high protein feeds, the animal's critical nutritive needs usually will be met.
Combinations of farm grains and most of the available high protein feeds
provide an adequate balance of the essential amino acids. Consequently,
mineral and vitamin supplements such as are available serve effectively in
correcting the nutritional inadequacies of such mixtures for swine.

C.   Some swine producers prefer to go even further toward dispensing
with the use of commercially prepared supplements in their ration

formulation. Not only do such producers use soybean oil meal or some other standard high protein feed as a source of supplemental protein, but they also use ground limestone or oystershell flour and defluorinated phosphate or steamed bone meal, along with trace-mineralized salt, for meeting their needs for supplemental minerals. Most such producers, however, still rely on commercially prepared vitamin supplements as a supplemental source of the critical vitamins.

In balancing swine rations as in balancing rations for certain other classes of livestock, no direct consideration is given to dry matter as such. It is simply understood that, for the most part at least, only feeds of low fiber content will be used. While under certain conditions some swine feeders still include limited amounts of alfalfa meal in their rations, most producers, in view of the relatively high cost of alfalfa meal and the nutritional adequacy of modern-day swine supplements, have dispensed with the use of this product in their feeding programs. Even when used, its level is usually restricted to not over about 10% of the ration for breeding stock and not over about 5% for growing-finishing animals. By so limiting the fiber content of the ration, problems associated with dry matter intake are for the most part avoided.

FIGURE 64.   Most market hogs today are self fed on either concrete or slotted floors. *(Courtesy of Ralston Purina Co., St. Louis, Mo.)*

In view of the above, the balancing of daily rations for swine usually entails certain assumptions and improvisions not encountered in balancing rations for other classes of livestock. Also, the ration-balancing procedure will vary, depending on the type of supplement used. For the sake of illustrating these points, let us proceed to consider the use of each type of supplement in the balancing of a daily ration for a 47.5 kg growing-finishing pig.

II.   Use of a "complete" supplement. When a complete supplement is used along with a grain, such as corn, then one needs to concern himself only with protein and energy in balancing the ration. When these two nutritive factors have been satisfied, other nutritive requirements also will have been met. The requirements of the above animal for TP and DE as obtained from Table 35 are as follows:

	TP kg	DE kcal
Daily requirements for a 47.5 kg growing-finishing pig	0.350	8250

In order to develop a combination of corn and supplement, the TP and DE contents of which would approximate the above requirements, the following steps are proposed.

Step 1. Determine the kg of corn necessary to supply the required amount of DE (8250 kcal) by dividing the latter amount by the kcal DE/kg of corn (3488). The answer is 2.365 kg. The TP and DE content of this amount of corn is then calculated as follows:

2.365 kg ground shelled corn	0.208	8249

Step 2. Calculate the amount of corn that must be replaced with 40% supplement to meet the protein requirement. This is accomplished by subtracting the amount of TP provided by 2.365 kg corn (0.208 kg) from the TP requirement (0.350 kg) and dividing the difference (0.142 kg) by the increase in TP brought about per kg of corn replaced by 40% supplement (0.40 − 0.088 = 0.312).

$$0.142 \text{ kg}/0.312 = 0.455 \text{ kg}$$

Step 3. Calculate the TP and DE contents of the ration revised accordingly.

1.910 kg ground shelled corn	0.168	6662
0.455 kg 40% supplement	0.182	1587*
TOTAL	0.350	8249

*No DE value is ordinarily available for commercial supplements. In arriving at this figure it is assumed that the DE value of the supplement is not greatly different from that of the corn.

(The above calculation could have been made also through the use of mathematical equations as described previously.)

It would appear from the above that 1.91 kg of ground shelled corn plus 0.455 kg of 40% complete swine supplement would meet the daily protein and energy needs of a 47.5 kg growing-finishing pig. Also, since a complete supplement normally contains supplemental levels of the critical minerals and vitamins, as well as of protein, it can be assumed that the pig's requirements for minerals and vitamins have also been met.

III.   Use of a "mineral and vitamin" supplement. Sometimes swine feeders rather than relying on a complete supplement as a source of supplemental protein will use some conventional high protein feed such as soybean oil meal for this purpose and then fall back on a commercially prepared "mineral and vitamin" supplement as a supplemental source of these nutrients. In such instances, the ration-balancing procedure is essentially the same as above except that soybean oil meal is used in the place of the complete supplement as a source of supplemental protein. A commercially prepared "mineral and vitamin" supplement is then added at the level recommended by the manufacturer to the daily allowance of corn and soybean oil meal to provide the needed minerals and vitamins. Usually the quantity of such supplements required amounts to less than 3% of the total ration. Consequently, their addition to the ration would not ordinarily necessitate any adjustment in the allowances of corn and soybean oil meal to stay within the acceptable range of the protein and energy requirements.

IV.   Use of a "vitamin" supplement. If only a vitamin supplement is to be used in balancing the ration, then the requirements for calcium and phosphorus, as well as for TP and DE, need to be listed and met with the feeds fed as follow:

	TP kg	DE kcal	Ca kg	P kg
Requirements of a 47.5 kg growing-finishing pig	0.350	8250	0.0125	0.0100

Step 1. Same as Step 1 above.

| 2.365 kg ground shelled corn | 0.208 | 8249 | 0.0008 | 0.0064 |

Step 2. Calculate the amount of corn to replace with 44% SOM to meet TP requirement $(0.350 - 0.208) \div (0.458 - 0.088) = 0.384$ kg.

1.981 kg ground shelled corn	0.174	6910	0.0006	0.0053
0.384 kg 44% SOM	0.176	1267	0.0013	0.0026
TOTAL	0.350	8177	0.0019	0.0079

Step 3. Calculate the amount of defluorinated phosphate to add to the ration to meet the phosphorus requirement: $(0.0100 - 0.0079) \div 0.18 = 0.012$ kg.

	TP *kg*	DE *kcal*	Ca *kg*	P *kg*
1.981 kg ground shelled corn	0.174	6910	0.0006	0.0053
0.384 kg 44% SOM	0.176	1267	0.0013	0.0026
0.012 kg defluorinated phosphate			0.0040	0.0022
TOTAL	0.350	8177	0.0059	0.0101

Step 4. Calculate the amount of ground limestone to add to ration to meet the calcium requirement: $(0.0125 - 0.0059) \div 0.3585 = 0.019$ kg.

	TP	DE	Ca	P
1.981 kg ground shelled corn	0.174	6910	0.0006	0.0053
0.384 kg 44% SOM	0.176	1267	0.0013	0.0026
0.012 kg defluorinated phosphate			0.0040	0.0022
0.019 kg ground limestone			0.0068	
TOTAL	0.350	8177	0.0127	0.0101
Percentage of requirement	100.0%	99.1%	101.6%	101.0%

To the above ration would then be added as much of the vitamin supplement as is recommended by the manufacturer. In most instances this will amount to less than 1% of the ration.

43 Some General Guides for Feeding Livestock

I.   **Beef cattle.**

A.   **For wintering dry brood cows in dry-lot.** Feed a little under 2% of the body weight daily as air-dry roughage (or equivalent silage) plus protein supplement if needed to balance the ration (seldom is over 0.5 kg [1.1 lb] per head daily of supplement required).

B.   **For stocker steers and heifers in dry-lot.** Feed a little over 2% of the body weight daily as air-dry roughage (or equivalent silage) plus protein supplement as needed to balance the ration (seldom is over 0.5 kg [1.1 lb] per head daily of supplement required).

C.   **For cows nursing calves in dry-lot.** Feed such cows about 50% more nutrients than dry cows. If roughage is low quality and/or poorly consumed, feed up to about 2 kg (4.4 lb) of grain and protein supplement per head daily.

D. **For fattening calves and yearlings.** Feed 2 - 2-1/2% of the body weight daily as grain and protein supplement and 1/2-1% of the body weight daily as air-dry roughage (or equivalent silage) along with minerals and vitamin A as needed. Use about 1 kg of protein supplement for each 8-12 kg of grain or 1 lb for each 8-12 lb.

E. **For mature bulls in dry-lot.** Feed 1 - 1-1/2% of the body weight daily as air-dry hay (or equivalent silage) plus concentrates [usually 1-3 kg (2.2-6.6 lb) per head daily] as needed to achieve the desired condition.

II. **Dairy cattle**

A. **For milking cows.** Feed 2% of the body weight daily as air-dry roughage (or equivalent silage) plus concentrates as follows:

    1 kg (lb) conc for each 4. kg (lb) low test milk
    1 kg (lb) conc for each 3.5 kg (lb) 4% milk
    1 kg (lb) conc for each 3. kg (lb) high test milk

Air-dry roughage may be reduced to as low as 1% of the body weight daily, but the concentrates will have to be increased accordingly.

B. **For growing dairy heifers.** Feed about the same as, or a little more liberally than, stocker beef heifers as outlined above.

C. **For mature dairy bulls in dry-lot.** Feed about the same as beef bulls as outlined above.

III. **Swine.**

A. **Feed growing-fattening pigs** as follows:

wt of pig in kg (lb)	Air-dry feed per head per day as % of body wt	% Crude Protein in ration
25 (55)	6.5	16.0
50 (110)	5.5	14.0
75 (165)	4.5	13.0
100 (220)	3.5	13.0

1. Feed corn (or a suitable substitute) along with a complete protein-mineral-vitamin supplement in such proportions as will meet the protein requirement,

or

Feed corn (or a suitable substitute) and soybean oil meal (or a suitable substitute) in such proportions as will meet the protein requirements, along with a mineral-vitamin supplement in such amounts as will provide the minimum requirements of the critical minerals and vitamins.

Fairly satisfactory substitutes for at least a part of the corn in pig rations are grain sorghum, wheat, barley, and/or oats.

Satisfactory substitutes for soybean oil meal, at least in part, are peanut oil meal, tankage, tankage with bone, meat scrap, meat and bone scrap, fish meal, and poultry by-product meal. Cottonseed meal (41%) may be used up to 9% of the ration.

2.   Some producers like to include around 2.5% dehydrated alfalfa meal in rations for growing-fattening pigs.

3.   Pigs are usually self-fed and when self-fed will consume feed at about the rates shown above.

B.   For sows nursing litters. Feed from 4.5 to 6.8 kg (10-15 lb) of feed per head daily (depending on the condition of the sow and the size of the litter) containing about 15% protein. Use corn (or a suitable substitute) along with a complete protein-mineral-vitamin supplement, or corn and soybean oil meal (or suitable substitutes) along with an appropriate mineral-vitamin supplement.

C.   For pregnant sows on either pasture or dry-lot. Feed from 1.6 to 2.0 kg (3.5-4.5 lb) of feed per head daily (depending on the condition of the sow) containing about 14% crude protein. Use corn (or a suitable substitute) along with a complete protein-mineral-vitamin supplement, or corn and soybean oil meal (or suitable substitutes) along with an appropriate mineral-vitamin supplement.

D.   Adult boars. Feed about the same as pregnant sows as outlined above.

IV.   **Horses, mules, and ponies.** Feed as follows:

PERCENTAGE OF BODY WEIGHT DAILY AS:

	Grain	Hay
Idle	0.00	1.75-2.00
Doing light work	0.75	1.25
Doing medium work	1.00	1.00
Doing heavy work	1.25	1.00

Little, if any, protein supplement will be required by mature animals, except for mares in lactation.

V.   **Sheep.**

    A.   **Breeding ewes.** Feed about 2.5-3.0% of body weight daily as hay (or equivalent silage) plus concentrates (up to 1 kg or 2.2 lb per head daily) as needed to balance the ration.

    B.   **Fattening lambs.** Feed about 1-1/2 - 2% of body weight daily each of concentrates and hay (or equivalent silage). Feeder lambs should be brought on to grain feeding gradually in order to avoid excessive losses from *enterotoxemia* (overeating disease).

44 Formulating Balanced Ration Mixtures— General

Most swine and certain other classes of livestock are fed on complete ration mixtures rather than specific daily amounts of certain feeds as arrived at by calculating a balanced daily ration. When complete ration mixtures are used, it is essential that such mixtures be formulated so as to meet the animal's minimum requirements for the various critical nutrients based on the amount of feed consumed. In order to calculate the concentration of any given nutrient needed in a ration mixture to meet an animal's daily requirement, it is necessary to know the animal's daily feed consumption. When animals are hand fed, this can be predetermined on the basis of calculated needs. When animals are self-fed, their anticipated feed consumption must be estimated. While self-fed animals eat all they want, an animal has certain physical and physiological limits to its capacity for feed consumption. This capacity is reasonably well established for most animals and can be obtained by referring to the dry matter column of the daily nutrient requirement table for the respective class of livestock. Once the anticipated daily feed consumption has been arrived at, the concentration of any given nutrient needed in this amount of feed to meet the animal's daily requirement can be calculated. These concentrations have been calculated for the various classes of livestock and are reported in the respective NRC tables on nutritive requirements.

Variations of several general procedures may be followed in formulating ration mixtures for use in livestock feeding.

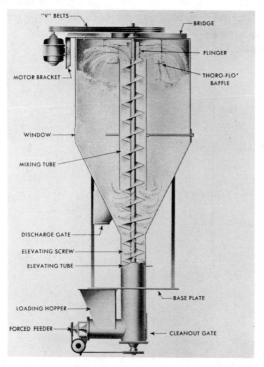

FIGURE 65.   Feed mixers are of two general types—horizontal and vertical. Shown is a drawing of a vertical type mixer with the different parts and interior action indicated. *(Courtesy of Sprout-Waldron, Muncy, Pa.)*

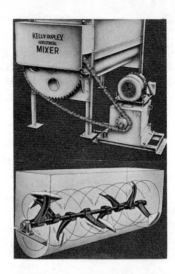

FIGURE 66.   Feed mixers are of two general types—horizontal and vertical. Shown is a typical horizontal type mixer with a schematic drawing of its interior action. *(Courtesy of The Duplex Mill and Mfg. Co., Springfield, Ohio)*

45 Formulating a Ration Mixture Based on a Balanced Daily Ration

I. In formulating a ration mixture by this procedure, one simply calculates a balanced daily ration as was done earlier for the 300 kg finishing yearling steer.

II. Once this has been done, the figures for this daily ration are converted to a percentage (per hundredweight) basis as follows:

$$\frac{\text{Wt of each feed in daily ration}}{\text{Total wt of daily ration}} \times 100 = \% \text{ of that feed in ration mix}$$

III. After the percentage or amount per Cwt of each feed in the ration has been determined, the amount per mix batch is calculated by multiplying the amount of each feed per Cwt by the number of Cwt per mix batch.

IV. This method as applied to the balanced daily ration developed earlier for a 300 kg finishing yearling steer using a restricted level of roughage is as follows:

	Amount in ration kg	% of ration	Amount per 1200 kg of mix kg	Amount per ton (2000 lb) of mix lb
Ground coastal bermudagrass hay	1.50	18.1	× 12 = 217.2	% × 20 = 362
Ground shelled corn	6.21	74.9	× 12 = 898.8	% × 20 = 1498
44% SOM	0.52	6.3	× 12 = 75.6	% × 20 = 126
Ground limestone	0.06	0.7	× 12 = 8.4	% × 20 = 14
TOTAL	8.29	100.0	1200.0	2000

46 Formulating Feed Mixtures by the Use of the Square Method

Frequently it is necessary to blend two or more feeds together into a mixture containing a certain definite percentage of some major nutritive factor. For this purpose a procedure generally referred to as the Square Method may be used.

I. **When only two feeds are involved.** Sometimes it is necessary to determine what combination of two feeds will give a mixture with a certain content of some particular nutrient. For example, a swine producer may need to know what combination of ground shelled corn and a 40% complete pig supplement will provide a mix suitable for self feeding to a group of pigs averaging about 27.5 kg (61 lb) in weight. From Table 32 it is determined that pigs of this weight should receive a 16% crude protein ration. What combination of corn and 40% supplement will provide a 16% crude protein mix may be quickly, easily, and precisely calculated by the use of the so-called Square Method. The steps in the use of the Square Method for the above purpose are as follows:

A.   Draw a square at the left side of the page.

B.   Insert the % crude protein desired in the final mixture (16%) in the middle of the square.

C.   Place "corn" with its % crude protein (8.8) on the upper left corner and "40% supplement" with its % crude protein (40.0) on the lower left corner. (For this method to work, one feed must be above the desired level of protein and the other below.)

D.   Subtract the % crude protein in corn (8.8) from the % crude protein desired in the mix (16.0), and place the difference (7.2) on the corner of the square diagonally opposite from the corn. This amount is supplement.

E.   Subtract the % crude protein desired in the mix (16.0) from the % crude protein in the supplement (40.0), and place the difference (24.0) on the corner of the square diagonally opposite from the supplement. This amount is corn.

F.   The above remainders represent the proportions of these two feeds which will provide a mix containing the desired % crude protein. The amounts are then converted to a percentage or a per hundredweight basis and then to other weight bases as desired for mixing purposes.

APPLICATION OF THE SQUARE METHOD AS OUTLINED ABOVE USING ONLY TWO FEEDS

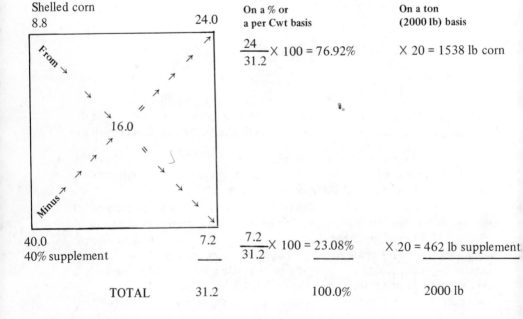

Shelled corn
8.8   24.0

On a % or
a per Cwt basis

$$\frac{24}{31.2} \times 100 = 76.92\%$$

On a ton
(2000 lb) basis

$$\times 20 = 1538 \text{ lb corn}$$

16.0

40.0   7.2
40% supplement

$$\frac{7.2}{31.2} \times 100 = 23.08\%$$

$$\times 20 = 462 \text{ lb supplement}$$

TOTAL   31.2   100.0%   2000 lb

II.   **When three or more feeds are involved.** Frequently it is desirable to use more than two feeds in formulating a feed mixture. For example, a swine producer may desire to use a mixture of corn, oats, and a 40% complete supplement in formulating a 14% crude protein mix for his pregnant sows. The following steps would be involved in the use of the Square Method in this connection.

    A.   Draw a square as in the previous example.

    B.   Place the % crude protein desired (14.0) in the middle of the square.

    C.   Separate the feeds into two groups, specify the proportion of each feed in each group, and calculate the weighted average % protein in each group. For this example let us assume that corn and oats were grouped together in the proportion of 2:1 with the supplement being used alone. The average % protein in the corn and oats must then be calculated as follows:

$$2 \times 8.8 \ = 17.6$$
$$1 \times 11.7 = \underline{11.7}$$
$$29.3/3 = 9.77\%$$

    D.   Place "2 corn + 1 oats" with its calculated % of crude protein (9.77) on the upper left corner of the square and "40% supplement" (40.00) on the lower left corner. (For the method to work, the figure on one left-hand corner of the square must be above and the one on the other below the level of protein desired.)

    E.   Subtract diagonally and proceed with the calculations as in the previous example.

    F.   Divide the final figure for "corn + oats" into 2/3 corn and 1/3 oats. (The proportion of each feed in each group must always be indicated initially and complied with in the final mixture.)

APPLICATION OF THE SQUARE METHOD AS
OUTLINED ABOVE USING THREE OR MORE FEEDS

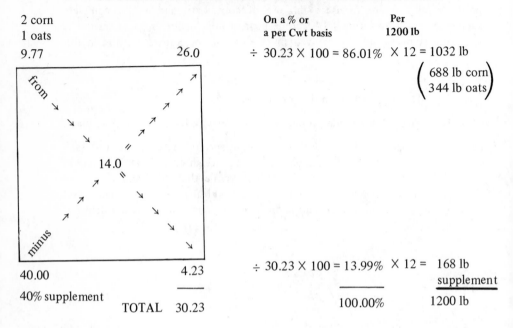

2 corn
1 oats
9.77

26.0

On a % or
a per Cwt basis

Per
1200 lb

$\div$ 30.23 $\times$ 100 = 86.01% $\times$ 12 = 1032 lb

$\left(\begin{array}{l} 688 \text{ lb corn} \\ 344 \text{ lb oats} \end{array}\right)$

14.0

40.00

4.23

40% supplement

TOTAL   30.23

$\div$ 30.23 $\times$ 100 = 13.99% $\times$ 12 =   168 lb
supplement

100.00%

1200 lb

III.   With a fixed percentage of one or more ration components. Sometimes a feeder desires to formulate a mixture containing a certain percentage of some nutrient, such as protein, but with a fixed percentage of one or more ration components.

A.   For example, a swine producer might wish to formulate a 14% crude protein mixture for pregnant sows using corn, oats, soybean oil meal, and a mineral and vitamin supplement. He wants to include in the mixture exactly 20% oats and 3.0% mineral and vitamin supplement. He then needs to know what combination of corn and soybean oil meal can be used to make up the other 77.0% of the mixture and give an overall mixture which contains exactly 14.0% crude protein. An adaptation of the Square Method may be used for this purpose.

B.   It is known that a crude protein level of 14.0% is desired for the overall mixture. This means that there is to be 14.0 lb of protein per 100 lb of mixture. Since 20 lb of each 100 lb of the mixture is oats (20.0%) then the oats in each 100 lb of the mixture would supply 2.34 lb of crude protein (11.7% of 20 lb). The mineral and vitamin supplement is essentially protein-free and so would contribute no protein of consequence to the mix. Hence, the oats (20.0 lb) and the mineral and vitamin supplement (3.0 lb) per 100 lb of mix would provide 2.34 lb crude protein. The remainder of the 14.0 lb of crude protein needed per 100 lb of mixture (14.00 lb − 2.34 lb = 11.66 lb) must then come from the 77.00 lb of corn and soybean oil meal per 100 lb of the overall mixture. In order to determine what combination of 77.0 lb of corn and soybean oil meal will provide the 11.66 lb of needed protein, an adaptation of the Square Method may be used.

C.   To do this, it is first necessary to calculate what % protein will be needed in the corn and soybean oil meal combination to provide 11.66 lb of protein per 77 lbs, as follows:

$$11.66 \div 77 \times 100 = 15.14$$

This figure is then used in conjunction with the Square Method as follows:

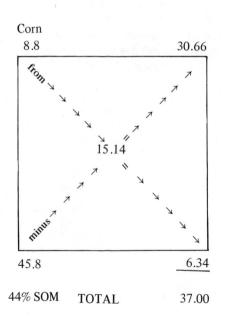

Corn
8.8   30.66

15.14

45.8   6.34

44% SOM   TOTAL   37.00

Per 77.0 parts
(calculated by ratio)

$$\frac{30.66}{37.0} = \frac{x}{77.0}$$

x = 63.81 parts corn

$$\frac{6.34}{37.0} = \frac{x}{77.0}$$

x = 13.19 parts SOM

Final ration mixture		Proof of 14% protein in final ration	
Ground oats	20.00%	× 0.117 =	2.34 lb
Mineral-vitamin supplement	3.00%	× 0.000 =	00.00 lb
Ground shelled corn	63.81%	× 0.088 =	5.62 lb
44% soybean oil meal	13.19%	× 0.458 =	6.04 lb
TOTAL	100.00%		14.00 lb (Protein per 100 lb mix)

IV. Use of the Square Method for formulating a balanced ration mixture for a finishing steer. Let us assume that we want to use the Square Method to develop a feed formula containing 12.0% crude protein, as fed basis, for a 300 kg finishing steer calf using ground shelled corn, ground coastal bermudagrass hay, 44% soybean oil meal, and supplemental minerals as needed.

> **A.** Let us first look at the crude protein content of the three basic feeds for grouping purposes.
>
> | | |
> |---|---|
> | Ground shelled corn | 8.8% |
> | Ground coastal bermudagrass hay | 9.0% |
> | 44% soybean oil meal | 45.8% |

Since shelled corn and coastal bermudagrass hay are similar in protein content, they would be placed together on one left-hand corner of the square with soybean oil meal on the other left-hand corner. Also, it is known from experience that one should use about 4 parts by weight of corn for each part by weight of hay in formulating a steer ration.

APPLICATION OF SQUARE METHOD IN
FORMULATING A STEER RATION

4 corn
1 hay
8.84

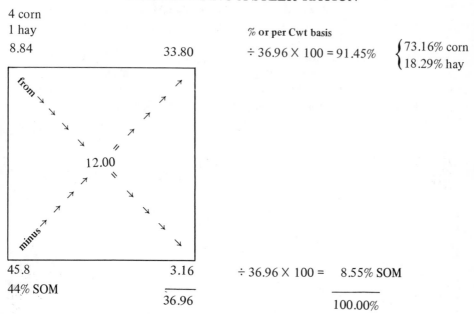

% or per Cwt basis

33.80   ÷ 36.96 × 100 = 91.45%   { 73.16% corn
                                        { 18.29% hay

12.00

45.8   3.16   ÷ 36.96 × 100 =   8.55% SOM
44% SOM   ‾‾‾‾
                        36.96   ‾‾‾‾‾‾‾‾
                                                     100.00%

Tentative mixture —

Ground shelled corn	73.16%
Ground coastal bermudagrass hay	18.29%
44% soybean oil meal	8.55%

B. The above tentative mixture is adequate from the standpoint of protein and energy because it was so developed. However, upon checking it for percentages of calcium and phosphorus it is found as shown below that it contains 0.11% calcium and 0.28% phosphorus on an as fed or air-dry basis.

	% Ca	Ca/Cwt mix	% P	P/Cwt mix
73.16% ground shelled corn	0.03	0.022	0.27	0.198
18.29% coastal bermudagrass hay	0.31	0.057	0.14	0.026
8.55% 44% SOM	0.32	0.027	0.67	0.057
		0.106		0.281

According to the NRC tables, the calcium and phosphorus requirements of this class of animal are 0.37% and 0.27%, respectively, dry basis, which are equivalent to 0.34% and 0.24%, respectively, air-dry (90% DM) basis.

From the above, it is apparent that while the above mixture is adequate in phosphorus (0.24% vs. 0.28%), it is quite low in calcium (0.34% vs. 0.11%).

This shortage of calcium might be corrected simply by adding 0.50% to 0.75% ground limestone to the mix on the basis of experience. The precise amount of ground limestone to be added, however, could be determined by the use of the Square Method as follows:

On a % or a per Cwt basis

Tentative mix

0.11   35.51   ÷ 35.74 × 100 = 99.36% tentative mix

0.34

35.85   0.23 ÷ 35.74 × 100 =   0.64% ground limestone

Ground limestone   100.00%

TOTAL   35.74

C.   Final ration mixture on a percent basis:

99.36 × 73.16 ÷ 100 =   72.69% ground shelled corn
99.36 × 18.29 ÷ 100 =   18.17% ground coastal bermudagrass hay
99.36 ×   8.55 ÷ 100 =   8.50% 44% soybean oil meal
                        0.64% ground limestone

TOTAL   100.00%

While the above final ration mixture, as calculated, will contain slightly less than the desired 12% crude protein, it should still be adequate for protein as well as for the other critical nutrients, except for possibly vitamin A. To be on the safe side, most cattle feeders today fortify their steer finishing rations at the rate of about 30,000 IU per steer per day.

47 Fortifying Steer Ration Mixtures with Vitamin A

I.   To accomplish a fortification of the above mixture with vitamin A, it would be assumed that each steer will consume the mixture at the rate of about 3% of its body weight daily. In other words, a 300 kg (661 lb) steer will consume approximately 9 kg (19.84 lb) of feed daily. Sufficient vitamin A supplement would then be added to provide 30,000 IU of vitamin A for each 9 kg (19.84 lb) of ration mixture.

II.   To illustrate the latter calculation, let us assume that the feeder's vitamin A supplement contains 4,000,000 IU of vitamin A per lb. If this amount (4,000,000) is divided by 30,000, the steer days of vitamin A per lb of supplement is obtained (133.3). By multiplying the steer days of vitamin A per lb of supplement (133.3) times the feed consumption per steer per day (19.84 lb), the amount of mix which 1 lb of supplement will fortify is obtained as follows:

$$\frac{\text{Vit A/lb suppl (4,000,000 IU)}}{\text{Vit A/steer/day (30,000)}} \times \text{Feed/steer/day (19.84 lb}$$

$$= \text{Lb mix fortified with 1 lb suppl (2645 lb)}$$

The lb of supplement required per ton of ration mix is then calculated by ratio as follows:

$$\frac{1}{2645} = \frac{x}{2000}$$

$$2645x = 2000$$

$$x = 0.756 \text{ (lb supplement per ton of mix)}$$

48 Use of Algebraic Equations in the Formulation of Feed Mixtures

Those who enjoy a mathematical approach to the solution of problems may find the use of algebraic equations preferable to the Square Method for formulating feed mixtures. In fact, appropriate algebraic methods may be used in the place of the Square Method in each of the preceding illustrations, as outlined below.

I. **When only two feeds are involved.** Algebraic equations may be used in the place of the Square Method for determining what combination of two different feeds will give a mixture containing a certain definite percentage of some particular nutrient. To illustrate, let us return to the swine feeder who desired to know what combination of ground shelled corn and 40% complete pig supplement would provide a mixture which contained exactly 16% crude protein. Through the use of the Square Method, it was determined that a mixture of 76.92% corn and 23.08% supplement would give the desired level

(16%) of protein. Algebraic equations can be used to make these same determinations as follows:

$$x = \text{lb corn per 100 lb mix}$$

$$y = \text{lb supplement per 100 lb mix}$$

$$x + y = 100 \text{ lb mix}$$

$$0.088x + 0.400y = 16.0 \text{ (lb protein/100 lb mix)}$$

(subtract) $0.088x + 0.088y = 8.8$

$$0.312y = 7.2$$

$$y = 7.2/0.312 = 23.08\% \text{ supplement in mix}$$

$$x = 100 - 23.08 = 76.92\% \text{ corn in mix}$$

II.   When three or more feeds are involved. To use algebraic equations for formulating a feed mixture containing a certain definite percentage of some particular nutrient using three or more feeds, it is first necessary, as with the Square Method, to separate the feeds into two logical groups, to specify the proportion of each feed in each group, and to calculate the weighted average % of the nutrient under consideration in each group. One is then ready to set up the appropriate equations and solve for the unknowns.

Using the same problem as was used in illustrating the Square Method in this connection (formulation of a 14% crude protein mix using corn and oats in the ratio of 2:1 along with a 40% supplement), one then would proceed as follows:

$$x = \text{lb corn + lb oats per 100 lb mix}$$

$$y = \text{lb supplement per 100 lb mix}$$

$$x + y = 100 \text{ lb of mix}$$

$$0.0977x + 0.4000y = 14.00 \text{ (lb protein/100 lb mix)}$$

(subtract) $0.0977x + 0.0977y = 9.77$

$$0.3023y = 4.23$$

$$y = 4.23/0.3023 = 13.99\% \text{ supplement in mix}$$

$$x = 100 - 13.99 = 86.01\% \text{ corn and oats (2:1) in mix}$$

III.   With a fixed percentage of one or more ration components. Let us assume that algebraic equations are to be used for solving the same problem as was used to illustrate the use of the Square Method in this connection (formulation of a 14% crude protein mix using corn and 44% soybean oil meal along with a fixed 20% oats and 3% mineral and vitamin supplement). As with the Square

Method, the amount of protein provided by the 20 lb of oats and 3 lb of mineral and vitamin supplement per 100 lb of mix must first be calculated (2.34 lb) and subtracted from the 14 lb desired in the overall mix (14.00 − 2.34 = 11.66 lb). This is the amount of protein which must come from the 77.0 lb (100 − 23) of corn and soybean oil meal per 100 lb of overall mix. The lb of corn and soybean oil meal, respectively, are then calculated by the use of appropriate equations as follow:

x = lb of corn per 77 lb of corn and SOM and per 100 lb of overall mix

y = lb of SOM per 77 lb of corn and SOM and per 100 lb of overall mix

$x + y$ = 77.0 lb of corn and SOM

$0.088x + 0.458y = 11.66$

(subtract) $0.088x + 0.088y =$ 6.78

$0.37y =$ 4.88

$y = 4.88/0.37 = 13.19$ lb SOM/77 lb corn and SOM

$x = 77 − 13.19 = 63.81$ lb corn/77 lb corn and SOM

Final ration mixture

Ground shelled corn	63.81%
44% soybean oil meal	13.19%
Ground oats	20.00%
Mineral and vitamin supplement	3.00%
TOTAL	100.00%

IV. When definite amounts of two nutrients from two feed sources are desired.
   Frequently it is desirable to calculate what combination of two feeds or two feed groups will provide the required amount of each of two different nutrients. Simultaneous equations are useful in making such determinations. For example, the daily nutrient requirements of a 300 kg finishing yearling steer are as listed below.

DM	TP	ME	Ca	P	Caro
kg	kg	Mcal	kg	kg	mg
8.3	0.92	21.7	0.029	0.021	46.0

   A.   In proceeding to balance a daily ration for such an animal using coastal bermudagrass hay, ground shelled corn, 44% soybean oil meal, and minerals, as a first step it might be desirable to determine what combination of air-dry roughage (as coastal bermudagrass hay) and air-dry concentrates (as ground shelled corn) would meet the steer's requirements for DM and ME. To use simultaneous equations for making this determination, one would proceed as follows:

x = kg of air-dry coastal bermudagrass hay containing 91.0% DM and 1.76 Mcal ME/kg

y = kg of air-dry ground shelled corn containing 86.0% DM and 2.83 Mcal ME/kg

(DM) $0.91x + 0.86y = 8.30$

(ME) $1.76x + 2.83y = 21.70$

$$1.76x + 1.66y = 16.05$$
$$1.17y = 5.65$$
$$y = 4.83 \text{ (kg corn, as fed)}$$

$$0.91x + (0.86 \times 4.83) = 8.30$$

$$x = \frac{8.30 - (0.86 \times 4.83)}{0.91} = 4.56 \text{ (kg hay, as fed)}$$

Proof of correct answers

$4.56 \times 91 \div 100 = 4.15$
$4.83 \times 86 \div 100 = 4.15$
TOTAL $= 8.30$ (kg DM required)

$4.56 \times 1.76 = 8.03$
$4.83 \times 2.83 = 13.67$
TOTAL$= 21.70$ (Mcal ME required)

The above method provides another alternative procedure for carrying out the initial phase of balancing a daily ration where both roughages and concentrates are being used on an as fed basis. It is especially useful when a high moisture roughage such as silage or haylage is to be fed.

B.   The above procedure also might be used in the determination of the amounts of corn and soybean oil meal that should be used to meet the daily requirements of a pig for protein and energy. For example, the daily nutrient requirements of a 47.5 kg growing-finishing pig are as follows:

TP kg	ME kcal	Ca kg	P kg	Caro mg
0.350	7920	0.0125	0.010	6.5

Simultaneous equations can be used to calculate the amounts of corn and soybean oil meal needed to provide the above amounts of TP and ME as follows:

$$x = \text{kg of corn}$$

$$y = \text{kg of soybean oil meal}$$

$$0.088x + 0.458y = 0.350$$

$$3275x + 2825y = 7920$$

$$3275x + 17045y = 13026$$

$$14220y = 5106$$

$$y = 0.3591 \text{ (kg 44\% SOM, as fed)}$$

$$x = \frac{0.350 - (0.458 \times 0.3591)}{0.088}$$

$$x = 2.108 \text{ (kg corn, as fed)}$$

Proof of correct answers

$$2.108 \times 0.088 = 0.186$$
$$0.359 \times 0.458 = \underline{0.164}$$

   TOTAL = 0.350 (kg TP required)

$$2.108 \times 3275 = 6904$$
$$0.359 \times 2825 = \underline{1014}$$

   TOTAL = 7918 (7920 kcal ME required)

The above amounts of corn and soybean oil meal could then be used as the basis for formulating a feed mix for 47.5 kg pigs as follows:

	Per hd per day	% or per Cwt	Per ton
Corn	2.108 kg	85.45	1709 lb
44% SOM	0.359 kg	14.55	291 lb
TOTAL	2.467 kg	100.00	2000 lb*

*To this around 50 to 60 lb of a mineral-vitamin supplement would need to be added to produce a balanced mixture.

49 Computerized Least-Cost Rations

A great deal of interest has been demonstrated by feed manufacturers and livestock feeders over the past few years in the computer formulation of so-called least-cost rations. Such rations differ from conventionally formulated rations only in that a computer is used in their formulation. As are conventionally formulated rations, computerized least-cost rations are supposedly balanced with respect to nutritional adequacy using the most economical, satisfactory sources available for supplying the various critical nutrients in the amounts needed. The advantage of the computer is that it will do almost instantly what would otherwise be a long, laborious, almost impossible task done by hand calculation.

I. **Basic requisites.** There are several basic requisites necessary for the computer formulation of least-cost rations. They are:

- Computer facilities
- Personnel trained in the use of computer facilities
- Information on nutritional requirements
- Information on feed suitability
- Information on feed composition
- Information on prevailing prices of available feeds.

A. Computer facilities and their use. Under prevailing conditions only the major feed companies and the larger livestock operations are in a position to justify having their own computer facilities and the personnel to operate them. On the other hand, anyone with a telephone can have computer services available by making the necessary arrangements with an organization or firm which is in a position to provide this type of service. Such organizations or firms ordinarily have the personnel required for establishing a working arrangement between the livestock feeder and the computer facility. The Agricultural Extension Services in most states are in a position to provide farmers with computer service.

While it is not in any way the intent of this book to provide basic instruction in computer operation, the author is pleased to provide such nutritional information as might be required by a computer operator for use in the computer formulation of least-cost rations for various classes of livestock.

B. Nutritional considerations. From the standpoint of nutritive requirements, such information takes on the form of so-called restrictions. These restrictions for the various critical nutrients are expressed as minimum and/or maximum values, depending on the nutrient under consideration. Generally speaking, the fewer the restrictions and the wider the restriction ranges consistent with ration quality, the better.

For certain critical nutrients such as salt, the trace minerals, and the vitamins, which are usually supplied at minimum allowance levels independent of the amounts of these nutrients provided by the major ration components, no formal restrictions are ordinarily provided. The rations are simply formulated with consideration being given to protein, energy, fat, fiber, calcium, and phosphorus, and as much of the trace minerals, the critical vitamins, and salt is added per ton as is required to meet the minimum requirements. In fact, sometimes the calcium and phosphorus restrictions are also omitted, and supplemental sources of these two nutrients are added as required based on the amounts of each supplied by the major ingredients as calculated by the computer.

C. Individual feed restrictions. In formulating rations, whether it be by conventional or computer procedures, it is frequently necessary or desirable to place certain limitations on the use of one or more potential ration components. These may be either minimum or maximum limitations, depending on the circumstances. For example, sometimes a feed may be used in the ration for a particular class of livestock up to a certain level, but it cannot be used in unlimited amounts without possible complications. In other instances, circumstances sometimes dictate that a certain minimum amount or some fixed amount of a feed be used in a ration. As with the various critical nutrients, such limitations are effected in computerized ration formulation through the use of appropriate restrictions. As indicated earlier, these restrictions may be in the form of minimum and/or maximum values. Should it be desirable to include a fixed level of some nutrient or feed in the ration, this would ordinarily be accomplished by simply using minimum and maximum restriction values of the same magnitude for the feed or nutrient in question.

D.   **Feed composition and prices.** Once appropriate restriction values have been established for the various critical nutrients and feeds, this information is fed into the computer along with information on the composition and price of the various alternative feeds available in the area. In this connection it should be remembered that the list of alternative feeds and ingredients must be of such a combination as will permit the computer to comply with the program restrictions on a reasonable basis in the formulation of the desired ration. In other words, a sufficient number of alternative sources of each critical nutrient should be provided to permit the computer considerable latitude in its selection process. Also, of course, the composition figures used for the various feeds and products under consideration must be accurate for the computer to turn out meaningful results.

For effective least-cost ration formulation, there must also be available reliable, realistic prices on potentially available feed ingredients. Such prices must, of course, be F.O.B. at the user's base of operation. Also, they must include the cost of any processing to which a feed must be subjected to put it in a form comparable to that of other alternative feeds. In addition, only products should be included in the list of alternative feed possibilities which can be handled with the user's existing facilities. Otherwise, the cost of providing any additional facilities needed to facilitate the use of such an ingredient should be given appropriate consideration.

II.   **Table of restrictions.** In the following table (Table 18) are listed some proposed restrictions for use in the computer formulation of rations for various classes of livestock. It should be noted in this connnection that under certain circumstances it might not be necessary to use all of the restrictions listed for a particular livestock class. In other instances, additional restrictions may need to be included. Also, in some instances, the restrictions are not hard and fast, and might be varied either up or down from that given, should circumstances seem to warrant it. While these restrictions are designed primarily for use in computer ration formulation, the student will frequently find these of assistance in connection with other feeds and feeding endeavors.

III.   Limitations to computer formulation. In the use of the computer for ration formulation, one can always be certain that a computerized least-cost ration is the cheapest balanced ration that can be formulated from the feeds available at the prices used. Also, a ration so formulated will always be balanced nutritionally insofar as the computer has been instructed with respect to nutritional considerations. It will not, however, always be the ration which will produce the greatest and/or the most efficient gains or production. There are some attributes of feeding value, such as palatability or acceptability, on which it is difficult to place a numerical value. In some instances, certain feeds tend to complement each other so as to effect a feeding value of the two feeds when fed together, which is above the average of the feeding value of the two feeds fed separately. Also, the feeding value of some feeds tends to vary with the level fed.

Furthermore, in the use of computerized least-cost rations for ruminants, considerable discretion will need to be exercised in switching from one ration to another. It is generally recognized that from two to three weeks are required for the rumen microflora to become completely adjusted to an entirely new ration, and something less than top performance is usually experienced during an adjustment period. Consequently, should the computer indicate a need for a major shift in the components of a ruminant ration, every effort should be made to accomplish such a shift in the form of a series of at least three or four minor ration adjustments extending over a period of about two weeks. Otherwise, varying degrees of various digestive disturbances and something less than top performance may be experienced.

However, in spite of the limitations in the use of computerized least-cost rations, the computer certainly is a tool which, if used discreetly, can do much toward minimizing feed costs for the livestock producer.

Table 18

SOME PROPOSED RESTRICTIONS FOR USE IN THE COMPUTER FORMULATION OF LEAST-COST RATIONS FOR SEVEN DIFFERENT CLASSES OF LIVESTOCK

PER 100 LB (AIR-DRY BASIS)

	Finishing steer calves		Finishing yrlg steers		Growing-fattening pigs (40-130 lb)		Growing-fattening pigs (130-220 lb)		Lactating sows		Lactating cows		Pleasure horses	
	Min	Max	Min	Max	Min	Max	Min	Max	Min	Max	Min	Max	Min	Max
Protein (use either)														
TP, lb	11.0	...	10.0	...	16-14	...	13.0	...	15.0	...	13.5	...	9.0	...
DP, lb	7.3	...	6.4	...	...	...	...	...	...	...	10.3	...	5.3	...
Energy (use any one)														
TDN, lb	72.0	...	72.0	...	75.0	...	75.0	...	75.0	...	58.5	...	56.1	...
DE, Mcal	...	...	...	...	150.0	...	150.0	...	150.0	...	118.4	...	112.2	...
ME, Mcal	118.0	...	118.0	...	144.0	...	144.0	...	144.0	...	93.9	...	...	...
$NE_{lact\ cows}$, Mcal	...	...	...	...	...	...	...	...	...	...	65.3	...	...	...
Fat	2.0	7.0	2.0	7.0	2.5	10.0	2.5	10.0	2.5	10.0	1.8	5.0	2.0	3.5
Crude fiber	7.0	...	7.0	...	...	5.0	...	5.0	...	8.0	11.7	...	11.0	...
Calcium	0.40	.80	0.35	0.70	0.65	0.80	0.50	0.70	0.75	0.90	0.43	0.86	0.30	0.60
Phosphorus	0.30	.40	0.25	0.35	0.50	0.65	0.40	0.50	0.50	0.75	0.32	0.43	0.20	0.30
Molasses (any type)	...	10.0	...	10.0	...	0.0	...	0.0	...	0.0	...	10.0	...	10.0
Dried citrus pulp	...	20.0	...	20.0	...	0.0	...	0.0	...	0.0	...	20.0	...	0.0
Dried beet pulp	...	15.0	...	15.0	...	0.0	...	0.0	...	0.0	...	15.0	...	...
Wheat	...	40.0	...	40.0	...	...	...	...	...	...	...	15.0	...	10.0
Wheat bran	...	10.0	...	10.0	...	0.0	...	0.0	...	10.0	...	25.0	...	25.0
Cottonseed meal	...	...	...	...	...	...	...	...	...	...	...	10.0	...	10.0
Corn gluten meal	...	...	...	...	...	9.0	...	9.0	...	9.0	...	...	...	...
Urea	...	1.0*	...	1.0*	...	0.0	...	0.0	...	0.0	...	1.0*	...	0.0

*Should be used at this level only after animal has been adapted to urea feeding.

50 Estimating Feed Requirements

In the development of plans for carrying on a livestock operation, it is essential that one be able to make fairly accurate estimates of feed requirements. Various approaches may be used for making such estimates, and the one to use will depend on the circumstances.

I.   One procedure which might be used is to develop a balanced daily ration such as might be used for a particular group of animals. The amounts of the various feeds making up this ration are then multiplied by the number of animals involved and in turn by the number of days the animals are to be fed on this ration. Each major change in the ration and/or in the number of animals to be fed will require a separate feed use period.

To illustrate this approach to estimating feed requirements, let us assume that a dairyman needs an estimate of his feed use for the coming year. It is his hope to keep an average of approximately 75 cows on the milking line during the year with an average production per cow of about 14,000 lb of around 4% milk. The cows will weigh on the average around 1100 lb. For about six months out of the year he plans to use an average daily ration of approximately as follows:

5.5 lb alfalfa hay	4.6 lb 44% soybean oil meal
33.0 lb corn silage	0.13 lb defluorinated phosphate
6.6 lb ground shelled corn	0.15 lb trace-mineralized salt
4.9 lb ground barley	0.03 lb ground limestone

For the other six months he is planning on using pasture to meet about half of his roughage needs. The daily feed needs per lactating cow during this period then would be the same as above except that the hay and silage requirements would be reduced by 50%.

In order to maintain a milking herd of 75 cows, the producer plans on having on hand throughout the year an average of approximately 15 dry cows to which he plans to feed about the same daily roughage allowance per cow as is fed to the milking cows but only one-third as much concentrates.

For the sake of simplifying this illustration it will be assumed that with this operation replacement females are brought in from an outside source.

The annual feed needs of the operation would then be calculated as follows:

	Alf Hay	Corn Sil	Sh Corn	Barley	44% SOM (Pounds)	Defl Phos	TM Salt	Grnd LS
Milking herd—6 mth, no past								
Per cow/day	5.5	33.0	6.6	4.9	4.6	0.13	0.15	0.03
For 75 cows, 183 days	75,488	452,925	90,585	67,253	63,135	1785	2059	412
Milking herd—6mths with past								
Per cow/day	2.75	16.5	6.6	4.9	4.6	0.13	0.15	0.03
For 75 cows, 183 days	37,744	226,462	90,585	67,253	63,135	1785	2059	412
Dry cows—6 mths, no past								
Per cow/day	5.5	33.0	2.2	1.6	1.5	0.04	0.05	0.01
For 15 cows, 183 days	15,098	90,585	6,039	4,392	4,118	110	137	27
Dry cows—6 mths with past								
Per cow/day	2.75	16.5	2.2	1.6	1.5	0.04	0.05	0.01
For 15 cows, 183 days	7,549	45,292	6,039	4,392	4,118	110	137	27
TOTAL	135,879	815,264	193,248	143,290	134,506	3790	4392	878
	67.9 tons	407.6 tons	3451 bu	2985 bu	67.3 tons	38 Cwt	44 Cwt	9 Cwt

How closely the above totals will approximate the producer's actual feed usage will to a large degree depend on how closely he follows his planned program of operation. Also, no doubt there will be miscellaneous feed ingredients required not covered in the above calculation. On the whole, however, such estimates should be adequate for developing crop production and feed purchasing plans and for developing an operating budget on which to base necessary financial arrangements.

II.   A somewhat different approach is sometimes used in estimating feed requirements for a steer finishing operation and other similar type enterprises. This approach, rather than being based on animal days involved and average daily feed consumption, is based on pounds of gain to be produced and the pounds of feed required per pound of gain. For example, a farmer may have on hand 100 head of steer calves which he would like to finish for market. The steers are currently averaging approximately 500 lb in weight, and he would like to put them on the market at about 900 lb. This means then that about 400 lb of gain is to be put on each steer or a total of 40,000 lb on the 100 head. He plans to self-feed these animals on a ration mixture as follows:

	%
Ground coastal bermudagrass hay	20.00
Ground shelled corn	70.00
44% soybean oil meal	9.00
Defluorinated phosphate	0.33
Ground limestone	0.33
Trace-mineralized salt	0.33
TOTAL	100.00

Experience has proven that steers of this age and weight will require over an average finishing period approximately 7 lb of the above ration per lb of gain. This means then that a total of about 280,000 lb feed of the above ingredient percentage composition would be required for finishing the 100 steers with amounts of the individual feeds as calculated below:

$$280,000 \times 0.20 \; = \; 56,000 \text{ lb hay (28 tons)}$$

$$280,000 \times 0.70 \; = 196,000 \text{ lb corn (3500 bu)}$$

$$280,000 \times 0.09 \; = \; 25,200 \text{ lb SOM (12.6 tons)}$$

$$280,000 \times 0.0033 = \quad 933 \text{ lb defluorinated phosphate (0.5 ton)}$$

$$280,000 \times 0.0033 = \quad 933 \text{ lb ground limestone (0.5 ton)}$$

$$280,000 \times 0.0033 = \quad 933 \text{ lb trace-mineralized salt (0.5 ton)}$$

With older steers and with steers carried to somewhat heavier weights, the feed requirement per lb of gain will increase to as much as 8 or 9 lb of air-dry feed per lb of gain, and possibly more. Reducing the roughage to concentrates ratio will tend to lower the feed required per lb of gain, whereas increasing the roughage to concentrates ratio will tend to increase the feed requirement per lb of gain. The above figures are based on the use of available growth-promoting products. If such products are not used, then feed requirements will need to be increased by about 10%.

III.   An approach similar to that used for steers above might also be used for estimating the feed requirements for a swine-feeding enterprise. Pigs being finished from weaning to market weight will on the average require approximately 3.25 lb of air-dry concentrates per lb gain.

IV.   In calculating feed requirements for pigs, as well as for steers, it should be pointed out that the feed required per unit of gain does not remain constant throughout the feeding period. In fact, the air-dry feed requirement per lb of gain will increase from the beginning of the finishing period to the end from around 2.5 up to about 4.0 lb for pigs and from about 5.5 lb up to around 8.5 lb for steer calves on a 20% roughage ration.

51 Weights, Measures, Volumes, and Capacities

I. **Bushel.**

A. A bushel is the volumetric equivalent of a cylinder 18.5 in. in diameter and 8 in. high. It is 2150.42 in.^{3} For practical purposes it is assumed to be 1.25 ft.^{3} Standard weights per bushel have been established for most grains, seeds, fruits, roots, etc.

B. **Standard weights per bushel (in lb).**

Wheat bran	20	
Peanuts in hull	22	
Cottonseed	32	
Oats grain	32	(actual wt will vary from 25-40)
Barley grain	48	(actual wt will vary from 44-50)
Shelled corn	56	

Sorghum grain	56	
Rye grain	56	
Wheat	60	
Soybeans	60	
Cowpeas	60	
Sweet potatoes	55	
Irish potatoes	60	
Ear corn	70	(is 2.5 ft^3 or amount required to shell out 56 lb.)
Snapped corn	80	(is 3.5 ft^3 or amount required to shell out 56 lb)
Water	77.6	(62.5 lb per ft^3)

II. Gallon (liquid).

    A. A liquid gallon is 231 in.^{3}

    B. **Standard weights per gallon** (in lb).

Vegetable oils	7.7
Water	8.34
Milk	8.6
Molasses	11.75

III. Miscellaneous.

    A. **Silage** weighs:

        30-40 lb per ft^3 in a horizontal silo
        40-50 lb per ft^3 in an upright silo

    B. **Baled hay.**

Approximate size of bale	=	14 in. × 18 in. × 36 in.
Approximate volume of bale	=	5.26 ft^3
Approximate weight per bale	=	60 lb
Weight per ft^3	=	11.4 lb

IV. Calculating volumes.

    A. **Rectangular** bins, etc.

$$\text{Volume} = \text{Length} \times \text{Width} \times \text{Depth}$$

        (All must be in a common unit of measure)

    B. **Cylindrical** bins, silos, tanks, drums, etc.

Volume = Area of circular base or end × Height or length of cylinder

$$A = \pi r^2$$
$$\pi = 3.1416$$
$$r = \text{½ diameter (or radius)}$$

C. **Cone-shaped** bin, etc.

Volume = Area of circular end × Height to point of cone × 1/3

V. **Calculating capacities.**

A. **Capacity in bushels.**

$$\frac{\text{Volume in ft}^3}{1.25} = \text{Capacity in bushels}$$

B. **Capacity in gallons.**

$$\frac{\text{Volume in in.}^3}{231} = \text{Capacity in gallons}$$

C. **Capacity in lb.**

No. of gallons × Wt per gal in lb = Capacity in lb

No. of bushels × Wt per bus in lb = Capacity in lb

No. of ft³ × Wt per ft³ in lb = Capacity in lb

VI. **Conversion of various measurements into the metric system.** See table of equivalents (Table 41).

52 Pastures— General

I. **Definition of pasture.** A pasture is an area of land on which there is a growth of forage on which designated livestock may graze at will. A good pasture is an area on which there is a growth of lush, green, nutritious, actively growing forage from which designated livestock can eat all they can consume in a relatively short period of time. Pastures vary greatly in quality, depending on type, growing conditions, and stage of maturity.

II. **Classification of pastures.** Pastures are classified on the basis of various considerations.

   A. **Legumes vs. nonlegumes.**

   1. A *legume* is a crop which has the capacity to harbor nitrifying bacteria in nodules on its roots and so can meet a part, if not all, of its own nitrogen needs.

2. A *nonlegume* is any crop which does not have the capacity of harboring nitrifying bacteria in nodules on its roots and so must depend on outside sources for its nitrogen supply.

B. **Annuals vs. perennials.**

1. An *annual* is a crop which must be propagated from seed each year it is grown.

2. A *perennial* is one which does not have to be reseeded each year but will reestablish itself from year to year from its roots.

C. **Summer vs. winter.**

1. A *summer pasture* is one which starts growing with the onset of warm weather in the spring and continues growing until halted by frost in the fall.

2. A *winter pasture* is one which starts active growth in the fall, remains alive during the winter, and makes rapid growth during the later winter and spring, with little if any growth during the summer.

D. **Temporary vs. permanent.**

1. A *temporary pasture* is one which is seeded on freshly cultivated soil for use through only one or a part of one grazing season. It will usually consist of an annual or a mixture of annuals.

2. A *permanent pasture* is one which, once established, remains as pasture for at least a period of years and in some instances continuously. It may consist of either perennials or reseeding annuals.

E. **Mixtures vs. pure seedings.**

1. A *pasture mixture* is a combination of two or more pasture crops on the same area. They are usually crops which complement each other from the standpoint of growth characteristics and/or nutritive value.

2. A *pure seeding* is a pasture which supposedly consists of only one species, although absolutely pure stands seldom are found.

F. Information on the classification of the more commonly grown pasture crops is summarized in Table 19.

53 Some Important Facts About Some of the More Important Pasture Crops

I. **Alfalfa.**

A. Alfalfa is a perennial summer legume which is widely grown as a pasture crop over much of the country above the southern coastal states.

B. No grazing crop is more palatable, more nutritious, or more productive than is alfalfa when grown under favorable conditions.

C. As pointed out previously, alfalfa requires a fertile, well-drained soil in order to thrive.

D. Pure stands of alfalfa are not ordinarily used as grazing for ruminants because of the danger of bloat.

E. To reduce the danger of bloat, alfalfa is usually seeded in combination with bromegrass, orchardgrass, timothy, and/or redtop when it is to be used for ruminant grazing.

    F.   Alfalfa sometimes suffers rather badly from winter kill under certain conditions.

    G.   Also, alfalfa suffers rather severely in some areas, especially through the South, from insects and/or diseases.

II.   **Bromegrass, orchardgrass, timothy, and redtop.**

    A.   These are all perennial summer grasses which are found primarily in the Corn Belt and the northeastern states.

    B.   As pasture forages these grasses are widely used in combinations with alfalfa and/or other legumes primarily to reduce the incidence of bloat among ruminant animals grazing the legumes.

    C.   These grasses are not widely used in either pure stands or combinations in the absence of legumes for grazing purposes.

III.   **Red clover and alsike clover.**

    A.   Red clover and alsike clover are both short-lived summer perennial legumes found primarily through the Corn Belt and in the northeastern states.

    B.   Both are widely used along with timothy and other grasses and legumes as components of rotational pastures where permanency is not a consideration.

    C.   As components of such pastures they are very palatable, nutritious, and productive.

    D.   While not as bad as alfalfa about causing bloat, both red clover and alsike clover in pure stands are subject to causing this trouble, and so both should be seeded with grasses when they are to be used for ruminant grazing.

    E.   Since both are relatively short-lived as perennials, other longer lasting legumes such as alfalfa and ladino clover should be used along with or in the place of these two clovers in permanent pasture mixtures.

IV.   **Kentucky bluegrass.**

    A.   Kentucky bluegrass is a rather short-growing perennial summer grass widely used for grazing throughout the Midwest and the northeastern states.

    B.   It is found to a more or less degree in most of the permanent pastures of these areas.

C.   Bluegrass, as it is commonly called, does best on fertile, well-drained soil which has been well limed or is naturally high in calcium content.

D.   It is very palatable and nutritious but is not a heavy producer except during the spring months.

E.   It is usually used in combination with certain short-growing legumes such as white or ladino clover to extend the grazing period.

V.   **Bermudagrass** (also called *common bermudagrass*).

A.   Bermudagrass is a close-growing perennial summer grass which over the years has been a widely prevalent weed crop in cotton and corn fields throughout the South, and as such fields have been diverted from row crops to grazing, bermudagrass has in most instances taken over.

B.   There are probably more acres of bermudagrass pasture in the South than any other type.

C.   Young, immature bermudagrass makes excellent pasture, but it drops rapidly in nutritive value with advancing maturity.

D.   Bermudagrass will respond to fertilizer but not to the degree that coastal bermudagrass will.

E.   It is relatively shallow rooted and not very drouth resistant.

F.   About 2-3 acres of bermudagrass with moderate fertilization will provide year-round feeding (pasture and hay) for one beef cow and her calf.

G.   Since bermudagrass is a nonlegume, it should be interseeded with legumes or else receive liberal applications of nitrogen fertilizer for good results.

V.   **Coastal bermudagrass.**

A.   Coastal bermudagrass is a hybrid bermudagrass developed by workers at the Georgia Coastal Plain Experiment Station during the 1940s.

B.   It ranks second in acreage only to bermudagrass as a grazing crop over most of the southern coastal states.

C.   Coastal bermudagrass will not survive the winters above the southern coastal states.

D.   It varies greatly in nutritive value during the course of the grazing season, depending on its degree of maturity, but it is very drouth resistant and has a high carrying capacity when properly fertilized.

E.   About one acre of coastal bermudagrass with moderately heavy fertilization will provide year-round feeding (grazing and hay) for one beef cow and her calf.

F.   A disadvantage of coastal bermudagrass is that it does not produce viable seed and must be propagated by the use of root stolons or sprigs. In a sense, however, this is an advantage since uncontrollable spread of the crop is thereby avoided.

G.   While very satisfactory as a grazing crop for beef brood herds, it is not regarded too highly as a grazing crop for fattening steers or producing dairy animals.

H.   Coastal bermudagrass, like bermudagrass, is a nonlegume and so is highly dependent on supplemental sources of nitrogen for good results.

I.   Animals on coastal bermudagrass pasture will do best if it is kept closely grazed.

VII. Bahiagrass.

A.   Bahiagrass is a medium-tall-growing perennial summer grass, extensive acreages of which are grown for grazing over much of the southern Coastal Plain.

B.   It will not survive the winter well above the Coastal Plain section.

C.   Bahiagrass will produce grazing essentially on a par with coastal bermudagrass under Coastal Plain conditions.

D.   It has the advantage over coastal bermudagrass in that it produces a viable seed and can be grown from seed.

E.   Like bermudagrass, it is a nonlegume and must be fertilized heavily with nitrogen for good results.

F.   Since bahiagrass does produce viable seed, a good portion of which are hard, its control sometimes poses a problem.

VIII. Fescue.

A.   Extensive acreages of various varieties of tall fescue, especially alta and Kentucky 31, are grown in many sections of the country above the southern Coastal Plain for grazing purposes.

B.   Fescue is a perennial winter grass, and stands of it are easily established from seed in those areas where it is adapted.

FIGURE 67. Beef brood cows on fescue pasture in mid-winter in Georgia, supplemented with hay. *(Courtesy of E.R. Beaty of the University of Georgia College of Agriculture Experiment Stations)*

    C.   Its major advantage is that it stays green throughout the winter in many sections and so is used extensively as a winter grazing crop, especially for beef brood cow herds and stocker animals.

    D.   Fescue is not good for milk production and so is not recommended for dairy herds.

    E.   Beef cows on fescue tend to be poor milk producers and as a result frequently wean lightweight calves.

    F.   Several problems have been associated with the use of fescue as pasture for beef cattle such as fescue foot, fat necrosis, grass tetany, and nitrate poisoning.

    G.   Fescue is sometimes grown in combination with ladino or other white clover which adds greatly to the pasture's productivity.

    H.   Animals on fescue grazing do best when the crop is kept closely grazed.

IX.   Dallisgrass.

    A.   Dallisgrass is a perennial summer grass which prevails to varying degrees in many permanent pastures over much of the Southeast.

    B.   It thrives especially well in lowland pastures, but its growth is not restricted to such areas.

C.   While it is sometimes included in seeding mixtures, its presence in the pasture sward is usually on a volunteer basis.

D.   As a pasture forage Dallisgrass would be regarded as about average in productivity and nutritive value.

E.   A fungus known as *ergot* will sometimes develop on Dallisgrass heads and, if eaten by cattle, may cause a condition known as *Dallisgrass poisoning*, which is a nonfatal staggering condition from which animals normally recover upon being removed from the pasture.

X.   **Oats, rye, barley and wheat.**

A.   Fall-seeded small grains are widely used in many sections of the country for grazing purposes.

B.   Since they remain more or less green and growing throughout the fall, winter, and spring, they are used primarily for providing winter grazing.

C.   The farther south the location, generally speaking, the greater the amount of winter grazing provided by these crops.

D.   To the extent that such grazing is available, it is generally very palatable and nutritious.

E.   Rye is usually the most productive of these crops during the winter months and in the colder sections, but it is not as palatable and nutritious as oats, wheat, and barley.

F.   When these crops are to be used strictly for grazing, they are sometimes seeded in combination with other crops such as ryegrass, crimson clover, and/or ladino clover.

G.   Oats, rye, barley, and wheat are all subject to causing grass tetany, especially if highly fertilized.

XI.   **Ryegrass.**

A.   Most ryegrass used for pasture is treated as a winter annual.

B.   It is widely used throughout the South and in the Pacific Coast states as a component of mixtures for winter grazing.

C.   It is usually used in combination with one or more of the small grains along with crimson and/or ladino clover in this connection.

D.   It is palatable, nutritious, and a fairly heavy producer, and it helps extend the length of the grazing period.

E.   It also adds strength to the pasture footing and helps to keep livestock from miring down during wet weather.

XII. Crimson clover.

A. Crimson clover is an annual winter legume used extensively in seeding mixtures for temporary winter grazing over much of the South.

B. Certain strains of crimson clover are reseeding and, if permitted to go to seed, will tend to reestablish themselves on permanent pasture areas each fall under southern conditions, thereby providing variable amounts of volunteer winter grazing.

C. Pure stands of lush crimson clover grazing will sometimes cause bloat in ruminant animals.

XIII. White clover.

A. White clover is a close-growing, fine-stemmed perennial legume found in most of the more humid sections of the country.

B. It has a good capacity for reseeding itself where conditions are not favorable for perennial growth.

C. In the North with ample moisture it continues active growth throughout the growing season but is killed back during the winter.

D. In the South it is usually killed back by the summer heat and drouth but remains more or less in active growth during the late fall, winter, and spring.

E. White clover is very palatable and nutritious and is an important component of many permanent pasture swards.

F. It is usually most prevalent and most productive in the lower, wetter areas within a pasture.

G. White clover is not very drouth resistant and usually ceases active growth during prolonged periods of hot, dry weather.

XIV. Ladino clover.

A. Ladino clover is a large-growing variety of white clover which was introduced into the U.S. from Italy several years ago.

B. It has essentially the same characteristics as a pasture forage as does white clover except that it is taller growing and is more productive.

C. It is widely used in combination with bromegrass and orchardgrass in the North and with tall fescue above the Coastal Plain in the South in the establishment of permanent pastures.

D. Being taller growing it is able to compete better than white clover in mixtures with tall-growing grasses.

    E.   It is seldom used in pure stand as grazing for ruminants because of the danger of bloat.

XV. **Annual lespedeza** (commonly referred to as *lespedeza*).

    A.   Lespedeza is a summer annual legume which is widely grown over much of the central and southern sections of the country.

    B.   There are several different types and varieties such as common, Korean, Kobe, and Tennessee 76.

    C.   While an annual, lespedeza effectively reseeds itself under most conditions, and one type or another is found in more or lesser degrees in most of the permanent pastures of those areas where it is adapted.

    D.   Lespedeza is quite palatable and very nutritious, but it is not a heavy producer.

    E.   It has a relatively short growing period and consequently should be used in combination with other crops for best results.

    F.   Lespedeza seed is very nutritious, and cattle grazing on lespedeza which has gone to seed tend to fatten quite readily.

    G.   Lespedeza is not normally a bloat producer.

XVI. **Sericea lespedeza** (commonly referred to as *sericea*).

    A.   Sericea is a perennial summer legume used extensively over the years in the South as a cover crop on land in the government soil bank.

    B.   Between that coming out of the soil bank and that seeded specifically for grazing purposes, there is a considerable acreage of sericea used for pasture, especially for beef cow herds, in the southeastern area.

    C.   Young, immature sericea is fairly palatable and nutritious, but as this crop matures, it becomes very unpalatable because of a buildup in its content of tannins.

    D.   For best results it should be grazed fairly closely or mowed frequently to keep down old growth and encourage new growth.

    E.   It is very drouth resistant and is frequently relied on by beef producers for grazing during periods of dry weather.

    F.   Sericea is one of few legumes which does not ordinarily cause any bloat trouble when grazed in pure stand by ruminant animals.

XVII. Kudzu.

    A.   Kudzu is a perennial summer leguminous vine sometimes used in the South for controlling erosion on steep land.

    B.   It is rather slow and tedious to establish since it is usually propagated from crowns.

    C.   Established areas of kudzu will provide excellent emergency grazing during periods of drouth.

    D.   It is quite palatable and nutritious but has a rather limited carrying capacity.

    E.   Heavy grazing during early spring is harmful to and may eliminate a stand of kudzu.

    F.   Kudzu is not normally a bloat producer.

FIGURE 68.   Arrowleaf clover in late spring in Georgia. Arrowleaf clover is a promising new forage crop for the Southeast. *(Courtesy of E.R. Beaty of the University of Georgia College of Agriculture Experiment Stations)*

XVIII. Arrowleaf clover.

    A.   Arrowleaf clover is a winter annual legume which is gaining popularity as an early spring grazing crop over much of the Southeast.

B.   Of the three strains available—Amclo, Yuchi, and Meechee—the first two are the most widely used.

C.   It is ordinarily seeded in the fall—October, November, and December.

D.   It requires a well-drained soil and does best on a fertile soil with a pH of 6.2-6.5.

E.   The type of seedbed is not too important so long as the soil surface is fairly free of vegetative material.

F.   Good stands of this crop have been obtained from top seeding closely grazed or burnt-over permanent pasture with 5-10 lbs of seed per acre.

G.   Arrowleaf clover seed should be inoculated with type O inoculant.

H.   It produces a high percentage of hard seeds and tends to reseed itself under favorable conditions.

I.   Arrowleaf clover makes its principal growth during March, April, May, and June.

J.   The Amclo strain matures about 30 days later than crimson clover and the Yuchi strain about 15-20 days later than the Amclo.

K.   It is a heavy producer of lush forage which is comparable to alfalfa in nutritive value.

L.   Cattle seem to relish it, but this is not true for horses.

M.   No trouble from bloat has been reported from grazing this group.

N.   While grown primarily for grazing purposes, surplus late growth arrowleaf clover is sometimes harvested for hay.

XIX. Millet and sudangrass.

A.   These are both rather coarse-growing summer annual grasses which are frequently planted for use as temporary summer grazing crops.

B.   Several new varieties and/or crosses of both have been developed over the past several years which are supposedly superior to the old varieties.

C.   Such crops have in the past most commonly been planted for use with dairy herds, but their use for finishing steers on pasture is on the increase in some sections.

D.   They are both usually regarded as emergency crops to be used for grazing if needed but, if not needed for grazing, harvested for hay or silage.

E.   Second-growth sudangrass following a severe drouth or an early frost will sometimes contain hydrocyanic acid which is quite poisonous to livestock.

XX.   Johnsongrass.

A.   Johnsongrass is a rank-growing perennial summer grass which occurs on a volunteer basis more or less throughout the southern states where it is one of the worst weeds of the row-crop farmer.

B.   While seldom, if ever, planted as a pasture crop, it is sometimes used for grazing purposes when available.

C.   If grazed before becoming too mature, it makes a fairly satisfactory grazing crop, but it is very low in protein.

D.   Continuous close grazing of Johnsongrass will tend to eliminate it from a pasture over a period of 2-3 years.

E.   As with sudangrass, second-growth Johnsongrass following a severe drouth or an early frost sometimes contains hydrocyanic acid which is very poisonous to livestock.

54 General Use of Pasture in Livestock Feeding

I. It is impossible to balance rations precisely for grazing animals.

   A. With a grazing program the amount of forage consumed cannot be precisely controlled or measured.

   B. Under grazing conditions it is impossible to know precisely the composition of the forage being consumed by the animal.

   1. Composition of forage varies during the course of the day—especially with respect to dry matter.

   2. Composition of forage varies from day to day, depending on weather conditions and stage of maturity.

   3. Composition of a forage species will vary, depending on the level of fertilization.

Table 19
CLASSIFICATION SUMMARY OF MAJOR PASTURE CROPS

	LEGUME	NON-LEGUME	ANNUAL	PERENNIAL	SUMMER	WINTER	TEMPORARY	PERMANENT
Bahigrass		X		X	X			X
Bluegrass		X		X	X			X
Bermudagrass		X		X	X			X
Bromegrass		X		X	X			X
Coastal bermuda-grass		X		X	X			X
Dallisgrass		X		X	X			X
Johnsongrass		X		X	X			X
Orchardgrass		X		X	X			X
Redtop		X		X	X			X
Tall fescue		X		X		X		X
Timothy		X		X	X			X
Barley, winter		X	X			X	X	
Millet		X	X		X		X	
Oats, winter		X	X			X	X	
Rye		X	X			X	X	
Ryegrass		X	X			X	X	
Sudangrass		X	X		X		X	
Wheat, winter		X	X			X	X	
Alfalfa	X			X	X			X
Alsike clover	X			X	X		X	
Kudzu	X			X	X			X
Ladino clover	X			X	X*	X*		X
Red clover	X		X	X	X		X	
Sericea	X			X	X			X
White clover	X			X	X*	X*		X
Crimson clover	X		X			X	X	
Lespedeza	X		X		X			X
Arrowleaf clover	X		X			X	X	

*Summer in the North, winter in the South.

4. Because of selective grazing on the part of the animal, it is impossible to know the exact nature of the forage being consumed by the animal.

C. In general, pasture forage dry matter has a composition and feeding value comparable to that of hay and/or silage.

1. Pasture is a roughage and should be so used in the feeding program.

2. In general, pasture may be used to replace the hay and/or silage in a balanced ration developed for dry-lot feeding.

3. Good lush pasture, in addition to replacing the hay and/or silage in a balanced ration for dry-lot feeding, may also reduce the amount of protein supplement required.

II. Use of pasture in feeding beef cattle.

A. Most of the feed for beef brood cows and growing stock may come from pasture.

1. If pasture is fair to good, only minerals will be required in addition to the pasture.

    a. Salt should always be provided. Trace-mineralized salt is probably best as insurance against a trace-mineral deficiency.

    b. In addition to salt, defluorinated phosphate, or steamed bone meal is usually provided on a free-choice basis.

    c. Some magnesium oxide may also be supplied during the late winter and early spring to prevent grass tetany.

2. If pasture is of poor quality, some supplemental protein and energy should be provided. This may be done by using

Cottonseed meal	Range cubes
Soybean meal	Protein blocks
"Hot meal"	Liquid supplements

3. Nursing calves on pasture are sometimes creep fed on a mixture of grain—most any combination available of corn, grain sorghum, oats, barley, and wheat.

4. Recently weaned calves on pasture should be supplemented with a limited amount of grain and protein supplement unless the pasture is of extremely good quality.

5. If the amount of pasture is inadequate, it should always be supplemented with hay, silage, and/or concentrates.

B. The extent to which pasture may serve effectively toward the ration of fattening cattle will depend on several factors.

1. **Age of cattle.** Older cattle will fatten on pasture much more readily than young cattle. Calves will fatten on pasture under only the most favorable conditions.

2. **Quality of pasture.** Pastures decrease in nutritive value with advancing maturity, so cattle generally do best and fatten most readily on young immature forage growth. To have such growth available over a period of time usually involves the use of temporary pastures.

3. **Freedom from diseases and parasites.** Animals afflicted with diseases and/or parasites (either internal and/or external) are inclined to be slow gainers and usually are unable to gain sufficiently on pasture to fatten.

4. **Environmental conditions.** The English breeds of cattle do not like to graze in the hot midday sun of the summer months. Also, pastures are becoming mature, and parasites are worst during

this period. Hence, if pasture is adequate, grazing cattle usually do best during the fall, winter, and spring months.

    5.   **Genetic potential of cattle to perform**. Cattle that have been bred to perform well, especially those selected to perform well on pasture, will come most nearly fattening effectively on pasture.

    6.   If the above mentioned factors are all favorable, cattle, including calves, may attain a low choice finish on pasture. However, at least some of the above factors are usually not favorable, and so pastures are not ordinarily relied on for producing choice grade slaughter cattle.

III.   **Pastures for feeding dairy cattle.**

    A.   *Good pasture* is satisfactory as the only feed, other than minerals, for feeding dry cows and growing dairy animals.

    B.   For milking cows good pasture may be used to replace all of the roughage and possibly a part of the concentrates, depending on:

        1.   **Quality of pasture.** Good pasture may be used as a greater portion of the ration than poor pasture without hurting production.

        2.   **Price of concentrates and milk.** The lower milk prices are and the higher concentrate prices are, the more economical it might be to use pasture in the feeding program for milking cows.

        3.   **Production potential of cows.** Cows with a low production potential may produce up to their potential on good pasture and limited grain whereas cows with high production potential will respond favorably to less pasture and more concentrates.

    C.   Cows on pasture are inclined to produce off-flavored milk.

        1.   This is especially true for pastures with wild onions or bitterweed. Such pastures should not be used for milking cows.

        2.   Most lush green pasture tends to impart varying degrees of off flavor to milk. Most such flavors can be avoided by keeping cows off of pasture for several hours prior to milking.

    D.   Lush green pasture will supply more protein than corn or sorghum silage and usually more than even good hay—hence, good pasture may reduce the amount of protein supplement that would otherwise be required.

    E.   Modern dairies are using pasture less and less in their feeding programs.

FIGURE 69. A dairy cow grazing on lush pasture. While pastures were formerly widely used in the feeding of the dairy herd, this is no longer true. Land prices in the vicinity of dairies are too high to permit its use for pasture. Also, pastures are too unreliable, too variable in quality, and too much of a source of off flavors in the milk to be looked upon with favor by many dairymen. *(Courtesy of Ralston Purina Co., St. Louis, Mo.)*

1. The price of the land on which most dairies are located is too high to justify its use for pasture.

2. The control of off flavors is too much of a problem.

3. Quality of pasture varies too greatly from week to week.

IV. **Pasture for sheep.**

A. Pasture may be used as the only feed, other than minerals, for all classes of sheep.

1. Sheep will come more closely to producing a top quality product (wool and fat market lambs) on pasture alone than will any other class of livestock.

2. If pasture is used as the only feed in a sheep program, it will need to be extra good at certain periods.

a. For the flushing period just prior to and during the breeding season to help bring ewes into heat and increase the number of multiple births.

b. For 2-3 weeks just prior to lambing to prevent a

condition known as *pregnancy disease* which is caused by too low a level of digestible carbohydrates.

    c.   During the nursing period—best for both ewes and lambs, but if the supply is short, good pasture may be used as a creep feed for lambs only.

    3.   All sheep and especially those on pasture must be subjected to a rigid program of internal parasite control.

V.   Pasture for horses.

    A.   Pasture may be used as the only feed, other than minerals, for all classes of nonworking horses.

    B.   Pasture may be used for working horses during nonworking hours if supplemented with grain at the rate of 3/4-1½ lb of grain per hundred lb of body weight, depending on the amount of work being done.

    C.   Hay is better than pasture for working horses. Hay will give a horse better wind and better endurance.

    D.   Lush pasture is not a desirable feed for saddle horses to be ridden on bridle paths or in the showring. It tends to produce a very loose, messy feces.

VI.   Pasture for hogs.

    A.   Up to about the early 1950s pasture was regarded as an essential factor in swine production.

    1.   Up to that time pasture was regarded as the most economical source of certain unidentified nutritive factors.

    2.   Up to then the use of freshly planted temporary pasture for sows with young pigs was regarded as the best way to keep internal parasite infestation of young pigs at a minimum.

    3.   Good pasture may be used to replace a varying proportion of the dry-lot concentrate mixture for swine depending on the class of animal:

        a.   10-20% of the concentrate feed for growing fattening pigs.

        b.   20-50% of the concentrate feed for sows nursing pigs.

        c.   50-75% of the concentrate feed for pregnant sows.

    4.   When pasture is to be used to replace a portion of the concentrate feed for swine, it is necessary to restrict the level of concentrates fed to the level desired by hand feeding. In other words, swine will not voluntarily reduce concentrate consumption on pasture if concentrates are self fed.

5. Most present-day, up-to-date swine operations do not use pastures for feeding lactating sows or growing-fattening pigs.

   a. With present-day knowledge of nutrition, pastures are not needed as a source of unidentified factors.

   b. It is becoming increasingly more difficult to produce parasite-free pigs on pasture. It can be done more effectively on concrete.

   c. At present-day prices, nutrients can be provided more cheaply through concentrates than with pasture.

6. Pasture may or may not be used for pregnant sows in modern-day swine operations. Some producers consider pastures desirable for this class of animals, but others believe that they can be handled more cheaply and just as effectively in dry-lot on dry feed.

55 Study Questions and Problems

I.   *Indicate by number whether each of the following feeds would be classified as:*

         1.   An air-dry energy feed.

         2.   An air-dry protein feed, animal origin.

         3.   An air-dry protein feed, plant origin.

         4.   An air-dry legume roughage.

         5.   An air-dry nonlegume hay.

         6.   An air-dry low-quality roughage.

         7.   An air-dry oil seed.

         8.   An air-dry mineral feed.

         9.   A high moisture feed.

1.	Peanut kernels.	*31.*	Ground limestone.
2.	Cottonseed meal.	*32.*	Meat scrap.
3.	Ground oystershells.	*33.*	Alfalfa meal.
4.	Meat and bone meal.	*34.*	Beet molasses.
5.	Dried skimmed milk.	*35.*	Alfalfa hay.
6.	Cowpea hay.	*36.*	Wheat grain.
7.	Grain sorghum grain.	*37.*	Lespedeza hay.
8.	Corn silage.	*38.*	Oat hulls.
9.	Defluorinated phosphate.	*39.*	Cottonseed.
10.	Tankage with bone.	*40.*	Millet hay.
11.	Fresh turnips.	*41.*	Soybean hay.
12.	Bermudagrass hay.	*42.*	Tankage.
13.	Red clover hay.	*43.*	Brewers dried grains.
14.	Shelled corn.	*44.*	Oats grain.
15.	Peanut hulls.	*45.*	Peanut oil meal
16.	Milo grain.	*46.*	Dried beet pulp.
17.	Ground corn cob.	*47.*	Cottonseed hulls.
18.	Orchardgrass pasture.	*48.*	Coastal bermudagrass haylage.
19.	Steamed bone meal.	*49.*	Fish meal.
20.	Soybean seed.	*50.*	Corn stalks.
21.	Fresh cow's milk.	*51.*	Dried citrus pulp.
22.	Dried bakery product.	*52.*	Corn gluten meal.
23.	Sericea hay.	*53.*	Ground snapped corn.
24.	Coastal bermuda greenchop.	*54.*	Oat hay.
25.	Poultry by-product meal.	*55.*	Rice hulls.
26.	Hominy feed.	*56.*	Johnsongrass hay.
27.	Bahiagrass hay.	*57.*	Barley grain.
28.	Oat straw.	*58.*	Wheat straw.
29.	Linseed meal.	*59.*	Feather meal.
30.	Fescue hay.	*60.*	Hegari grain.

II. *Indicate by number whether each of the compositions below is the composition of —*

1. An air-dry energy feed.

2. An air-dry protein feed, animal origin.

3. An air-dry protein feed, plant origin.

4. An air-dry legume roughage.

5. An air-dry nonlegume hay.

6. An air-dry low-quality roughage.

7. An air-dry oil seed.

8. An air-dry mineral product.

9. A molasses.

10. A high moisture grain.

11. A haylage.

12. A silage.

13. A fresh forage.

14. A wet by-product.

15. A root crop.

16. Fresh whole or skimmed milk.

	DM %	ASH %	FIBER %	EE %	PROT %	TDN %	CA %	P %	CARO mg/kg
1.	94.2	24.9	2.5	9.4	54.9	62.0	8.49	4.18	*
2.	95.7	80.4	1.9	1.9	7.1	*	30.92	14.01	*
3.	89.0	3.0	5.3	1.7	11.6	71.9	0.07	0.40	*
4.	86.9	6.0	36.2	1.7	3.6	38.2	0.31	0.09	*
5.	11.3	0.5	3.4	0.2	1.3	7.2	0.10	0.01	*
6.	91.0	4.5	29.6	1.5	9.0	49.1	0.31	0.14	179.3
7.	12.9	1.2	1.2	0.2	1.3	10.6	0.04	0.04	114.9
8.	9.3	0.7	0.0	0.1	3.3	8.5	0.13	0.10	*
9.	12.6	0.7	0.0	3.6	3.5	16.1	0.12	0.09	0.9
10.	17.6	1.7	4.9	0.6	3.0	11.8	0.24	0.05	50.0
11.	90.0	7.1	29.1	2.3	13.2	52.7	1.30	0.20	16.1
12.	75.0	1.0	1.7	3.3	7.7	68.0	0.03	0.24	3.6
13.	50.0	4.9	15.3	0.9	8.5	26.8	0.71	0.10	35.7
14.	92.7	3.5	16.9	22.9	23.1	86.6	0.14	0.68	*
15.	92.7	6.1	10.9	5.6	41.4	73.6	0.19	1.09	*
16.	85.4	1.5	7.1	3.4	8.0	72.6	0.04	0.23	3.5
17.	91.4	19.0	0.6	9.8	60.4	67.8	5.14	2.91	*
18.	23.8	1.1	3.8	1.5	5.5	15.9	0.07	0.12	*
19.	99.9	96.8	0.0	0.0	0.0	0.0	35.85	0.02	0.0
20.	91.5	4.4	59.8	1.2	6.6	16.7	0.25	0.06	*
21.	91.5	4.5	13.1	1.2	47.4	70.3	0.20	0.65	*
22.	94.8	2.4	2.8	47.7	28.4	131.1	0.06	0.43	*
23.	23.1	1.1	0.6	0.1	2.2	18.5	0.01	0.05	*
24.	30.7	2.3	7.3	0.9	3.0	17.6	0.08	0.06	5.0
25.	88.5	2.1	2.3	3.1	8.9	71.3	0.03	0.29	*
26.	28.2	2.2	6.6	0.7	1.9	16.1	0.06	0.09	7.3
27.	67.7	5.4	0.0	0.2	5.7	52.5	1.20	0.12	*
28.	88.6	4.5	30.2	2.3	6.3	48.8	0.36	0.15	47.3
29.	87.8	6.3	38.3	1.4	3.2	42.8	0.14	0.07	*
30.	17.7	2.1	2.8	0.6	5.0	11.8	0.25	0.09	26.3

*None for all practical purposes.

III. *Match by number the materials listed below with the following compositions (as fed basis).*

	DM %	ASH %	FIBER %	EE %	PROT %	TDN %	CA %	P %	CARO mg/kg
1.	86.0	1.1	2.0	3.8	8.8	78.0	0.03	0.27	4.1
2.	89.0	3.0	5.3	1.7	11.6	71.9	0.07	0.40	*
3.	88.9	3.4	10.6	4.5	11.7	66.3	0.09	0.33	*
4.	77.5	8.9	0.0	0.1	6.6	61.1	0.12	0.03	*
5.	75.0	1.0	1.7	3.3	7.7	68.0	0.03	0.24	3.6
6.	99.0	0.0	0.0	0.0	281.2	*	0.0	0.0	0.0
7.	94.6	3.7	0.6	2.9	87.4	62.2	0.42	0.51	*
8.	93.7	21.2	1.8	9.0	59.2	65.5	6.43	3.39	*
9.	94.1	28.0	2.4	12.4	47.1	62.7	11.47	5.25	*
10.	91.3	5.9	2.9	1.1	50.8	76.3	0.29	0.65	*
11.	92.7	6.1	10.9	5.6	41.4	73.6	0.19	1.09	*
12.	90.9	4.9	5.3	17.4	37.9	83.1	0.24	0.58	*
13.	95.7	80.4	1.9	1.9	7.1	*	30.92	14.01	*
14.	99.8	?	0.0	0.0	0.0	0.0	33.00	18.00	0.0
15.	99.0	96.8	0.0	0.0	0.0	0.0	35.82	0.02	0.0
16.	0.0	0.0	0.0	0.0	0.0	0.0	0.0	0.0	0.0
17.	99.0	99.0	0.0	0.0	0.0	0.0	0.0	0.0	0.0
18.	99.0	*	0.0	0.0	0.0	93.0	0.0	0.0	0.0
19.	99.9	0.0	0.0	99.9	0.0	0.0	0.0	0.0	0.0
20.	99.0	0.0	0.0	99.0	0.0	172.8	0.0	0.0	0.0
21.	92.2	9.7	22.9	2.9	17.9	56.7	1.36	0.28	88.0
22.	91.4	9.0	28.0	1.7	15.5	49.1	1.29	0.19	65.0
23.	91.2	7.5	26.8	1.7	7.2	44.7	0.37	0.19	88.3
24.	90.2	2.6	40.6	1.4	4.0	46.5	0.13	0.06	*
25.	27.8	1.9	7.6	1.2	2.2	19.7	0.08	0.06	4.4
26.	25.9	2.4	5.5	1.1	5.7	16.8	0.44	0.08	62.6
27.	23.1	1.1	0.6	0.1	2.2	18.5	0.01	0.05	*
28.	12.6	0.7	0.0	3.6	3.5	16.1	0.12	0.09	0.9
29.	9.3	0.7	0.0	0.1	3.3	8.5	0.13	0.10	*
30.	50.0	4.9	15.3	0.9	8.5	26.8	0.71	0.10	35.7

*None for all practical purposes.

1. Feather meal.
2. Oats grain.
3. Defluorinated phosphate.
4. Soybean seed.
5. Bermudagrass hay.
6. Fresh cow's milk.
7. Urea.
8. Alfalfa pasture.
9. Yellow shelled corn.
10. Dehydrated alfalfa meal.
11. Cottonseed hulls.
12. Fresh skimmed milk.
13. Pure glucose.
14. White potatoes.

15. Water.
16. Tankage.
17. Mineral oil.
18. Ground limestone.
19. Steamed bone meal.
20. Cane molasses.
21. Tankage with bone.
22. Barley grain.

23. 49% soybean oil meal.
24. Alfalfa hay.
25. 41% cottonseed meal.
26. High moisture shelled corn.
27. Alfalfa haylage.
28. Corn silage.
29. Corn oil.
30. Sand.

IV. *Calculate the percentage of crude protein in the following feed mixture:*

Ground shelled corn	400 lb
Ground barley grain	300 lb
Cane molasses	100 lb
44% soybean oil meal	200 lb
Defluorinated phosphate	10 lb
Trace-mineralized salt	20 lb
Urea (45% N)	30 lb

SOLUTION:

	% PROT (FROM TABLE 37)	LB PROT PER LB FEED		LB FEED IN MIX		TOTAL LB PROT FROM EACH FEED
Corn	8.8	0.088	X	400	=	35.2
Barley	11.6	0.116	X	300	=	34.8
Molasses	4.5	0.045	X	100	=	4.5
SOM	45.8	0.458	X	200	=	91.6
Defluorinated phosphate	0.0	0.000	X	10	=	0.0
TM salt	0.0	0.000	X	20	=	0.0
Urea	281.2	2.812	X	30	=	84.4
TOTAL				1060		250.5

$$\frac{250.5}{1060} \times 100 = 23.63\% \ (answer)$$

V. *Calculate the percentage crude fiber in the feed mixture in IV.*

ANSWER   3.39%.

VI. *Calculate the percentage of calcium in the following feed mixture:*

Ground shelled corn	550 lb
41% cottonseed meal	50 lb

Ground oats	400 lb
Steamed bone meal	5 lb
Ground limestone	5 lb
Trace-mineralized salt	5 lb
Urea (45% N)	25 lb

SOLUTION:

	% Ca (FROM TABLE 37)	LB Ca PER LB OF FEED	LB FEED IN MIX	TOTAL LB Ca FROM EACH FEED
Corn	0.03	0.0003	550	0.165
Cottonseed meal	0.16	0.0016	50	0.080
Oats	0.09	0.0009	400	0.360
Bone meal	30.92	0.3092	5	1.546
Limestone	35.85	0.3585	5	1.792
TM salt	0.0	0.0000	5	0.000
Urea	0.0	0.0000	25	0.000
TOTAL			1040	3.943

$$\frac{3.943}{1040} \times 100 = 0.38\% \ (answer)$$

VII. *Calculate the percentage of phosphorus in the feed mixture in VI.*

ANSWER 0.39%.

VIII. *Calculate the percentage of crude protein in a mixture of 1,900 lb cane molasses and 100 lb urea (45% N).*

ANSWER 18.34%.

IX. *A farmer has a silo 20 ft in diameter and 60 ft high.*

1. How many tons of silage can be stored in this silo?

2. How many pounds of shelled corn could be stored in this silo?

SOLUTION:

1. Tons of silage that can be stored in silo.

$$\text{Volume} = \text{Area of base} \times \text{height}$$
$$V = \pi r^2 \times h$$
$$V = 3.1416 \times 10^2 \times 60$$
$$V = 18849.6 \ ft^3$$

Approximate weight of silage $- 45 \ lb/ft^3$

$$\frac{(18,849.6)\,(45)}{2000} = 424.1 \text{ tons } (answer)$$

2.   Pounds of shelled corn that can be stored in silo.

$$\frac{\text{Volume in ft}^3}{\text{Ft}^3 \text{ per bu}} \times \text{Wt per bu shelled corn} = \text{Capacity of silo}$$
$$\text{in lb of shelled corn}$$

$$\frac{18849.6}{1.25} \times 56 = 844,462 \text{ lb } (answer)$$

X.   *A farmer has a silo 14 ft in diameter and 30 ft high.*

  1.   What is the volume of the silo in cubic feet?

  2.   What is the capacity of the silo in lb of barley grain?

ANSWERS:

  1.   4,618.2 ft³

  2.   177,337 lb

XI.   *A metal drum is 22 in. in diameter and 34 in. high.*

  1.   What is the capacity of the drum in gallons?

  2.   What is the capacity of the drum in bushels?

SOLUTION:

  1.   Capacity of drum in gallons.

$$\text{Cap in gal} = \frac{\pi r^2 \times h}{231 \text{ (in.}^3 \text{ per gal)}}$$

$$= \frac{3.1416 \times 11^2 \times 34}{231}$$

$$= 55.95 \text{ gal } (answer)$$

  2.   Capacity of drum in bushels.

$$\text{Cap in bu} = \frac{\pi r^2 \times h}{2150 \text{ (in.}^3 \text{ per bu)}}$$

$$= \frac{3.1416 \times 11^2 \times 34}{2150}$$

$$= 6.01 \text{ bu } (answer)$$

XII.   *Calculate the pounds of cane molasses that can be stored in a horizontal tank 16 ft long and 6 ft in diameter.*

SOLUTION:

$$\begin{aligned} \text{Volume} &= \text{Area of end} \times \text{length} \\ &= \pi r^2 \times \text{length} \\ &= 3.1416 \times 3^2 \times 16 \\ &= 452.4 \text{ ft}^3 \end{aligned}$$

$$\text{Ft}^3 \times \text{gal per ft}^3 \times \text{Wt per gal of molasses} = \text{Total lb of molasses}$$
$$452.4 \times 7.48 \times 11.75 = 39{,}761 \text{ lb } (\textit{answer})$$

or

$$\frac{\text{Ft}^3 \times \text{in.}^3 \text{ per ft}^3}{\text{In.}^3 \text{ per gal}} \times \text{Wt per gal of molasses} = \text{Total lb of molasses}$$

$$\frac{452.4 \times 1728}{231} \times 11.75 = 39{,}764 \text{ lb } (\textit{answer})$$

XIII. *A farmer has a circular feed bin with a cone-shaped bottom. The bin is 69 in. in diameter and has an overall depth of 88 in. The cone section has a depth of 56 in. What is the capacity of the bin in bushels?*

SOLUTION:

Volume of cylindrical section.

$$\begin{aligned} V &= \pi r^2 \times \text{Depth} \\ &= 3.1416 \times 34.5^2 \times (88 - 56) \\ &= 3.1416 \times 1190.25 \times 32 \\ &= 119{,}657.3 \text{ in.}^3 \end{aligned}$$

Volume of cone section.

$$\begin{aligned} V &= \pi r^2 \times \text{Depth} \times 1/3 \\ &= 3.1416 \times 1190.25 \times 56 \times 1/3 \\ &= 69800.1 \text{ in.}^3 \end{aligned}$$

Capacity of bin in bu.

$$\begin{aligned} C &= \frac{\text{Total volume in in.}^3}{\text{In.}^3 \text{ in a bu}} \\ &= \frac{119{,}657.3 + 69{,}800.1}{2150.42} \\ &= 88.1 \text{ bu } (\textit{answer}) \end{aligned}$$

XIV. *A farmer has a circular grain storage bin with a cone-shaped roof. The bin is 18 ft in diameter and has a height to the eave of 128 in. and to the peak of the cone of 164 in. What is the capacity of the bin in bu when filled to the peak of the cone?*

ANSWER   2375.8 bu.

XV. *A farmer has a circular grain bin 20 ft in diameter filled to a height of 30 ft with snapped corn. Calculate the lb of grain, cobs, and shucks in the bin.*

SOLUTION:

First, calculate the volume of the bin in ft^3.

$$V = \pi r^2 \times h$$
$$= 3.1416 \times 10^2 \times 30$$
$$= 9424.8 \text{ ft}^3$$

Next, calculate the bu of snapped corn in bin.

$$\frac{\text{Ft}^3 \text{ in bin}}{\text{Ft}^3 \text{ per bu snapped corn}} = \text{bu snapped corn}$$

$$\frac{9424.8}{3.5} = 2{,}692.8 \text{ bu}$$

Next, calculate the lb of snapped corn in the bin.

$$2692.8 \text{ bu} \times 80 = 215{,}424 \text{ lb snapped corn}$$

Next, calculate the lb of grain, cobs, and shucks.
$$70\% \text{ of } 215{,}424 \quad = \quad 150{,}798$$
$$17.5\% \text{ of } 215{,}424 \quad = \quad 37{,}699$$
$$12.5\% \text{ of } 215{,}424 \quad = \quad 26{,}928$$

XVI. *A farmer has a watering tank which is 2 ft high, 3 ft wide, and 10 ft long, and has circular ends. What is the capacity of the tank in gallons?*

SOLUTION:

To solve, divide the tank into two sections. Consider the two ends as one section — a cylinder with a depth of 2 ft and a radius one-half the width of the tank ($\frac{1}{2} \times$ 3 ft = 1.5 ft).

Consider the remainder of the tank as the other section — a rectangular-shaped section which is 2 ft high, 3 ft wide, and 7 ft (10 − 3) long.

Combine the two sections to arrive at total capacity.

ANSWER: 420 gal.

XVII. *A farmer has a feed tub which is 11 in. deep and has a diameter of 16.5 in. at the bottom and 20 in. at the top. What is the capacity of the tub in gallons? In bushels?*

SOLUTION:

To solve, consider the tub to be the top section of a cone from which the bottom section (a smaller cone) has been removed.

First, calculate the height of the two cones by the use of a ratio.

$$x : 20 \; = (x - 11) : 16.5$$
$$16.5x \; = 20x - 220$$
$$3.5x \; = 220$$
$$x \; = 62.86 \text{ in. (height of larger cone)}$$
$$x - 11 \; = 51.86 \text{ in. (height of smaller cone)}$$

Next, calculate volume of the two cones.

$$3.1416 \times 8.25^2 \times 51.86 \times \frac{1}{3} = 3696.3 \text{ in.}^3 \text{ in smaller cone}$$

$$3.1416 \times 10^2 \times 62.86 \times \frac{1}{3} = 6582.7 \text{ in.}^3 \text{ in larger cone}$$

The difference in the volume of the two cones equals the volume of the tub:

$$6582.7 - 3696.3 = 2886.4 \text{ in.}^3$$
$$= 12.5 \text{ gal}$$
$$= 1.34 \text{ bu} \quad (answers)$$

XVIII. *Using ground shelled corn and 44% soybean oil meal, formulate a 1,000 lb, 16% crude protein mix.*

SOLUTION:

USE OF THE SQUARE METHOD

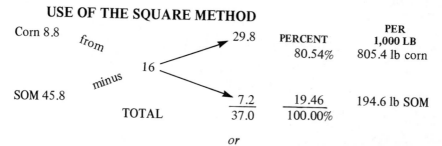

			PERCENT	PER 1,000 LB
Corn 8.8	*from*	29.8	80.54%	805.4 lb corn
	16			
SOM 45.8	*minus*	7.2	19.46	194.6 lb SOM
	TOTAL	37.0	100.00%	

or

USE OF ALGEBRAIC EQUATIONS

$$x = \text{Percent of corn in mix}$$
$$y = \text{Percent of SOM in mix}$$

$$0.088x + 0.458y = 16.0$$
$$0.088x + 0.088y = 8.8$$
$$0.37\ y = 7.2$$
$$y = 19.46\%\ \text{SOM in mix}$$
$$x = 100 - y = 80.54\%\ \text{corn in mix}$$

XIX. *Calculate on a ton basis what combination of cane molasses and urea (45% N) would provide a 20% crude protein product.*

SOLUTION:

USING THE SQUARE METHOD		PER 100 LB	PER 2000 LB

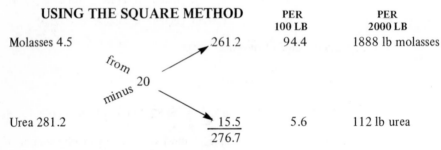

Molasses 4.5   261.2   94.4   1888 lb molasses

Urea 281.2   15.5   5.6   112 lb urea
                                276.7

USING EQUATIONS

ANSWER: Same as above.

XX. *A farmer wants to add enough urea (45% N) to some sorghum forage (1.9% crude protein) as the silo is filled to produce a silage containing 3.5% crude protein. How many lb of the urea should be added per ton of fresh forage?*

SOLUTION:

USING EQUATIONS

Let $x = \%$ urea

Let $y = \%$ forage

$$2.812x + 0.019y = 3.5$$
$$0.019x + 0.019y = 1.9$$
$$2.793x + 0.0y = 1.6$$

$$x = 0.573 \, (\% \text{ urea})$$
$$y = 100 - x = 99.427 \, (\% \text{ forage})$$

If 0.573 lb of urea is added per 99.427 lb of forage, then 11.53 lb (*answer*) of urea would be added per 2,000 lb of forage.

USING THE SQUARE METHOD

ANSWER: Same as above.

XXI. *A farmer is making silage from overripe corn containing 47% dry matter. He wants to add sufficient water to the corn forage as it enters the silo to bring the dry matter content of the silage down to 33%. How much water should be added per ton of corn forage? (Use both the Square Method and equations.)*

ANSWER: 848.5 or 101.7 gal.

XXII. *Using a mixture of 2 parts shelled corn and 1 part oats in conjunction with 44% soybean oil meal, formulate a 1,200 lb, 15% crude protein mix for lactating sows.*

SOLUTION:

First, calculate the percentage of protein in corn-oats mixture.

$$2 \times 8.8 = 17.6$$
$$1 \times 11.7 = \underline{11.7}$$
$$29.3 \div 3 = 9.77\%$$

Then proceed with equations as follow:

Let x = % of corn-oats mixture in overall mix

Let y = % SOM in overall mix

$$0.0977x + 0.4580y = 15.0$$
$$0.0977x + 0.0977y = 9.77$$
$$0.3603y = 5.23$$

$$y = 14.52\% \text{ SOM}$$

$$x = 100 - y = 85.48\% \text{ corn-oats mixture}$$

Per 1200 lb

$$14.52 \times 12 = 174.2 \text{ lb SOM}$$
$$85.48 \times 12 = 1025.8 \text{ lb corn and oats} \quad \left\{ \begin{array}{l} 683.9 \text{ lb of corn} \\ 341.9 \text{ lb of oats} \end{array} \right.$$

Calculate using the Square Method.

ANSWER: Same as above.

XXIII.   *Using the following feeds*
1.   Ground grain sorghum grain.
2.   Brewers dried grains.
3.   41% cottonseed meal.
4.   Ground oats grain.
5.   44% soybean oil meal.

Calculate a satisfactory 2,000 lb, 16% crude protein concentrate mix for a dairy cow.

SOLUTION:

Use either Square Method or equations to calculate. An innumerable number of satisfactory combinations possible.

XXIV.   *The daily protein and metabolizable energy requirements of a 300 kg yearling finishing steer are 0.92 kg and 21.7 Mcal, respectively. Upon subtracting the protein and energy of 1.5 kg of coastal bermudagrass hay to be fed to the steer, there still remains a need for 0.785 kg of protein and 19.06 Mcal ME to be derived from the concentrates — ground shelled corn and 41% cottonseed meal. Using simultaneous equations, calculate the amount of each of these two feeds which should be fed to meet the protein and ME needs.*

SOLUTION:

$$\text{Let } x = \text{Amount of corn}$$
$$\text{Let } y = \text{Amount of CSM}$$

Then

$$
\begin{aligned}
0.088x + 0.414y &= 0.785 \\
2.83x + 2.37y &= 19.06 \\
2.83x + 13.31y &= 25.25 \\
-10.94y &= -6.19 \\
y &= 0.566 \text{ kg 41\% CSM}
\end{aligned}
$$

$$
\begin{aligned}
0.088x + (0.414 \times 0.566) &= 0.785 \\
0.088x + 0.234 &= 0.785 \\
0.088x &= 0.551 \\
x &= 6.26 \text{ kg sh corn}
\end{aligned}
$$

XXV. *If the following feeds can be bought at the prices indicated, which would be the best buy as a source of TDN for fattening cattle.*

Shelled corn	$1.96 per bu
Barley grain	1.68 per bu
Milo grain	2.96 per cwt
Oats grain	1.05 per bu
Cane molasses	.36 per gal

SOLUTION:

	COST PER LB	COST PER CWT	TDN PER CWT	COST PER LB TDN
Corn	3.50¢	$3.50	78.0	4.49¢
Barley	3.50	3.50	71.9	4.87
Milo	2.96	2.96	71.3	4.15
Oats	3.28	3.28	66.3	4.95
Molasses	3.06	3.06	73.9	4.14

ANSWER:

Cane molasses is actually the cheapest source of TDN. However, it cannot be used at over about 10-15% of the ration for best results. Also, molasses requires special equipment for storing and mixing. Consequently, either milo or a combination of milo and molasses would be the best buy.

XXVI. *The following feeds are available at the prices indicated. Which would be the cheapest satisfactory source of protein for growing-fattening pigs?*

44% soybean oil meal	$183.00 per ton
49% soybean oil meal	205.00 per ton
36% cottonseed meal	158.00 per ton
41% cottonseed meal	173.00 per ton
Peanut oil meal	200.00 per ton
Meat scrap	220.00 per ton

SOLUTION:

	COST PER LB	COST PER CWT	PROTEIN PER CWT	COST PER LB PROTEIN
44% SOM	9.15¢	$ 9.15	45.8	19.98¢
49% SOM	10.25	10.25	50.8	20.18
36% CSM	7.90	7.90	35.9	22.01
41% CSM	8.65	8.65	41.4	20.89
POM	10.00	10.00	47.4	21.09
Scrap	11.00	11.00	54.9	20.03

ANSWER:

From the above it is apparent tht 44% SOM is the cheapest source of protein under the prevailing prices. Since soybean oil meal is a very satisfactory protein supplement for growing-fattening pigs, it would be the best buy.

XXVII. *Using the Petersen method, calculate the nutritive worth of ground cottonseed when shelled corn is $2.00 per bu and 44% soybean oil meal is $100.00 per ton.*

SOLUTION:

$$\text{Price of corn per ton} = \frac{2.00}{56} \times 2000 = 71.43$$

CORN AND SOM CONSTANTS FROM TABLE 17		PRICE OF CORN AND SOM PER TON	
0.904	X	$ 71.43	= $64.57
0.224	X	100.00	= 22.40
			$86.97 per ton
			(*answer*)

XXVIII. *A farmer plans to feed out 60 head of steer calves from an average initial weight of 500 lb to an average finished weight of 1,000 lb using ground earcorn, 41% cottonseed meal, and minerals. Calculate the amount of each feed he will need to carry out this operation.*

SOLUTION:

Approximate ration mix to be fed.

Ground earcorn	90.0%
41% cottonseed meal	9.0%
Defluorinated phosphate	0.33%
Ground limestone	0.33%
Trace-mineralized salt	0.33%

Total gain to be produced.

$$60 \times 500 = 30,000$$

Total feed to be needed.

lb gain $\times$ feed/lb gain = total feed needed

$$30,000 \times 7.5 = 225,000 \text{ lb}$$

Amounts of individual feeds needed.

90% of 225,000 = 202,500 lb earcorn

9% of 225,000 = 20,250 lb 41% CSM

0.33% of 225,000 = 750 lb defluorinated phosphate

= 750 lb ground limestone

= 750 lb TM salt

XXIX. *A farmer has 50 head of dry, pregnant beef cows averaging approximately 1,000 lb in weight which he plans to carry through a 100-day wintering period in dry-lot. Calculate the amount of feed he will need for this purpose.*

SOLUTION:

First, balance a daily ration for a 1,000 lb dry, pregnant beef cow, and then multiply each item in the ration by the number of cows (50) and then by the number of days (100).

56 Tables on Nutrient Requirements

Following are several tables which list the nutrient requirements of various classes of beef cattle, dairy cattle, horses, sheep, and swine, respectively. These requirements are those published by the National Research Council of the National Academy of Sciences and are presented here with the Academy's permission. The specific publications from which the respective sets of requirements were obtained are as listed below:

- *NUTRIENT REQUIREMENTS OF BEEF CATTLE*, Fourth Revised Edition, Publication ISBN 0-309-01754-8, Committee on Animal Nutrition, National Academy of Sciences — National Research Council, Washington, D.C. 1970.

- *NUTRIENT REQUIREMENTS OF DAIRY CATTLE*, Fourth Revised Edition, Publication ISBN 0-309-01916-8, Committee on Animal Nutrition, National Academy of Sciences — National Research Council, Washington, D.C. 1971.

- *NUTRIENT REQUIREMENTS OF HORSES*, Third Revised Edition, Publication ISBN 0-309-02045-X, Committee on Animal Nutrition, National Academy of Sciences — National Research Council, Washington, D.C. 1973.

- *NUTRIENT REQUIREMENTS OF SHEEP*, Fourth Revised Edition, Publication ISBN 0-309-01693-2, Committee on Animal Nutrition, National Academy of Sciences — National Research Council, Washington, D.C. 1968.

- *NUTRIENT REQUIREMENTS OF SWINE*, Seventh Revised Edition, Publication ISBN 0-309-02140-5, Committee on Animal Nutrition, National Academy of Sciences — National Research Council, Washington, D.C. 1973.

The requirements for the different livestock species as presented in the following tables are basically the same as those given in the above publications. In a few instances, however, to facilitate their use in this book, some of the figures have been transposed from one base to another, such as kilograms to pounds, grams to kilograms, vitamin A to carotene, and digestible energy to TDN. In each case standard conversion values have been used.

Table 20.
NUTRIENT REQUIREMENTS OF BEEF CATTLE
(DAILY NUTRIENTS PER ANIMAL)

BODY WEIGHT kg(lb)	AVERAGE DAILY GAIN kg(lb)	DAILY DRY MATTER PER ANIMAL[a] kg(lb)	TOTAL PROTEIN kg(lb)	DIGESTIBLE PROTEIN kg(lb)	ENERGY ME[b] Mcal	ENERGY TDN[c] kg(lb)	Ca kg(lb)	P kg(lb)	CAROTENE mg	VITAMIN A thousands IU
Finishing Steer Calves										
150(331)	0.90(1.98)	3.5(7.7)	0.45(0.99)	0.30(0.66)	9.9	2.7(6.0)	0.021(0.046)	0.015(0.033)	19.5	7.8
200(441)	1.00(2.20)	5.0(11.0)	0.61(1.35)	0.41(0.90)	13.4	3.7(8.2)	0.023(0.051)	0.017(0.037)	27.5	11.0
300(661)	1.10(2.43)	7.1(15.7)	0.87(1.92)	0.58(1.28)	19.0	5.3(11.7)	0.026(0.057)	0.019(0.042)	39.5	15.8
400(882)	1.10(2.43)	8.8(19.4)	0.98(2.16)	0.62(1.37)	23.5	6.5(14.3)	0.025(0.055)	0.020(0.044)	49.0	19.6
450(992)	1.05(2.32)	9.4(20.7)	1.04(2.29)	0.67(1.48)	25.1	6.9(15.2)	0.021(0.046)	0.021(0.046)	52.0	20.8
Finishing Yearling Steers										
250(551)	1.30(2.87)	7.2(15.9)	0.80(1.76)	0.51(1.12)	18.8	5.2(11.5)	0.029(0.064)	0.020(0.044)	40.0	16.0
300(661)	1.30(2.87)	8.3(18.3)	0.922(2.03)	0.62(1.37)	21.7	6.0(13.2)	0.029(0.064)	0.021(0.046)	46.0	18.4
400(882)	1.30(2.87)	10.3(22.7)	1.14(2.51)	0.73(1.61)	26.9	7.4(16.3)	0.028(0.062)	0.023(0.051)	57.0	22.8
500(1102)	1.20(2.65)	11.5(25.4)	1.28(2.82)	0.82(1.81)	30.0	8.3(18.3)	0.026(0.057)	0.026(0.057)	64.0	25.6
Finishing Two-Year-Old Steers										
350(772)	1.40(3.09)	10.3(22.7)	1.14(2.51)	0.73(1.61)	26.4	7.3(16.1)	0.030(0.066)	0.024(0.053)	57.0	22.8
400(882)	1.40(3.09)	11.3(24.9)	1.25(2.76)	0.80(1.76)	28.9	8.0(17.6)	0.030(0.066)	0.025(0.055)	63.0	25.2
500(1102)	1.40(3.09)	13.4(29.5)	1.49(3.29)	0.95(2.09)	34.3	9.5(20.9)	0.030(0.066)	0.030(0.066)	74.5	29.8
550(1213)	1.30(2.87)	13.7(30.2)	1.52(3.35)	0.97(2.14)	35.1	9.7(21.4)	0.030(0.066)	0.030(0.066)	76.0	30.4
Finishing Heifer Calves										
150(331)	0.80(1.76)	3.5(7.7)	0.45(0.99)	0.30(0.66)	9.9	2.7(6.0)	0.018(0.040)	0.013(0.029)	19.5	7.8
200(441)	0.90(1.98)	5.0(11.0)	0.61(1.35)	0.41(0.90)	13.4	3.7(8.2)	0.021(0.046)	0.015(0.033)	27.5	11.0
300(661)	1.00(2.20)	7.3(16.1)	0.89(1.96)	0.59(1.30)	19.5	5.4(11.9)	0.023(0.051)	0.018(0.040)	40.5	16.2
400(882)	0.95(2.09)	8.7(19.2)	0.97(2.14)	0.62(1.37)	23.2	6.4(14.1)	0.023(0.051)	0.019(0.042)	48.5	19.4
Finishing Yearling Heifers										
250(551)	1.20(2.65)	7.6(16.8)	0.84(1.85)	0.54(1.19)	19.8	5.5(12.1)	0.027(0.060)	0.020(0.044)	42.0	16.8
300(661)	1.20(2.65)	8.6(19.0)	0.95(2.09)	0.61(1.35)	22.4	6.2(13.7)	0.027(0.060)	0.020(0.044)	48.0	19.2
400(882)	1.20(2.65)	10.7(23.6)	1.19(2.62)	0.76(1.68)	27.9	7.7(17.0)	0.030(0.066)	0.024(0.053)	59.5	23.8
450(992)	1.10(2.43)	11.0(24.3)	1.22(2.69)	0.78(1.72)	28.7	7.9(17.4)	0.024(0.053)	0.024(0.053)	61.0	24.4

Table 20 (Cont.)

BODY WEIGHT kg(lb)	AVERAGE DAILY GAIN kg(lb)	DAILY DRY MATTER PER ANIMAL[a] kg(lb)	TOTAL PROTEIN kg(lb)	DIGESTIBLE PROTEIN kg(lb)	ENERGY		Ca kg(lb)	P kg(lb)	CAROTENE mg	VITAMIN A thousands IU
					ME[b] Mcal	TDN[c] kg(lb)				
Growing Steers										
150(331)	0.00(0.00)	2.7(6.0)	0.21(0.46)	0.11(0.24)	5.6	1.5(3.3)	0.005(0.011)	0.005(0.011)	15.0	6.0
	0.25(0.55)	3.1(6.8)	0.34(0.75)	0.22(0.49)	7.1	2.0(4.4)	0.008(0.018)	0.007(0.015)	17.0	6.8
	0.50(1.10)	3.2(7.1)	0.39(0.86)	0.26(0.57)	8.4	2.3(5.1)	0.012(0.026)	0.010(0.022)	17.5	7.0
	0.75(1.65)	3.2(7.1)	0.43(0.95)	0.29(0.64)	9.0	2.5(5.5)	0.017(0.037)	0.013(0.029)	17.5	7.0
200(441)	0.00(0.00)	3.3(7.3)	0.26(0.57)	0.14(0.31)	6.8	1.9(4.2)	0.006(0.013)	0.006(0.013)	18.5	7.4
	0.25(0.55)	4.5(9.9)	0.45(0.99)	0.27(0.60)	9.3	2.6(5.7)	0.008(0.018)	0.008(0.018)	25.0	10.0
	0.50(1.10)	4.9(10.8)	0.54(1.19)	0.35(0.77)	11.2	3.1(6.8)	0.013(0.029)	0.010(0.022)	27.0	10.8
	0.75(1.65)	5.0(11.0)	0.56(1.23)	0.36(0.79)	12.5	3.5(7.7)	0.018(0.040)	0.014(0.031)	28.0	11.2
300(661)	0.00(0.00)	4.5(9.9)	0.35(0.77)	0.19(0.42)	9.3	2.6(5.7)	0.008(0.018)	0.008(0.018)	25.0	10.0
	0.25(0.55)	6.1(13.5)	0.54(1.19)	0.32(0.71)	12.6	3.5(7.7)	0.011(0.024)	0.011(0.024)	34.0	13.6
	0.50(1.10)	7.7(17.0)	0.77(1.70)	0.47(1.04)	15.9	4.4(9.7)	0.014(0.031)	0.014(0.031)	43.5	17.4
	0.75(1.65)	8.0(17.6)	0.89(1.96)	0.57(1.26)	18.2	5.0(11.0)	0.017(0.037)	0.015(0.033)	44.5	17.8
400(882)	0.00(0.00)	5.6(12.3)	0.44(0.97)	0.24(0.53)	11.5	3.2(7.1)	0.010(0.022)	0.010(0.022)	31.0	12.4
	0.25(0.55)	7.7(17.0)	0.64(1.41)	0.35(0.77)	15.9	4.4(9.7)	0.014(0.031)	0.014(0.031)	43.0	17.2
	0.50(1.10)	9.7(21.4)	0.86(1.90)	0.50(1.10)	20.0	5.5(12.1)	0.017(0.037)	0.017(0.037)	54.0	21.6
	0.75(1.65)	9.9(21.8)	0.88(1.94)	0.51(1.12)	22.6	6.3(13.9)	0.018(0.040)	0.018(0.040)	55.0	22.0
Growing Heifers										
150(331)	0.00(0.00)	2.7(6.0)	0.21(0.46)	0.11(0.24)	5.6	1.5(3.3)	0.005(0.011)	0.005(0.011)	15.0	6.0
	0.25(0.55)	3.2(7.1)	0.36(0.79)	0.23(0.51)	7.3	2.0(4.4)	0.008(0.018)	0.007(0.015)	17.5	7.0
	0.50(1.10)	3.2(7.1)	0.39(0.86)	0.26(0.57)	8.4	2.3(5.1)	0.012(0.026)	0.010(0.022)	18.0	7.2
	0.75(1.65)	3.3(7.3)	0.44(0.97)	0.30(0.66)	9.3	2.6(5.7)	0.017(0.037)	0.013(0.029)	18.5	7.4
200(441)	0.00(0.00)	3.3(7.3)	0.26(0.57)	0.14(0.31)	6.8	1.9(4.2)	0.006(0.013)	0.006(0.013)	18.5	7.4
	0.25(0.55)	4.6(10.1)	0.46(1.01)	0.28(0.62)	9.5	2.6(5.7)	0.008(0.018)	0.008(0.018)	25.5	10.2
	0.500(1.10)	5.0(11.0)	0.56(1.23)	0.36(0.79)	11.4	3.2(7.1)	0.013(0.029)	0.010(0.022)	28.0	11.2
	0.75(1.65)	5.4(11.9)	0.60(1.32)	0.38(0.84)	13.5	3.7(8.2)	0.018(0.040)	0.014(0.031)	30.0	12.0

Table 20 (Cont.)

BODY WEIGHT kg(lb)	AVERAGE DAILY GAIN kg(lb)	DAILY DRY MATTER PER ANIMAL[a] kg(lb)	TOTAL PROTEIN kg(lb)	DIGESTIBLE PROTEIN kg(lb)	ENERGY ME[b] Mcal	ENERGY TDN[c] kg(lb)	Ca kg(lb)	P kg(lb)	CAROTENE mg	VITAMIN A thousands IU
300(661)	0.00(0.00)	4.5(9.9)	0.35(0.77)	0.19(0.42)	9.3	2.6(5.7)	0.008(0.018)	0.008(0.018)	25.0	10.0
	0.25(0.55)	6.2(13.7)	0.55(1.21)	0.32(0.71)	12.8	3.5(7.7)	0.011(0.024)	0.011(0.024)	34.5	13.8
	0.50(1.10)	8.2(18.1)	0.82(1.81)	0.50(1.10)	16.9	4.7(10.4)	0.015(0.033)	0.015(0.033)	45.5	18.2
	0.75(1.65)	8.6(19.0)	0.95(2.09)	0.61(1.35)	19.6	5.4(11.9)	0.017(0.037)	0.015(0.033)	47.5	19.0
400(882)	0.00(0.00)	5.6(12.3)	0.44(0.97)	0.24(0.53)	11.5	3.2(7.1)	0.010(0.022)	0.010(0.022)	31.0	12.4
	0.25(0.55)	7.7(17.0)	0.64(1.41)	0.35(0.77)	15.9	4.4(9.7)	0.014(0.031)	0.014(0.031)	43.0	17.2
	0.50(1.10)	10.2(22.5)	0.91(2.01)	0.53(1.17)	21.0	5.8(12.8)	0.018(0.040)	0.018(0.040)	56.5	22.6
	0.75(1.65)	10.6(23.4)	0.94(2.07)	0.55(1.21)	24.2	6.7(14.8)	0.019(0.042)	0.019(0.042)	59.0	23.6
Dry Pregnant Mature Cows										
350(772)	—	5.8(12.8)	0.34(0.75)	0.16(0.35)	10.3	2.8(6.2)	0.009(0.020)	0.009(0.020)	35.0	14.0
400(882)	—	6.4(14.1)	0.38(0.84)	0.18(0.40)	11.5	3.2(7.1)	0.010(0.022)	0.010(0.022)	38.8	15.5
450(992)	—	6.8(15.0)	0.40(0.88)	0.19(0.42)	12.4	3.4(7.5)	0.012(0.026)	0.012(0.026)	42.0	16.8
500(1102)	—	7.6(16.8)	0.44(0.97)	0.21(0.46)	13.6	3.8(8.4)	0.012(0.026)	0.012(0.026)	45.5	18.2
550(1213)	—	8.0(17.6)	0.47(1.04)	0.22(0.49)	14.4	4.0(8.8)	0.012(0.026)	0.012(0.026)	48.8	19.5
600(1323)	—	8.6(19.0)	0.50(1.10)	0.24(0.53)	15.5	4.3(9.5)	0.013(0.029)	0.013(0.029)	52.0	20.8
Cows Nursing Calves, First 3-4 Months Postpartum										
350(772)	—	8.6(19.0)	0.79(1.74)	0.46(1.01)	17.7	4.9(10.8)	0.025(0.055)	0.020(0.044)	83.0	33.2
400(882)	—	9.3(20.5)	0.86(1.90)	0.50(1.10)	19.2	5.3(11.7)	0.026(0.057)	0.021(0.046)	90.0	36.0
450(992)	—	9.9(21.8)	0.91(2.01)	0.53(1.17)	20.4	5.6(12.3)	0.028(0.062)	0.022(0.049)	96.2	38.5
500(1102)	—	10.5(23.2)	0.97(2.14)	0.57(1.26)	21.6	6.0(13.2)	0.028(0.062)	0.023(0.051)	102.5	41.0
Bulls, Growth and Maintenance (Moderate Activity)										
300(661)	1.00(2.20)	8.7(19.2)	1.21(2.67)	0.84(1.85)	20.4	5.6(12.3)	0.023(0.051)	0.018(0.040)	85.0	34.0
400(882)	0.90(1.98)	10.0(22.0)	1.33(2.93)	0.90(1.98)	23.5	6.5(14.3)	0.019(0.042)	0.018(0.040)	97.0	38.8
500(1102)	0.70(1.54)	12.0(26.5)	1.60(3.53)	1.08(2.38)	25.8	7.1(15.7)	0.021(0.046)	0.021(0.046)	116.5	46.6
600(1323)	0.50(1.10)	11.6(25.6)	1.42(3.13)	0.94(2.07)	24.9	6.9(15.2)	0.021(0.046)	0.021(0.046)	113.0	45.2
700(1543)	0.30(0.66)	12.7(28.0)	1.41(3.11)	0.90(1.98)	26.2	7.2(15.9)	0.023(0.051)	0.023(0.051)	123.5	49.4
800(1764)	0.00(0.00)	9.9(21.8)	0.99(2.18)	0.60(1.32)	20.4	5.6(12.3)	0.018(0.040)	0.018(0.040)	96.2	38.5
900(1984)	0.00(0.00)	10.7(23.6)	1.07(2.36)	0.65(1.43)	22.0	6.1(13.4)	0.019(0.042)	0.019(0.042)	104.0	41.6

ᵃFeed intake was calculated from the NE requirements and average NE values for the kind of ration being fed.

ᵇME requirements for growing and finishing cattle were calculated from the NE_m and NE_{gain} requirements for weights and rates of gain.

ᶜTDN was calculated from ME by assuming 3.6155 kcal of ME per g of TDN.

TABLE 21
NUTRIENT REQUIREMENTS OF BEEF CATTLE
(NUTRIENT CONCENTRATION IN RATION DRY MATTER)

BODY WEIGHT kg (lb)	DAILY AVERAGE GAIN kg (lb)	DAILY DRY MATTER PER ANIMAL[a] kg (lb)	TOTAL PROTEIN %	DIGESTIBLE PROTEIN %	ENERGY ME[b] Mcal/kg	ENERGY TDN[c] %	Ca %	P %	CAROTENE mg/kg	VITAMIN A thousands IU/kg
Finishing Steer Calves										
150(331)	0.90(1.98)	3.5(7.7)	12.8	8.6	2.82	78	0.60	0.43	5.5	2.2
200(441)	1.00(2.20)	5.0(11.0)	12.2	8.1	2.67	74	0.46	0.34	5.5	2.2
300(661)	1.10(2.43)	7.1(15.7)	12.2	8.1	2.67	74	0.37	0.27	5.5	2.2
400(882)	1.10(2.43)	8.8(19.4)	11.1	7.1	2.67	74	0.28	0.23	5.5	2.2
450(992)	1.05(2.31)	9.4(20.7)	11.1	7.1	2.67	74	0.22	0.22	5.5	2.2
Finishing Yearling Steers										
250(551)	1.30(2.87)	7.2(15.9)	11.1	7.1	2.61	72	0.40	0.28	5.5	2.2
300(661)	1.30(2.87)	8.3(18.3)	11.1	7.1	2.61	72	0.35	0.25	5.5	2.2
400(882)	1.30(2.87)	10.3(22.7)	11.1	7.1	2.61	72	0.27	0.22	5.5	2.2
500(1,102)	1.20(2.65)	11.5(25.4)	11.1	7.1	2.61	72	0.23	0.22	5.5	2.2
Finishing Two-Year-Old Steers										
350(772)	1.40(3.09)	10.3(22.7)	11.1	7.1	2.56	71	0.29	0.22	5.5	2.2
400(882)	1.40(3.09)	11.3(24.9)	11.1	7.1	2.56	71	0.27	0.22	5.5	2.2
500(1,102)	1.40(3.09)	13.4(29.5)	11.1	7.1	2.56	71	0.22	0.22	5.5	2.2
550(1,213)	1.30(2.87)	13.7(30.2)	11.1	7.1	2.56	71	0.22	0.22	5.5	2.2
Finishing Heifer Calves										
150(331)	0.80(1.76)	3.5(7.7)	12.8	8.6	2.82	78	0.51	0.37	5.5	2.2
200(441)	0.90(1.98)	5.0(11.0)	12.2	8.1	2.67	74	0.42	0.30	5.5	2.2
300(661)	1.00(2.20)	7.3(16.1)	12.2	8.1	2.67	74	0.31	0.25	5.5	2.2
400(882)	0.95(2.09)	8.7(19.2)	11.1	7.1	2.67	74	0.26	0.22	5.5	2.2

TABLE 21 (Continued)

BODY WEIGHT kg (lb)	DAILY AVERAGE GAIN kg (lb)	DAILY DRY MATTER PER ANIMAL[a] kg (lb)	TOTAL PROTEIN %	DIGESTIBLE PROTEIN %	ENERGY ME[b] Mcal/kg	ENERGY TDN[c] %	Ca %	P %	CAROTENE mg/kg	VITAMIN A thousands IU/kg
Finishing Yearling Heifers										
250(551)	1.20(2.65)	7.6(16.8)	11.1	7.1	2.61	72	0.36	0.26	5.5	2.2
300(661)	1.20(2.65)	8.6(19.0)	11.1	7.1	2.61	72	0.31	0.23	5.5	2.2
400(882)	1.20(2.65)	10.7(23.6)	11.1	7.1	2.61	72	0.28	0.22	5.5	2.2
450(992)	1.10(2.43)	11.0(24.3)	11.1	7.1	2.61	72	0.22	0.22	5.5	2.2
Growing Steers										
150(331)	0.00(0.00)	2.7(6.0)	7.8	4.2	2.06	57	0.19	0.19	5.5	2.2
	0.25(0.55)	3.1(6.8)	11.1	7.1	2.28	63	0.26	0.23	5.5	2.2
	0.50(1.10)	3.2(7.1)	12.2	8.1	2.61	72	0.38	0.31	5.5	2.2
	0.75(1.65)	3.2(7.1)	13.3	9.0	2.82	78	0.53	0.41	5.5	2.2
200(441)	0.00(0.00)	3.3(7.3)	7.8	4.2	2.06	57	0.18	0.18	5.5	2.2
	0.25(0.55)	4.5(9.9)	10.0	6.1	2.06	57	0.18	0.18	5.5	2.2
	0.50(1.10)	4.9(10.8)	11.1	7.1	2.28	63	0.27	0.20	5.5	2.2
	0.75(1.65)	5.0(11.0)	11.1	7.1	2.50	69	0.36	0.28	5.5	2.2
300(661)	0.00(0.00)	4.5(9.9)	7.8	4.2	2.06	57	0.18	0.18	5.5	2.2
	0.25(0.55)	6.1(13.4)	8.9	5.2	2.06	57	0.18	0.18	5.5	2.2
	0.50(1.10)	7.7(17.0)	10.0	6.1	2.06	57	0.18	0.18	5.5	2.2
	0.75(1.65)	8.0(17.6)	11.1	7.1	2.28	63	0.21	0.18	5.5	2.2
400(882)	0.00(0.00)	5.6(12.3)	7.8	4.2	2.06	57	0.18	0.18	5.5	2.2
	0.25(0.55)	7.7(17.0)	8.3	4.6	2.06	57	0.18	0.18	5.5	2.2
	0.50(1.10)	9.7(21.4)	8.9	5.2	2.06	57	0.18	0.18	5.5	2.2
	0.75(1.65)	9.9(21.8)	8.9	5.2	2.28	63	0.18	0.18	5.5	2.2

TABLE 21 (Continued)

BODY WEIGHT kg(lb)	DAILY AVERAGE GAIN kg/lb	DAILY DRY MATTER PER ANIMAL[a] kg(lb)	TOTAL PROTEIN %	DIGESTIBLE PROTEIN %	ENERGY ME[b] Mcal/kg	ENERGY TDN[c] %	Ca %	P %	CAROTENE mg/kg	VITAMIN A thousands IU/kg
Growing Heifers										
150(331)	0.00(0.00)	2.7(6.0)	7.8	4.2	2.06	57	0.19	0.19	5.5	2.2
	0.25(0.55)	3.2(7.1)	11.1	7.1	2.28	63	0.25	0.22	5.5	2.2
	0.50(1.10)	3.2(7.1)	12.2	8.1	2.61	72	0.38	0.31	5.5	2.2
	0.75(1.65)	3.3(7.3)	13.3	9.0	2.82	78	0.52	0.39	5.5	2.2
200(441)	0.00(0.00)	3.3(7.3)	7.8	4.2	2.06	57	0.18	0.18	5.5	2.2
	0.25(0.55)	4.6(10.1)	10.0	6.1	2.06	57	0.18	0.18	5.5	2.2
	0.50(1.10)	5.0(11.0)	11.1	7.1	2.28	63	0.26	0.20	5.5	2.2
	0.75(1.65)	5.4(11.9)	11.1	7.1	2.50	69	0.33	0.26	5.5	2.2
300(661)	0.00(0.00)	4.5(9.9)	7.8	4.2	2.06	57	0.18	0.18	5.5	2.2
	0.25(0.55)	6.2(13.7)	8.9	5.2	2.06	57	0.18	0.18	5.5	2.2
	0.50(1.10)	8.2(18.1)	10.0	6.1	2.06	57	0.18	0.18	5.5	2.2
	0.75(1.65)	8.6(19.0)	11.1	7.1	2.28	63	0.18	0.18	5.5	2.2
400(882)	0.00(0.00)	5.6(12.3)	7.8	4.2	2.06	57	0.18	0.18	5.5	2.2
	0.25(0.55)	7.7(17.0)	8.3	4.6	2.06	57	0.18	0.18	5.5	2.2
	0.50(1.10)	10.2(22.5)	8.9	5.2	2.06	57	0.18	0.18	5.5	2.2
	0.75(1.65)	10.6(23.4)	8.9	5.2	2.28	63	0.18	0.18	5.5	2.2
Dry Pregnant Mature Cows										
350(772)	—	5.8(12.8)	5.9	2.8	1.80	50	0.16	0.16	6.1	2.4
400(882)	—	6.4(14.1)	5.9	2.8	1.80	50	0.16	0.16	6.1	2.4
450(992)	—	6.8(15.0)	5.9	2.8	1.80	50	0.16	0.16	6.1	2.4

TABLE 21 (Continued)

BODY WEIGHT kg(lb)	DAILY AVERAGE GAIN kg(lb)	DAILY DRY MATTER PER ANIMAL[a] kg(lb)	TOTAL PROTEIN %	DIGESTIBLE PROTEIN %	ENERGY		Ca %	P %	CAROTENE mg/kg	VITAMIN A thousands IU/kg
					ME[b] Mcal/kg	TDN[c] %				
500(1102)	—	7.6(16.8)	5.9	2.8	1.80	50	0.16	0.16	6.1	2.4
550(1213)	—	8.0(17.6)	5.9	2.8	1.80	50	0.16	0.16	6.1	2.4
600(1323)	—	8.6(19.0)	5.9	2.8	1.80	50	0.16	0.16	6.1	2.4
Cows Nursing Calves, First 3-4 Months Postpartum										
350(772)	—	8.6(19.0)	9.2	5.4	2.06	57	0.29	0.23	9.7	3.9
400(882)	—	9.3(20.5)	9.2	5.4	2.06	57	0.28	0.23	9.7	3.9
450(992)	—	9.9(21.8)	9.2	5.4	2.06	57	0.28	0.22	9.7	3.9
500(1102)	—	10.5(23.1)	9.2	5.4	2.06	57	0.27	0.22	9.7	3.9
Bulls, Growth and Maintenance (Moderate Activity)										
300(661)	1.00(2.20)	8.7(19.2)	13.9	9.6	2.35	65	0.26	0.21	9.7	3.9
400(882)	0.90(1.98)	10.0(22.0)	13.3	9.0	2.35	65	0.19	0.18	9.7	3.9
500(1102)	0.70(1.54)	12.0(26.5)	13.3	9.0	2.15	60	0.18	0.18	9.7	3.9
600(1323)	0.50(1.10)	11.6(25.6)	12.2	8.1	2.15	60	0.18	0.18	9.7	3.9
700(1543)	0.30(0.66)	12.7(28.0)	11.1	7.1	2.06	57	0.18	0.18	9.7	3.9
800(1764)	0.00(0.00)	9.9(21.8)	10.0	6.1	2.06	57	0.18	0.18	9.7	3.9
900(1984)	0.00(0.00)	10.7(23.6)	10.0	6.1	2.06	57	0.18	0.18	9.7	3.9

[a]Feed intake was calculated from the NE requirements and average NE values for the kind of ration being fed.

[b]ME requirements for growing and finishing cattle were calculated from the NE_m and NE_{gain} requirements for weights and rates of gain (Table 22).

[c]TDN was calculated from ME by assuming 3.6155 kcal of ME per g of TDN.

TABLE 22
NET ENERGY REQUIREMENTS OF GROWING AND FINISHING BEEF CATTLE
(MEGACALORIES PER ANIMAL PER DAY)

BODY WEIGHT (kg):	100	150	200	250	300	350	400	450	500
NE_m REQUIRED:	2.43	3.30	4.10	4.84	5.55	6.24	6.89	7.52	8.14
DAILY GAIN (kg)				NE_{gain} REQUIRED					
Steers									
0.1	0.17	0.23	0.28	0.34	0.39	0.43	0.48	0.52	0.56
0.2	0.34	0.46	0.57	0.68	0.78	0.88	0.97	1.06	1.14
0.3	0.52	0.70	0.87	1.03	1.18	1.33	1.47	1.61	1.74
0.4	0.70	0.95	1.18	1.40	1.60	1.80	1.99	2.17	2.34
0.5	0.89	1.20	1.49	1.77	2.02	2.27	2.51	2.74	2.97
0.6	1.08	1.46	1.81	2.15	2.46	2.76	3.05	3.33	3.60
0.7	1.27	1.73	2.14	2.53	2.90	3.26	3.60	3.93	4.25
0.8	1.47	2.00	2.47	2.93	3.36	3.77	4.17	4.55	4.92
0.9	1.67	2.27	2.81	3.34	3.82	4.29	4.74	5.18	5.60
1.0	1.88	2.55	3.16	3.75	4.29	4.82	5.33	5.82	6.29
1.1	2.09	2.84	3.52	4.17	4.78	5.37	5.93	6.47	7.00
1.2	2.31	3.13	3.88	4.60	5.27	5.92	6.55	7.14	7.73
1.3	2.53	3.43	4.26	5.04	5.77	6.49	7.17	7.82	8.46
1.4	2.76	3.74	4.63	5.49	6.29	7.06	7.81	8.52	9.22
1.5	2.99	4.05	5.02	5.95	6.81	7.65	8.46	9.23	9.98
Heifers									
0.1	0.18	0.25	0.30	0.36	0.41	0.46	0.51	0.56	0.61
0.2	0.37	0.50	0.62	0.74	0.84	0.95	1.05	1.14	1.24
0.3	0.57	0.77	0.95	1.13	1.29	1.45	1.60	1.75	1.90
0.4	0.77	1.05	1.30	1.54	1.76	1.98	2.18	2.39	2.58
0.5	0.99	1.34	1.66	1.96	2.25	2.53	2.79	3.05	3.30
0.6	1.21	1.64	2.03	2.40	2.75	3.09	3.41	3.73	4.03
0.7	1.44	1.95	2.42	2.85	3.27	3.68	4.06	4.44	4.80
0.8	1.67	2.28	2.81	3.33	3.82	4.28	4.73	5.17	5.59
0.9	1.92	2.60	3.23	3.81	4.37	4.91	5.42	5.93	6.41
1.0	2.17	2.94	3.65	4.32	4.95	5.56	6.14	6.71	7.26
1.1	2.43	3.30	4.09	4.84	5.55	6.23	6.88	7.52	8.13
1.2	2.70	3.66	4.55	5.37	6.16	6.92	7.64	8.35	9.03
1.3	2.98	4.04	5.01	5.92	6.79	7.63	8.42	9.21	9.96
1.4	3.26	4.42	5.49	6.49	7.44	8.36	9.23	10.09	10.91
1.5	3.56	4.82	5.98	7.07	8.11	9.11	10.06	11.00	11.90

TABLE 23
DAILY NUTRIENT REQUIREMENTS OF DAIRY CATTLE

BODY WEIGHT kg[a]	DAILY GAIN kg[a]	DRY FEED kg[a]	PROTEIN TOTAL kg[a]	PROTEIN DIGESTIBLE kg[a]	ENERGY NEm Mcal	ENERGY NEgain Mcal	ENERGY DE Mcal	ENERGY ME Mcal	ENERGY TDN kg[a]	Ca kg[a]	P kg[a]	CAROTENE mg	VITAMIN A 1000 IU	VITAMIN D IU
Growing Heifers (Large Breeds)														
40	0.20	0.5[b]	0.110	0.100	0.9	0.4	2.2	1.8	0.5	0.0022	0.0017	4.2	1.7	265
45	0.30	0.6[b]	0.135	0.120	1.1	0.5	2.6	2.1	0.6	0.0032	0.0025	4.8	1.9	300
55	0.40	1.2	0.180	0.145	1.3	0.6	4.0	3.3	0.9	0.0045	0.0035	5.8	2.3	360
75	0.75	2.1	0.330	0.245	1.5	0.9	6.6	5.4	1.5	0.0091	0.0070	7.9	3.2	495
100	0.75	2.9	0.370	0.260	2.0	1.1	8.8	7.2	2.0	0.0109	0.0084	11	4	660
150	0.75	4.1	0.435	0.295	3.1	1.5	11.9	9.8	2.7	0.015	0.012	16	6	990
200	0.75	5.3	0.500	0.330	4.1	1.8	15.0	12.3	3.4	0.018	0.014	21	8	1320
250	0.75	6.5	0.570	0.365	4.8	2.2	17.6	14.4	4.0	0.021	0.016	26	10	–
300	0.75	7.5	0.640	0.395	5.6	2.5	19.8	16.2	4.5	0.024	0.018	32	13	–
350	0.75	8.4	0.715	0.430	6.2	2.8	21.6	17.7	4.9	0.025	0.019	37	15	–
400	0.75	9.3	0.800	0.465	6.9	3.1	22.9	18.8	5.2	0.026	0.020	42	17	–
450	0.70	9.5	0.885	0.495	7.5	3.1	23.4	19.2	5.3	0.027	0.021	48	19	–
500	0.60	9.5	0.935	0.505	8.1	2.9	23.4	19.2	5.3	0.027	0.021	53	21	–
550	0.40	8.9	0.915	0.475	8.7	2.0	22.0	18.0	5.0	0.026	0.020	58	23	–
600	0.15	8.6	0.810	0.405	9.3	0.7	19.0	15.5	4.3	0.024	0.018	64	26	–
Growing Heifers (Small Breeds)														
20	0.10	0.3[b]	0.065	0.060	0.6	0.2	1.3	1.1	0.3	0.0011	0.0008	2.1	0.8	130
25	0.15	0.4[b]	0.090	0.080	0.8	0.3	1.8	1.5	0.4	0.0015	0.0011	2.6	1.0	165
35	0.30	0.8	0.135	0.110	0.9	0.5	2.6	2.1	0.6	0.0032	0.0025	3.7	1.5	230
50	0.50	1.2	0.215	0.160	1.0	0.9	4.0	3.3	0.9	0.0049	0.0038	5.3	2.1	330
75	0.55	1.7	0.275	0.190	1.5	1.0	5.3	4.3	1.2	0.0070	0.0054	7.9	3.2	495
100	0.55	2.4	0.320	0.210	2.1	1.1	7.1	5.8	1.6	0.009	0.007	11	4	660
150	0.55	3.6	0.390	0.245	3.7	1.3	10.1	8.3	2.3	0.012	0.009	16	6	990
200	0.55	4.8	0.465	0.280	4.1	1.6	12.8	10.5	2.9	0.015	0.011	21	8	1320
250	0.55	6.1	0.550	0.320	4.8	1.9	15.4	12.6	3.5	0.017	0.013	26	10	–
300	0.50	6.8	0.590	0.330	5.6	2.0	16.7	13.7	3.8	0.019	0.014	32	13	–

TABLE 23 (Continued)

BODY WEIGHT kg[a]	DAILY GAIN kg[a]	DRY FEED kg[a]	PROTEIN TOTAL kg[a]	PROTEIN DIGESTIBLE kg[a]	ENERGY NE_m Mcal	ENERGY NE_gain Mcal	ENERGY DE Mcal	ENERGY ME Mcal	ENERGY TDN kg[a]	Ca kg[a]	P kg[a]	CAROTENE mg	VITAMIN A 1000 IU	VITAMIN D IU
350	0.35	6.6	0.585	0.315	6.2	1.5	16.3	13.4	3.7	0.019	0.014	37	15	—
400	0.15	6.4	0.555	0.290	6.9	0.7	15.9	13.0	3.6	0.019	0.014	42	17	—
450	0.05	6.1	0.580	0.290	7.5	0.5	15.0	12.3	3.4	0.019	0.014	48	19	—
Growing Bulls (Large Breeds)														
40	0.20	0.5[b]	0.110	0.100	0.9	0.4	2.2	1.8	0.5	0.0022	0.0017	4.2	1.7	265
45	0.30	0.6[b]	0.135	0.120	1.1	0.5	2.6	2.1	0.6	0.0032	0.0025	4.8	1.9	300
55	0.40	1.2	0.180	0.145	1.3	0.6	4.0	3.3	0.9	0.0045	0.0035	5.8	2.3	360
75	0.80	2.1	0.345	0.255	1.6	1.0	6.6	5.4	1.5	0.0097	0.0075	7.9	3.2	495
100	1.00	3.2	0.455	0.320	2.1	1.3	9.7	8.0	2.2	0.013	0.010	11.0	4.0	660
150	1.00	4.5	0.520	0.355	3.2	1.8	13.2	10.8	3.0	0.018	0.014	16	6	990
200	1.00	5.9	0.595	0.390	4.5	2.2	16.7	13.7	3.8	0.021	0.016	21	8	1320
250	1.00	7.3	0.670	0.430	6.0	2.7	19.8	16.3	4.5	0.024	0.018	26	10	—
300	1.00	8.7	0.745	0.465	7.2	3.0	22.9	18.8	5.2	0.027	0.020	32	13	—
350	1.00	10.2	0.830	0.500	8.1	3.4	26.0	21.3	5.9	0.029	0.022	37	15	—
400	1.00	11.8	0.930	0.540	9.0	3.8	29.1	23.8	6.6	0.030	0.023	42	17	—
450	1.00	12.5	1.055	0.590	9.8	4.1	30.8	25.3	7.0	0.030	0.023	48	19	—
500	0.90	13.0	1.110	0.610	10.6	4.0	32.2	26.4	7.3	0.030	0.023	53	21	—
550	0.80	13.8	1.160	0.625	11.4	3.8	33.9	27.8	7.7	0.030	0.023	58	23	—
600	0.70	13.8	1.190	0.630	12.1	3.5	33.9	27.8	7.7	0.030	0.023	64	26	—
650	0.60	13.6	1.220	0.635	12.9	3.2	33.5	27.5	7.6	0.030	0.023	69	28	—
700	0.50	13.4	1.235	0.630	13.6	2.8	33.1	27.1	7.5	0.030	0.023	74	30	—
750	0.40	13.2	1.240	0.620	14.4	2.3	32.6	26.8	7.4	0.030	0.023	79	32	—
800	0.25	12.7	1.165	0.570	15.1	1.4	31.3	25.7	7.1	0.030	0.023	85	34	—
850	0.10	12.1	1.060	0.510	15.7	0.6	30.0	24.5	6.8	0.030	0.023	90	36	—
Growing Bulls (Small Breeds)														
20	0.10	0.3[b]	0.065	0.060	0.6	0.2	1.3	1.1	0.3	0.0011	0.0008	2.1	0.8	130
25	0.15	0.4[b]	0.090	0.080	0.8	0.3	1.8	1.5	0.4	0.0015	0.0011	2.6	1.0	165
35	0.30	0.8	0.135	0.110	0.9	0.5	2.6	2.2	0.6	0.0032	0.0025	3.7	1.5	230
50	0.65	1.4	0.265	0.200	1.0	1.1	4.4	3.6	1.0	0.0065	0.0050	5.3	2.1	330
75	0.75	2.0	0.345	0.240	1.5	1.3	6.2	5.1	1.4	0.0084	0.0065	7.9	3.2	495

Body wt (kg)														
100	0.75	2.8	0.390	0.255	2.1	1.6	8.4	6.9	1.9	0.011	0.008	11	4	660
150	0.75	4.3	0.460	0.295	3.1	1.9	11.9	9.8	2.7	0.015	0.011	16	6	990
200	0.75	5.7	0.530	0.330	4.5	2.3	15.0	12.3	3.4	0.018	0.014	21	8	1320
250	0.75	7.0	0.610	0.365	6.0	2.7	17.6	14.5	4.0	0.021	0.016	26	10	—
300	0.75	8.2	0.680	0.395	7.2	3.1	20.3	16.6	4.6	0.023	0.017	32	13	—
350	0.75	9.3	0.760	0.430	8.1	3.4	22.9	18.8	5.2	0.024	0.018	37	15	—
400	0.70	10.2	0.820	0.450	8.9	3.6	25.1	20.6	5.7	0.025	0.019	42	17	—
450	0.60	10.4	0.875	0.465	9.8	3.3	25.6	20.9	5.8	0.026	0.020	48	19	—
500	0.40	10.0	0.885	0.455	10.6	2.3	24.7	20.2	5.6	0.026	0.020	53	21	—
550	0.25	10.0	0.845	0.420	11.4	1.4	24.7	20.2	5.6	0.025	0.019	58	23	—
600	0.10	9.8	0.800	0.385	12.1	0.6	24.2	19.9	5.5	0.024	0.018	64	26	—
Veal Calves														
35	0.50	0.7[b]	0.155	0.130	1.0	0.8	3.1	2.5	0.7	0.0030	0.0023	3.7	1.5	230
40	0.80	1.1[b]	0.240	0.205	1.5	1.4	4.8	4.0	1.1	0.0048	0.0037	5.3	2.1	330
75	1.00	1.4[b]	0.310	0.260	1.9	1.8	6.2	5.1	1.4	0.0079	0.0059	7.9	3.2	495
100	1.15	1.7[b]	0.375	0.320	2.3	2.2	7.5	6.1	1.7	0.0111	0.0080	11.0	4.0	660
150	1.30	2.4[b]	0.485	0.410	3.0	3.0	10.6	8.7	2.4	0.0160	0.0110	16.0	6.0	990
Maintenance of Mature Breeding Bulls														
500	—	8.3	0.640	0.300	9.5	—	20.3	16.6	4.6	0.020	0.015	53	21	—
600	—	9.6	0.735	0.345	10.8	—	23.8	19.5	5.4	0.022	0.017	64	26	—
700	—	10.9	0.830	0.390	12.3	—	26.9	22.1	6.1	0.025	0.019	74	30	—
800	—	12.0	0.915	0.430	13.9	—	29.5	24.2	6.7	0.027	0.021	85	34	—
900	—	13.1	1.000	0.470	15.2	—	32.2	26.4	7.3	0.030	0.023	95	38	—
1000	—	14.1	1.075	0.505	16.9	—	34.8	28.6	7.9	0.032	0.025	106	42	—
1100	—	15.1	1.160	0.545	18.2	—	37.0	30.4	8.4	0.035	0.027	117	47	—
1200	—	16.1	1.235	0.580	19.5	—	39.7	32.5	9.0	0.038	0.029	127	51	—
1300	—	17.1	1.310	0.615	20.7	—	42.3	34.7	9.6	0.040	0.031	138	55	—
1400	—	18.1	1.380	0.650	21.9	—	44.5	39.8	10.1	0.043	0.033	148	59	—

[a]To convert kg to lb, multiply by 2.205.
[b]Based on milk replacer.

Table 24

DAILY NUTRIENT REQUIREMENTS OF LACTATING DAIRY CATTLE

BODY WEIGHT kg(lb)	DRY FEED kg(lb)	PROTEIN		ENERGY				Ca kg(lb)	P kg(lb)	CAROTENE mg	VITA- MIN A 1000 IU
		TOTAL kg(lb)	DIGEST-IBLE kg(lb)	NE LACTAT-ING COWS Mcal[a]	DE Mcal	ME Mcal	TDN kg(lb)				
Maintenance of Mature Lactating Cows[b]											
350(772)	5.0(11.0)	0.468(1.03)	0.220(0.49)	6.9	12.3	10.1	2.8(6.2)	0.014(0.031)	0.011(0.024)	37	15
400(882)	5.5(12.1)	0.521(1.15)	0.245(0.54)	7.6	13.6	11.2	3.1(6.8)	0.017(0.037)	0.013(0.029)	42	17
450(992)	6.0(13.2)	0.585(1.29)	0.275(0.61)	8.3	15.0	12.3	3.4(7.5)	0.018(0.040)	0.014(0.031)	48	19
500(1102)	6.5(14.3)	0.638(1.41)	0.300(0.66)	9.0	16.3	13.4	3.7(8.2)	0.020(0.044)	0.015(0.033)	53	21
550(1213)	7.0(15.4)	0.691(1.52)	0.325(0.72)	9.6	17.6	14.4	4.0(8.8)	0.021(0.046)	0.016(0.035)	58	23
600(1323)	7.5(16.5)	0.734(1.62)	0.345(0.76)	10.3	18.9	15.5	4.2(9.3)	0.022(0.049)	0.017(0.037)	64	26
650(1433)	8.0(17.6)	0.776(1.71)	0.365(0.80)	10.9	19.8	16.2	4.5(9.9)	0.023(0.051)	0.018(0.040)	69	28
700(1543)	8.5(18.7)	0.830(1.83)	0.390(0.86)	11.6	21.1	17.3	4.8(10.6)	0.025(0.055)	0.019(0.042)	74	30
750(1654)	9.0(19.8)	0.872(1.92)	0.410(0.90)	12.2	22.0	18.0	5.0(11.0)	0.026(0.057)	0.020(0.044)	79	32
800(1764)	9.5(20.9)	0.915(2.02)	0.430(0.95)	12.8	23.3	19.1	5.3(11.7)	0.027(0.060)	0.021(0.046)	85	34
Maintenance and Pregnancy (Last 2 Months of Gestation)											
350(772)	6.4(14.1)	0.570(1.26)	0.315(0.69)	8.7	15.8	13.0	3.6(7.9)	0.021(0.046)	0.016(0.035)	67	27
400(882)	7.2(15.9)	0.650(1.43)	0.355(0.78)	9.7	17.2	14.1	4.0(8.8)	0.023(0.051)	0.018(0.040)	76	30
450(992)	7.9(17.4)	0.730(1.61)	0.400(0.88)	10.7	19.4	15.9	4.4(9.7)	0.026(0.057)	0.020(0.044)	86	34
500(1102)	8.6(19.0)	0.780(1.72)	0.430(0.95)	11.6	21.1	17.3	4.8(10.6)	0.029(0.064)	0.022(0.049)	95	38
550(1213)	9.3(20.5)	0.850(1.87)	0.465(1.03)	12.6	22.9	18.8	5.2(11.5)	0.031(0.068)	0.024(0.053)	105	42
600(1323)	10.0(22.0)	0.910(2.01)	0.500(1.10)	13.5	24.6	20.2	5.6(12.3)	0.034(0.075)	0.026(0.057)	114	46
650(1433)	10.6(23.4)	0.960(2.12)	0.530(1.17)	14.4	26.4	21.6	6.0(13.2)	0.036(0.079)	0.028(0.062)	124	50
700(1543)	11.3(24.9)	1.000(2.20)	0.555(1.22)	15.3	27.7	22.7	6.3(13.9)	0.039(0.086)	0.030(0.066)	133	53
750(1654)	12.0(26.5)	1.080(2.38)	0.595(1.31)	16.2	29.5	24.2	6.7(14.8)	0.042(0.093)	0.032(0.071)	143	57
800(1764)	12.6(27.8)	1.150(2.54)	0.630(1.39)	17.0	31.2	25.6	7.1(15.7)	0.044(0.097)	0.034(0.075)	152	61
Milk Production (Nutrients Required per kg(lb) of Milk[c])											
% Fat											
2.5(2.5)		0.066(0.066)	0.042(0.042)	0.59(0.27)	1.12(0.51)	0.91(0.41)	0.255(0.255)	0.0024(0.0024)	0.0017(0.0017)		
3.0(3.0)		0.070(0.070)	0.045(0.045)	0.64(0.29)	1.23(0.56)	0.99(0.45)	0.280(0.280)	0.0025(0.0025)	0.0018(0.0018)		
3.5(3.5)		0.074(0.074)	0.048(0.048)	0.69(0.31)	1.34(0.61)	1.06(0.48)	0.305(0.305)	0.0027(0.0026)	0.0019(0.0019)		
4.0(4.0)		0.078(0.078)	0.051(0.051)	0.74(0.34)	1.46(0.66)	1.13(0.51)	0.330(0.330)	0.0028(0.0028)	0.0020(0.0020)		
4.5(4.5)		0.082(0.082)	0.054(0.054)	0.78(0.35)	1.57(0.71)	1.21(0.55)	0.355(0.355)	0.0029(0.0029)	0.0021(0.0021)		
5.0(5.0)		0.086(0.086)	0.056(0.056)	0.83(0.38)	1.68(0.76)	1.28(0.58)	0.380(0.380)	0.0030(0.0030)	0.0022(0.0022)		
5.5(5.5)		0.090(0.090)	0.058(0.058)	0.88(0.40)	1.79(0.81)	1.36(0.62)	0.405(0.405)	0.0031(0.0031)	0.0023(0.0023)		
6.0(6.0)		0.094(0.094)	0.060(0.060)	0.93(0.42)	1.90(0.86)	1.43(0.65)	0.430(0.430)		0.0024(0.0024)		

[a]The energy requirements for maintenance, reproduction, and milk production of lactating cows are expressed in terms of $NE_{lactating}$ cows.

[b]Maintenance of lactating cows = 0.085 Mcal $NE_{lactating}$ cows/$kg^{3/4}$. To allow for growth, add 20 percent to the maintenance allowance during the first lactation and 10 percent during the second lactation.

[c]The energy requirement is presented as the actual amount required with no adjustment to compensate for any reduction in feed value at high levels of feed intake. To account for depressions in digestibility, which occur at high planes of nutrition with certain types of rations, such as corn silage, coarse textured grains or forages with high cell-wall content (e.g., Bermuda grass, sorghum, etc.), an increase of 3 percent should be allowed for each 10 kg of milk produced above 20 kg/day.

TABLE 25
NUTRIENT CONTENT OF RATIONS FOR DAIRY CATTLE

QUANTITY PER KG OF DRY MATTER

NUTRIENTS	CALF MILK REPLACER[a]		CALF STARTER		HEIFER GROWER RATION		DRY COW RATION		LACTATING COW RATIONS — DAILY MILK PRODUCTION <20 kg		20-30 kg		>30 kg		MATURE BULL RATION	
	Mn	Mx	Mn	Mx	Mn	Mx	Mn	Mx	Mn	Mx	Mn	Mx	Mn	Mx	Mn	Mx
Protein (g)	220.0		160.0		100.0		85		140		150		160		77	
Digestible (g)	200.0		120.0		62.0		51		105		114		123		36	
Energy, Mcal																
Digestible (DE)	4.2		3.2		2.9		2.3		2.7		2.9		3.1		2.5	
Metabolizable (ME)	3.4		2.6		2.4		1.9		2.1		2.3		2.5		2.0	
NE_m	2.4		1.8		1.7		1.1									
NE_{gain}	1.5		1.2		1.1											
$NE_{lactation}$									1.4		1.6		1.8			
TDN (g)	950		720		660		530		600		650		700		560	
Ether extract (g)	100.0		25		20		20		20		20		20		20	
Crude fiber (g)	0	30		150		150		150		130		130		130	150	
Calcium (g)	5.5		4.1		3.4		3.4		4.3		4.7		5.3		2.4	
Phosphorus (g)	4.2		3.2		2.6		2.6		3.3		3.5		3.9		1.8	
Magnesium (g)	0.6		0.7		0.8		0.8		1.0		1.0		1.0		0.8	
Potassium (g)	7.0		7.0		7.0		7.0		7.0		7.0		7.0		7.0	
Sodium, (g)	1.0		1.0		1.0		1.0		1.8		1.8		1.8		1.0	
Sodium chloride (g)	2.5		2.5		2.5		2.5		4.5		4.5		4.5		2.5	
Sulfur (g)	2.0		2.0		2.0		2.0		2.0		2.0		2.0		2.0	
Iron (mg)	100.0		100.0		100.0		100.0		100.0		100.0		100.0		100.0	
Cobalt (mg)	0.1	10	0.1	10	0.1	10	0.1	10	0.1	10	0.1	10	0.1	10	0.1	10
Copper (mg)	10.0	100	10.0	100	10.0	100	10.0	100	10.0	100	10.0	100	10.0	100	10.0	100
Manganese (mg)	20.0		20.0		20.0		20.0		20.0		20.0		20.0		20.0	
Zinc (mg)	40.0	500	40.0	500	40.0	500	40.0	1000	40.0	1000	40.0	1000	40.0	1000	40.0	1000
Iodine (mg)	0.1		0.1		0.1		0.6		0.6		0.6		0.6		0.1	
Molybdenum (mg)		6		6		6		6		6		6		6		6
Fluorine (mg)		40		30		30		40		40		40		40		40
Selenium (mg)	0.1	5	0.1	5	0.1	5	0.1	5	0.1	5	0.1	5	0.1	5	0.1	5
Carotene (mg)	9.5		4.2		4.0		8.0		8.0		8.0		8.0		8.0	
Vitamin A equiv. (IU)	3800		1600		1500		3200		3200		3200		3200		3200	
Vitamin D (IU)	600		250		250		300		300		300		300		300	
Vitamin E (mg)	300															

[a]The following minimum quantities of B-complex vitamins are suggested for milk replacers: niacin, 2.6 mg; pantothenic acid, 13 mg; riboflavin, 6.5 mg; pyridoxine, 6.5 mg; thiamine, 6.5 mg; folic acid, 0.5 mg; biotin, 0.1 mg; vitamin B_{12}, 0.07 mg; choline, 2.6 g. It appears that adequate amounts of these vitamins are furnished when calves have functional rumens (usually at 6 weeks of age) by a combination of rumen synthesis and natural feedstuffs.

TABLE 26
NUTRIENT REQUIREMENTS OF MATURE HORSES, PREGNANT MARES, AND LACTATING MARES
(DAILY NUTRIENTS PER ANIMAL)

BODY WEIGHT kg (lb)	DRY MATTER[a] kg (lb)	DIGESTIBLE ENERGY Mcal	TDN kg (lb)	PROTEIN kg (lb)	DIGESTIBLE PROTEIN kg (lb)	VITAMIN A thousands IU[b]	CAROTENE mg	Ca kg (lb)	P kg (lb)
Mature Horses at Rest (maintenance)									
200(441)	3.00(6.6)	8.24	1.87(4.1)	0.300(0.66)	0.160(0.35)	5.0	12.5	0.0080(0.018)	0.0060(0.013)
400(882)	5.04(11.1)	13.86	3.14(6.9)	0.505(1.11)	0.268(0.59)	10.0	25.0	0.0160(0.035)	0.0120(0.026)
500(1102)	5.96(13.1)	16.39	3.72(8.2)	0.597(1.32)	0.317(0.70)	12.5	31.2	0.0200(0.044)	0.0150(0.033)
600(1323)	6.83(15.1)	18.79	4.26(9.4)	0.684(1.51)	0.364(0.80)	15.0	37.5	0.0240(0.053)	0.0180(0.040)
Mature Horses at Light Work (2 hr/day)									
200(441)	3.80(8.4)	10.44	2.37(5.2)	0.383(0.84)	0.202(0.45)	5.0	12.5	0.0080(0.018)	0.0060(0.013)
400(882)	6.68(14.7)	18.36	4.16(9.2)	0.672(1.48)	0.355(0.78)	10.0	25.0	0.0160(0.035)	0.0120(0.026)
500(1102)	7.96(17.6)	21.89	4.96(10.9)	0.803(1.77)	0.424(0.93)	12.5	31.2	0.0200(0.044)	0.0150(0.033)
600(1323)	9.23(20.4)	25.39	5.76(12.7)	0.930(2.05)	0.491(1.08)	15.0	37.5	0.0240(0.053)	0.0180(0.040)
Mature Horses at Medium Work (2 hr/day)									
200(441)	4.79(10.6)	13.16	2.98(6.6)	0.483(1.07)	0.255(0.56)	5.0	12.5	0.0092(0.020)	0.0070(0.015)
400(882)	8.65(19.1)	23.80	5.40(11.9)	0.871(1.92)	0.460(1.01)	10.0	25.0	0.0172(0.038)	0.0130(0.029)
500(1102)	10.43(23.0)	28.69	6.51(14.3)	1.047(2.31)	0.553(1.22)	12.5	31.2	0.0212(0.047)	0.0160(0.035)
600(1323)	12.22(26.9)	33.55	7.61(16.8)	1.229(2.71)	0.649(1.43)	15.0	37.5	0.0252(0.056)	0.0190(0.042)
Mares, Last 90 Days of Pregnancy									
200(441)	3.16(7.0)	8.70	1.97(4.3)	0.364(0.80)	0.216(0.48)	10.0	25.0	0.0104(0.023)	0.0080(0.018)
400(882)	5.41(11.9)	14.88	3.37(7.4)	0.613(1.35)	0.375(0.83)	20.0	50.0	0.0195(0.043)	0.0150(0.033)
500(1102)	6.31(13.9)	17.35	3.93(8.7)	0.725(1.60)	0.434(0.96)	25.0	62.5	0.0240(0.053)	0.0180(0.040)
600(1323)	7.25(16.0)	19.95	4.52(10.0)	0.837(1.85)	0.502(1.11)	30.0	75.0	0.0280(0.062)	0.0210(0.046)
Mares, Peak of Lactation									
200(441)	5.54(12.2)	15.24	3.46(7.6)	0.750(1.65)	0.480(1.06)	10.0	25.0	0.0340(0.075)	0.0234(0.052)
400(882)	8.91(19.6)	24.39	5.53(12.2)	1.181(2.60)	0.748(1.65)	20.0	50.0	0.0420(0.093)	0.0356(0.078)
500(1102)	10.04(22.1)	27.62	6.26(13.8)	1.317(2.90)	0.829(1.83)	25.0	62.5	0.0470(0.104)	0.0386(0.085)
600(1323)	10.92(24.1)	30.02	6.80(15.0)	1.404(3.10)	0.876(1.93)	30.0	75.0	0.0510(0.112)	0.0390(0.086)

[a]Assume 2.75 Mcal of digestible energy per kg of 100 percent dry feed.
[b]One mg of beta-carotene equals 400 IU of vitamin A.

TABLE 27
NUTRIENT REQUIREMENTS OF GROWING HORSES
(DAILY NUTRIENTS PER ANIMAL)

AGE months	BODY WEIGHT kg (lb)	PERCENTAGE OF MATURE WEIGHT	DAILY GAIN kg (lb)	DRY MATTER[a] kg (lb)	DIGESTIBLE ENERGY Mcal	TDN kg (lb)	PROTEIN kg (lb)	DIGESTIBLE PROTEIN kg (lb)	VITAMIN A thousands IU[b]	CAROTENE mg	Ca kg (lb)	P kg (lb)
200 kg (441 lb) Mature Weight												
3	50(110)	25.0	0.70(1.54)	2.94(6.48)	7.43	1.68(3.7)	0.526(1.16)	0.383(0.84)	2.0	5.0	0.0174(0.038)	0.0109(0.024)
6	90(198)	45.0	0.50(1.10)	3.10(6.84)	8.53	1.95(4.3)	0.462(1.02)	0.315(0.69)	3.6	9.0	0.0166(0.037)	0.0104(0.023)
12	135(298)	67.5	0.20(0.44)	2.89(6.37)	7.95	1.81(4.0)	0.338(0.75)	0.206(0.45)	5.4	13.4	0.0120(0.026)	0.0075(0.017)
18	165(364)	82.5	0.10(0.22)	2.94(6.48)	8.08	1.81(4.0)	0.314(0.69)	0.181(0.40)	6.6	16.5	0.0104(0.023)	0.0065(0.014)
42	200(441)	100.0	0 0	3.00(6.62)	8.24	1.86(4.1)	0.300(0.66)	0.160(0.35)	5.0	12.5	0.0080(0.018)	0.0060(0.013)
400 kg (882 lb) Mature Weight												
3	85(187)	21.3	1.00(2.20)	3.80(8.38)	10.44	2.36(5.2)	0.741(1.63)	0.553(1.22)	3.4	8.5	0.0261(0.058)	0.0164(0.036)
6	170(375)	42.5	0.65(1.43)	4.51(9.94)	12.41	2.81(6.2)	0.640(1.41)	0.430(0.95)	6.8	17.0	0.0350(0.077)	0.0219(0.048)
12	260(573)	65.0	0.40(0.88)	4.96(10.94)	13.63	3.08(6.8)	0.600(1.32)	0.370(0.82)	10.4	26.0	0.0220(0.049)	0.0148(0.033)
18	330(728)	82.5	0.25(0.55)	5.13(11.31)	14.10	3.22(7.1)	0.575(1.27)	0.339(0.75)	14.2	35.5	0.0190(0.042)	0.0138(0.030)
42	400(882)	100.0	0 0	5.04(11.11)	13.86	3.13(6.9)	0.505(1.11)	0.268(0.59)	10.0	25.0	0.0160(0.035)	0.0120(0.026)
500 kg (1102 lb) Mature Weight												
3	110(243)	22.0	1.10(2.43)	4.39(9.68)	12.07	2.72(6.0)	0.834(1.84)	0.618(1.36)	4.4	11.0	0.0305(0.067)	0.0191(0.042)
6	225(496)	45.0	0.80(1.76)	5.60(12.35)	15.40	3.49(7.7)	0.800(1.76)	0.536(1.18)	9.0	22.5	0.0460(0.101)	0.0287(0.063)
12	325(717)	65.0	0.55(1.21)	6.11(13.47)	16.81	3.81(8.4)	0.750(1.65)	0.472(1.04)	11.0	27.5	0.0260(0.057)	0.0174(0.038)
18	400(882)	80.0	0.35(0.77)	6.24(13.76)	17.16	3.90(8.6)	0.700(1.54)	0.418(0.92)	16.0	40.0	0.0230(0.051)	0.0161(0.036)
42	500(1102)	100.0	0 0	5.96(13.14)	16.39	3.72(8.2)	0.597(1.32)	0.317(0.70)	12.5	31.2	0.0200(0.044)	0.0150(0.033)
600 kg (1323 lb) Mature Weight												
3	140(309)	23.3	1.25(2.76)	5.15(11.36)	14.15	3.22(7.1)	0.958(2.11)	0.705(1.55)	5.6	14.0	0.0520(0.115)	0.0322(0.071)
6	265(584)	44.2	0.85(1.87)	6.26(13.80)	17.21	3.90(8.6)	0.870(1.92)	0.582(1.28)	10.6	26.5	0.0512(0.113)	0.0320(0.071)
12	385(849)	64.1	0.60(1.32)	6.86(15.13)	18.86	4.26(9.4)	0.837(1.85)	0.524(1.16)	15.4	38.5	0.0329(0.073)	0.0206(0.045)
18	480(1058)	80.0	0.35(0.77)	6.98(15.39)	19.20	4.35(9.6)	0.775(1.71)	0.458(1.01)	19.2	48.0	0.0313(0.069)	0.0196(0.043)
42	600(1323)	100.0	0 0	6.83(15.06)	18.79	4.26(9.4)	0.684(1.51)	0.364(0.80)	15.0	37.5	0.0240(0.053)	0.0180(0.040)

[a]Assume 2.75 Mcal of digestible energy per kg of 100 percent dry feed.
[b]One mg of beta-carotene equals 400 IU of vitamin A.

TABLE 28
NUTRIENT REQUIREMENTS OF MATURE HORSES, PREGNANT MARES, AND LACTATING MARES
(NUTRIENT CONCENTRATION IN RATION DRY MATTER)

BODY WEIGHT kg (lb)	DAILY FEED[a] (DRY BASIS) PER ANIMAL kg (lb)	PERCENTAGE OF LIVE WEIGHT	DIGESTIBLE ENERGY Mcal	TDN %	PROTEIN %	DIGESTIBLE PROTEIN %	Ca %	P %	CAROTENE mg/kg	VITAMIN A 1000 IU/kg
Mature Horses at Rest (maintenance)										
200(441)	3.00(6.6)	1.5	2.75	62.4	10.0	5.3	0.26	0.20	4.18	1.67
400(882)	5.04(11.1)	1.3	2.75	62.4	10.0	5.3	0.31	0.24	4.95	1.98
500(1102)	5.96(13.1)	1.2	2.75	62.4	10.0	5.3	0.33	0.25	5.25	2.10
600(1323)	6.83(15.1)	1.1	2.75	62.4	10.0	5.3	0.35	0.26	5.42	2.17
Mature Horses at Light Work (2 hr/day)										
200(441)	3.80(8.4)	1.9	2.75	62.4	10.0	5.3	0.21	0.15	3.30	1.32
400(882)	6.68(14.7)	1.7	2.75	62.4	10.0	5.3	0.24	0.18	3.80	1.50
500(1102)	7.96(17.6)	1.6	2.75	62.4	10.0	5.3	0.25	0.18	3.93	1.57
600(1323)	9.23(20.4)	1.5	2.75	62.4	10.0	5.3	0.26	0.19	4.08	1.63
Mature Horses at Medium Work (2 hr/day)										
200(441)	4.79(10.6)	2.4	2.75	62.4	10.0	5.3	0.19	0.14	2.60	1.04
400(882)	8.65(19.1)	2.2	2.75	62.4	10.0	5.3	0.20	0.15	2.90	1.16
500(1102)	10.43(23.0)	2.1	2.75	62.4	10.0	5.3	0.20	0.15	3.00	1.20
600(1323)	12.22(26.9)	2.0	2.75	62.4	10.0	5.3	0.20	0.15	3.08	1.23
Mares, Last 90 Days of Pregnancy										
200(441)	3.16(7.0)	1.6	2.75	62.4	11.5	6.9	0.33	0.25	7.90	3.16
400(882)	5.41(11.9)	1.4	2.75	62.4	11.5	6.9	0.36	0.28	9.25	3.70
500(1102)	5.31(11.7)	1.3	2.75	62.4	11.5	6.9	0.38	0.29	9.90	3.96
600(1323)	7.25(16.0)	1.2	2.75	62.4	11.5	6.9	0.39	0.29	10.35	4.14
Mares, Peak of Lactation										
200(441)	5.54(12.2)	2.8	2.75	62.4	13.5	8.7	0.61	0.41	4.50	1.80
400(882)	8.91(19.6)	2.2	2.75	62.4	13.3	8.4	0.47	0.40	5.60	2.24
500(1102)	10.04(22.1)	2.0	2.75	62.4	13.1	8.3	0.47	0.37	6.22	2.49
600(1323)	10.92(24.1)	1.8	2.75	62.4	12.9	8.0	0.47	0.36	6.88	2.75

[a]Assume 2.75 Mcal of digestible energy per kg of ration dry matter.

TABLE 29
NUTRIENT REQUIREMENTS OF GROWING HORSES
(NUTRIENT CONCENTRATION IN RATION DRY MATTER)

AGE months	BODY WEIGHT kg (lb)	PERCENTAGE OF MATURE WEIGHT	DAILY FEED[a] (DRY BASIS) DAILY GAIN kg (lb)	PER ANIMAL kg (lb)	PERCENTAGE OF LIVE WEIGHT	PERCENTAGE OF OR AMOUNT PER KG OF RATION DRY MATTER DIGESTIBLE ENERGY Mcal	TDN %	DIGESTIBLE PROTEIN %	PROTEIN %	Ca %	P %	CAROTENE mg/kg	VITAMIN A 1000 IU/kg
200 kg (441 lb) Mature Weight													
3	50(110)	25.0	0.70(1.54)	2.94(6.5)	5.9	2.75	62.4	17.9	13.0	0.59	0.37	1.70	0.68
6	90(198)	45.0	0.50(1.10)	3.10(6.8)	3.4	2.75	62.4	14.9	10.2	0.53	0.34	2.90	1.16
12	135(298)	67.5	0.20(0.44)	2.89(6.4)	2.1	2.75	62.4	11.7	7.1	0.41	0.25	4.68	1.87
18	165(364)	82.5	0.10(0.22)	2.94(6.5)	1.8	2.75	62.4	10.7	6.2	0.35	0.22	5.60	2.24
42	200(441)	100	0 0	3.00(6.6)	1.5	2.75	62.4	10.0	5.3	0.29	0.20	4.18	1.67
400 kg (882 lb) Mature Weight													
3	85(187)	21.3	1.00(2.20)	3.80(8.4)	4.5	2.75	62.4	19.5	14.6	0.68	0.43	2.22	0.89
6	170(375)	42.5	0.65(1.43)	4.51(9.9)	2.7	2.75	62.4	14.2	9.5	0.78	0.48	3.78	1.51
12	260(573)	65.0	0.40(0.88)	4.96(10.9)	1.9	2.75	62.4	12.1	7.5	0.45	0.30	5.25	2.10
18	330(728)	82.5	0.25(0.55)	5.13(11.3)	1.6	2.75	62.4	11.2	6.6	0.37	0.27	6.92	2.77
42	400(882)	100	0 0	5.04(11.1)	1.3	2.75	62.4	10.0	5.3	0.32	0.24	4.95	1.98
500 kg (1102 lb) Mature Weight													
3	110(243)	22.0	1.10(2.43)	4.39(9.7)	4.0	2.75	62.4	19.0	14.1	0.69	0.44	2.50	1.00
6	225(496)	45.0	0.80(1.76)	5.60(12.3)	2.5	2.75	62.4	14.3	9.6	0.82	0.51	4.02	1.51
12	324(717)	65.0	0.55(1.21)	6.11(13.5)	1.9	2.75	62.4	12.3	7.7	0.43	0.28	4.50	1.80
18	400(882)	80.0	0.35(0.77)	6.24(13.8)	1.6	2.75	62.4	11.3	6.7	0.37	0.26	6.40	2.56
42	500(1102)	100	0 0	5.96(13.1)	1.2	2.75	62.4	10.0	5.3	0.34	0.25	5.25	2.10
600 kg (1323 lb) Mature Weight													
3	140(309)	23.3	1.25(2.76)	5.15(11.4)	3.7	2.75	62.4	18.6	13.7	1.01	0.63	2.72	1.09
6	265(584)	44.2	0.85(1.87)	6.26(13.8)	2.4	2.75	62.4	13.9	9.2	0.81	0.51	4.22	1.69
12	385(849)	64.1	0.60(1.32)	6.86(15.1)	1.8	2.75	62.4	12.2	7.6	0.48	0.30	5.60	2.24
18	480(1058)	80.0	0.35(0.77)	6.98(15.4)	1.5	2.75	62.4	11.1	6.6	0.45	0.28	6.88	2.75
42	600(1323)	100	0 0	6.83(15.1)	1.1	2.75	62.4	10.0	5.3	0.35	0.26	5.50	2.20

[a] Assume 2.75 Mcal of digestible energy per kg of 100 percent dry feed.

Table 30
DAILY NUTRIENT REQUIREMENTS OF SHEEP
(DAILY NUTRIENTS PER ANIMAL)

Body Weight kg(lb)	Dry Matter kg(lb)	TDN kg(lb)	DE[a] Mcal	Protein kg(lb)	DP[b] kg(lb)	Ca gm (lb)	P gm (lb)	Salt gm	Carotene mg	Vit. A mcg[c]	Vit. A IU	Vit. D IU
EWES												
Nonlactating and First 15 weeks of Gestation												
45(99)	1.08(2.38)	0.59(1.30)	2.6	0.095(0.21)	0.054(0.12)	3.2(0.007)	2.5(0.006)	9.0	1.7	280	935	250
54(119)	1.26(2.78)	0.68(1.50)	3.0	0.109(0.24)	0.059(0.13)	3.3(0.007)	2.6(0.006)	10.0	2.0	300	1100	300
64(141)	1.35(2.98)	0.77(1.70)	3.4	0.122(0.27)	0.068(0.15)	3.4(0.008)	2.7(0.006)	11.0	2.4	396	1320	350
73(161)	1.53(3.37)	0.86(1.80)	3.8	0.136(0.30)	0.073(0.16)	3.5(0.008)	2.8(0.006)	12.0	2.7	446	1485	400
Last 6 Weeks of Gestation												
45(99)	1.53(3.37)	0.91(2.00)	4.0	0.145(0.32)	0.082(0.18)	4.2(0.009)	3.1(0.007)	10.0	5.8	696	2320	250
54(119)	1.71(3.77)	1.00(2.20)	4.4	0.154(0.34)	0.086(0.19)	4.4(0.010)	3.3(0.007)	11.0	6.8	816	2720	300
64(141)	1.89(4.17)	1.09(2.40)	4.8	0.163(0.36)	0.091(0.20)	4.6(0.010)	3.5(0.008)	12.0	7.9	948	3160	350
73(161)	1.98(4.37)	1.13(2.49)	5.0	0.168(0.37)	0.091(0.20)	4.8(0.011)	3.7(0.008)	13.0	9.1	1092	3640	400
First 8 to 10 Weeks of Lactation												
45(99)	1.89(4.17)	1.24(2.73)	5.4	0.181(0.40)	0.100(0.22)	6.2(0.014)	4.6(0.010)	11.0	5.8	696	2320	250
54(119)	2.07(4.56)	1.33(2.93)	5.8	0.190(0.42)	0.104(0.23)	6.5(0.014)	4.8(0.011)	12.0	6.8	816	2720	300
64(141)	2.25(4.96)	1.40(3.09)	6.2	0.200(0.44)	0.109(0.24)	6.8(0.015)	5.0(0.011)	13.0	7.9	948	3160	350
73(161)	2.34(5.16)	1.43(3.15)	6.2	0.209(0.46)	0.113(0.25)	7.1(0.016)	5.2(0.012)	14.0	9.1	1092	3640	400
Last 12 to 14 Weeks of Lactation												
45(99)	1.53(3.37)	0.91(2.00)	4.0	0.145(0.32)	0.082(0.18)	4.6(0.010)	3.4(0.008)	10.0	5.8	696	2320	250
54(119)	1.71(3.77)	1.00(2.20)	4.4	0.154(0.34)	0.086(0.19)	4.8(0.011)	3.6(0.008)	11.0	6.8	816	2720	300
64(141)	1.89(4.17)	1.09(2.40)	4.8	0.163(0.36)	0.091(0.20)	5.0(0.011)	3.8(0.008)	12.0	7.9	948	3160	350
73(161)	1.98(4.37)	1.13(2.49)	5.0	0.168(0.37)	0.091(0.20)	5.2(0.012)	4.0(0.009)	13.0	9.1	1092	3640	400

Table 30 (Cont.)

Body Weight kg(lb)	Dry Matter kg(lb)	TDN kg(lb)	DE[a] Mcal	Protein kg(lb)	DP[b] kg(lb)	Ca gm(lb)	P gm(lb)	Salt gm	Carotene mg	Vit. A mcg[c]	Vit. A IU	Vit. D IU
Replacement Lambs and Yearlings												
27(59)	1.08(2.38)	0.68(1.50)	3.0	0.136(0.30)	0.073(0.16)	2.9(0.006)	2.6(0.006)	8.0	1.7	230	765	150
36(79)	1.26(2.78)	0.73(1.61)	3.2	0.127(0.28)	0.068(0.15)	3.0(0.007)	2.7(0.006)	9.0	2.3	310	1065	200
45(99)	1.35(2.98)	0.77(1.70)	3.4	0.118(0.26)	0.064(0.14)	3.1(0.007)	2.8(0.006)	10.0	2.8	378	1260	250
54(119)	1.35(2.98)	0.77(1.70)	3.4	0.109(0.24)	0.059(0.13)	3.2(0.007)	2.9(0.007)	11.0	3.4	459	1530	300
RAMS												
Lambs and Yearlings												
36(79)	1.26(2.78)	0.91(2.01)	4.0	0.145(0.32)	0.082(0.18)	3.0(0.007)	2.7(0.006)	9.0	2.3	310	1035	200
45(99)	1.53(3.37)	0.95(2.09)	4.2	0.145(0.32)	0.082(0.18)	3.1(0.007)	2.8(0.006)	10.0	2.8	378	1260	250
54(119)	1.71(3.77)	0.95(2.09)	4.2	0.145(0.32)	0.082(0.18)	3.2(0.007)	2.9(0.007)	11.0	3.4	459	1530	300
64(141)	1.89(4.17)	1.04(2.29)	4.6	0.145(0.32)	0.082(0.18)	3.3(0.007)	3.0(0.007)	11.0	4.0	540	1800	350
73(161)	1.98(4.37)	1.09(2.40)	4.8	0.145(0.32)	0.082(0.18)	3.4(0.008)	3.1(1.007)	12.0	4.5	608	2025	400
LAMBS												
Fattening (Gaining 0.35 to 0.45 lb/hd/day)												
27(59)	1.08(2.38)	0.68(1.50)	3.0	0.145(0.32)	0.082(0.18)	2.9(0.007)	2.6(0.006)	8.0	1.0	165	500	150
32(71)	1.26(2.78)	0.82(1.81)	3.6	0.154(0.34)	0.086(0.19)	2.9(0.007)	2.6(0.006)	8.0	1.2	198	660	175
36(79)	1.35(2.98)	0.95(2.09)	4.2	0.163(0.36)	0.091(0.20)	3.0(0.007)	2.7(0.006)	9.0	1.4	231	770	200
41(90)	1.53(3.37)	1.04(2.29)	4.6	0.163(0.36)	0.091(0.20)	3.0(0.007)	2.7(0.006)	9.0	1.5	248	825	225
45(99)	1.62(3.57)	1.09(2.40)	4.8	0.163(0.36)	0.091(0.20)	3.1(0.007)	2.8(0.006)	10.0	1.7	280	935	250

TABLE 31

NUTRIENT REQUIREMENTS OF SHEEP IN PERCENTAGE OR AMOUNT PER KG OF TOTAL RATION (BASED ON AIR-DRY FEED CONTAINING 90 PERCENT DRY MATTER)

| | | DAILY FEED | | | | | | | | | | | | |
| | | PER CENTAGE OF OR AMOUNT PER KG OF AIR-DRY FEED | | | | | | | | | | | | |
BODY WEIGHT kg (lb)	DAILY GAIN OR LOSS gm (lb)	PER ANIMAL kg (lb)	% LIVE-WEIGHT	TDN %	DE[a] Mcal	PROTEIN %	Dp[b] %	Ca %	P %	SALT %	CAROTENE mg	VIT. A mcg[c]	VIT. A IU	VIT. D IU
EWES														
Nonlactating and First 15 Weeks of Gestation														
45(99)	32(0.07)	1.2(2.6)	2.6	50	2.2	8.0	4.4	0.27	0.21	0.8	0.7	108	360	96
54(119)	32(0.07)	1.4(3.1)	2.5	50	2.2	8.0	4.4	0.24	0.19	0.7	0.7	110	367	100
64(141)	32(0.07)	1.5(3.3)	2.4	50	2.2	8.0	4.4	0.22	0.17	0.7	0.7	116	388	103
73(161)	32(0.07)	1.7(3.7)	2.4	50	2.2	8.0	4.4	0.20	0.16	0.7	0.7	117	391	105
Last 6 Weeks of Gestation														
45(99)	168(0.37)	1.7(3.7)	3.8	52	2.3	8.4	4.6	0.24	0.18	0.6	1.5	183	610	66
54(119)	168(0.37)	1.9(4.2)	3.5	52	2.3	8.2	4.5	0.23	0.17	0.6	1.6	194	648	71
64(141)	168(0.37)	2.1(4.6)	3.3	52	2.3	8.0	4.4	0.22	0.16	0.6	1.7	206	687	76
73(161)	168(0.37)	2.2(4.9)	3.0	52	2.3	7.8	4.3	0.22	0.16	0.6	1.8	228	758	83
First 8 to 10 Weeks of Lactation														
45(99)	-36(-0.08)	2.1(4.6)	4.6	59	2.6	8.7	4.8	0.30	0.22	0.5	1.3	151	504	54
54(119)	-36(-0.08)	2.3(5.1)	4.2	58	2.6	8.4	4.6	0.28	0.21	0.5	1.4	163	544	60
64(141)	-36(-0.08)	2.5(5.5)	3.9	56	2.5	8.0	4.4	0.27	0.20	0.5	1.5	172	574	64
73(161)	-36(-0.08)	2.6(5.7)	3.6	55	2.4	8.0	4.4	0.27	0.20	0.5	1.6	192	638	70
Last 12 to 14 Weeks of Lactation														
45(99)	32(0.07)	1.7(3.7)	3.8	52	2.3	8.4	4.6	0.26	0.20	0.6	1.5	183	610	66
54(119)	32(0.07)	1.9(4.2)	3.5	52	2.3	8.2	4.5	0.25	0.19	0.6	1.6	194	648	71
64(141)	32(0.07)	2.1(4.6)	3.3	52	2.3	8.0	4.4	0.24	0.18	0.6	1.7	206	687	76
73(161)	32(0.07)	2.2(4.9)	3.0	52	2.3	7.8	4.3	0.24	0.18	0.6	1.9	228	758	83

TABLE 31 (Continued)

BODY WEIGHT kg (lb)	DAILY GAIN OR LOSS gm (lb)	PER ANIMAL kg (lb)	% LIVE-WEIGHT	TDN %	DE^a Mcal	PROTEIN %	DP^b %	Ca %	P %	SALT %	CAROTENE mg	VIT. A mcg^c	VIT. A IU	VIT. D IU
Replacement Lambs and Yearlings														
27(59)	136(0.30)	1.2(2.6)	4.5	55	2.4	11.0	6.0	0.21	0.19	0.6	0.6	85	283	50
36(79)	91(0.20)	1.4(3.1)	4.0	50	2.2	8.7	4.8	0.20	0.18	0.6	0.7	97	323	62
45(99)	64(0.14)	1.5(3.3)	3.4	50	2.2	7.6	4.2	0.20	0.18	0.6	0.8	111	370	74
54(119)	32(0.07)	1.5(3.3)	2.8	50	2.2	7.0	3.9	0.20	0.18	0.7	1.0	135	450	88
RAMS														
Lambs and Yearlings														
36(79)	181(0.40)	1.4(3.1)	4.0	62	2.7	10.0	5.5	0.20	0.18	0.6	0.7	97	323	62
45(99)	136(0.30)	1.7(3.7)	3.7	57	2.5	8.6	4.7	0.18	0.16	0.6	0.8	102	340	68
54(119)	91(0.20)	1.9(4.2)	3.5	50	2.2	7.6	4.2	0.17	0.15	0.6	0.8	109	364	71
64(141)	45(0.10)	2.1(4.6)	3.3	50	2.2	6.9	3.8	0.16	0.14	0.5	0.9	117	391	76
73(161)	45(0.10)	2.2(4.9)	3.0	50	2.2	6.6	3.6	0.15	0.14	0.5	0.9	127	422	83
LAMBS														
Fattening														
27(59)	159(0.35)	1.2(2.6)	4.5	55	2.5	12.0	6.6	0.23	0.21	0.6	0.4	61	204	56
32(71)	181(0.40)	1.4(3.1)	4.4	58	2.6	11.0	6.1	0.21	0.18	0.6	0.4	64	213	57
36(79)	204(0.45)	1.5(3.3)	4.3	62	2.8	10.7	5.9	0.19	0.18	0.6	0.4	68	226	59
41(90)	204(0.45)	1.7(3.7)	4.2	62	2.7	9.5	5.3	0.18	0.16	0.6	0.4	70	230	61
45(99)	181(0.40)	1.8(4.0)	3.9	62	2.7	9.4	5.2	0.18	0.16	0.6	0.4	72	240	64

[a] 1 kg TDN = 4.4 Mcal DE (digestible energy).

[b] DP = digestible protein.

[c] Vitamin A alcohol, 0.3 mcg is equivalent to 1 IU of vitamin A activity. If vitamin A acetate is used, the vitamin A alcohol should be multiplied by 1.15 to obtain equivalent vitamin A activity. The comparable multiplier for vitamin A palmitate is 1.83.

TABLE 32
NUTRIENT REQUIREMENTS OF GROWING SWINE FED *AD LIBITUM:* PERCENTAGE OR AMOUNT PER KILOGRAM OF AIR-DRY DIET

	5-10	10-20	20-35	35-60	60-100
LIVEWEIGHT (kg):	5-10	10-20	20-35	35-60	60-100
(lb):	11-22	22-44	44-77	77-132	132-220
DAILY GAIN (kg):	0.30	0.50	0.60	0.75	0.90
(lb):	0.66	1.10	1.32	1.65	1.98
NUTRIENTS			*REQUIREMENTS*		
Energy and Protein					
TDN (%)	79.4	79.4	74.8	74.8	74.8
Digestible energy[a] (kcal)	3,500	3,500	3,300	3,300	3,300
Metabolizable energy[a] (kcal)	3,360	3,360	3,170	3,170	3,170
Crude protein[b] (%)	22	18	16	14	13
Inorganic Nutrients (%)					
Calcium	0.80	0.65	0.65	0.50	0.50
Phosphorus	0.60	0.50	0.50	0.40	0.40
Sodium	—	0.10	0.10	—	—
Chlorine	—	0.13	0.13	—	—
Vitamins					
Beta-carotene (mg)	4.4	3.5	2.6	2.6	2.6
Vitamin A (IU)	2,200	1,750	1,300	1,300	1,300
Vitamin D (IU)	220	200	200	125	125
Vitamin E (mg)	11	11	11	11	11
Thiamine (mg)	1.3	1.1	1.1	1.1	1.1
Riboflavin (mg)	3.0	3.0	2.6	2.2	2.2
Niacin[c] (mg)	22.0	18.0	14.0	10.0	10.0
Pantothenic acid (mg)	13.0	11.0	11.0	11.0	11.0
Vitamin B_6 (mg)	1.5	1.5	1.1	—	—
Choline (mg)	1,100	900	—	—	—
Vitamin B_{12} (μg)	22	15	11	11	11

TABLE 32 (Continued)

NUTRIENTS   REQUIREMENTS

Amino Acids (%)

Arginine	0.28	0.23	0.20	0.18	0.16
Histidine	0.25	0.20	0.18	0.16	0.15
Isoleucine	0.69	0.56	0.50	0.44	0.41
Leucine	0.83	0.68	0.60	0.52	0.48
Lysine	0.96	0.79	0.70	0.61	0.57
Methionine + cystine[d]	0.69	0.56	0.50	0.44	0.41
Phenylalanine + tyrosine[e]	0.69	0.56	0.50	0.44	0.41
Threonine	0.62	0.51	0.45	0.39	0.37
Tryptophan	0.18	0.15	0.13	0.11	0.11
Valine	0.69	0.56	0.50	0.44	0.41

[a]These suggested energy levels are derived from corn-based diets. When barley or medium- or low-energy grains are fed, these energy levels will not be met. Formulations based on barley or similar grains are satisfactory for pigs weighing 20-100 kg, but feed conversion will normally be reduced with the lower-energy diets.

[b]Approximate protein levels required to meet the essential amino acid needs. If cereal grains other than corn are used, an increase of 1 or 2 percent of protein may be required.

[c]It is assumed that all the niacin in the cereal grains and their by-products is in a bound form and thus is largely unavailable.

[d]Methionine can fulfill the total requirement; cystine can meet at least 50 percent of the total requirement.

[e]Phenylalanine can fulill the total requirement; tyrosine can fulfill 30 percent of the total requirement.

TABLE 33
DAILY NUTRIENT REQUIREMENTS OF GROWING SWINE

LIVEWEIGHT (kg):	7.5	15	27.5	47.5	80
(lb):	17	33	61	105	176
FEED INTAKE (AIR-DRY) (kg):	0.600	1.250	1.700	2.500	3.500
(lb):	1.32	2.76	3.75	5.51	7.72
NUTRIENTS		*REQUIREMENTS*			
Dry Matter (kg)	0.54(1.19 lb)	1.12(2.47 lb)	1.53(3.37 lb)	2.25(4.96 lb)	3.15(6.95 lb)
Energy and Protein[a,b]					
TDN (kg)	0.48(1.06 lb)	0.99(2.18 lb)	1.27(2.80 lb)	1.87(4.12 lb)	2.62(5.78 lb)
Digestible energy (kcal)	2,100	4,370	5,610	8,250	11,550
Metabolizable energy (kcal)	2,020	4,200	5,390	7,920	11,090
Crude protein (kg)	0.132(0.291 lb)	0.225(0.496 lb)	0.272(0.600 lb)	0.350(0.772 lb)	0.455(1.00 lb)
Inorganic Nutrients					
Calcium (kg)	0.0048(0.0105 lb)	0.0081(0.0179 lb)	0.0110(0.0243 lb)	0.0125(0.0276 lb)	0.0175(0.0386 lb)
Phosphorus (kg)	0.0036(0.0079 lb)	0.0063(0.0139 lb)	0.0085(0.0189 lb)	0.0100(0.0220 lb)	0.0140(0.0309 lb)
Sodium (kg)	–	0.0013	0.0017	–	–
Chlorine (kg)	–	0.0016	0.0022	–	–
Vitamins					
Beta-carotene[c] (mg)	2.6	4.4	4.4	6.5	9.1
Vitamin A (IU)	1,300	2,200	2,200	3,250	4,550
Vitamin D (IU)	132	250	340	312	437
Vitamin E (mg)	6.6	13.8	18.7	27.5	38.5
Thiamine (mg)	0.8	1.4	1.9	2.8	3.9
Riboflavin (mg)	1.8	3.8	4.4	5.5	7.7
Niacin[d] (mg)	13.2	22.5	23.8	25.0	35.0
Pantothenic acid (mg)	7.8	13.8	18.7	27.5	38.5
Vitamin B_6 (mg)	0.9	1.9	1.9	–	–
Choline (mg)	660	1,125	–	–	–
Vitamin B_{12} (μg)	13.2	18.8	18.7	27.5	38.5

NUTRIENTS

Amino Acids (g)		REQUIREMENTS		
Arginine	1.6	2.8	3.4	4.4	5.7
Histidine	1.5	2.5	3.1	3.9	5.1
Isoleucine	4.1	7.0	8.5	10.9	14.2
Leucine	5.0	8.4	10.2	13.1	17.1
Lysine	5.8	9.8	11.9	15.3	19.9
Methionine + cystine[e]	4.1	7.0	8.5	10.9	14.2
Phenylalanine + tyrosine[f]	4.1	7.0	8.5	10.9	14.2
Threonine	3.7	6.3	7.6	9.8	12.8
Tryptophan	1.1	1.8	2.2	2.8	3.7
Valine	4.1	7.0	8.5	10.9	14.2

[a]These suggested energy levels are derived from corn-based diets. When barley or medium- or low-energy grains are fed, these energy levels will not be met. Formulations based on barley or similar grains are satisfactory for pigs weighing 20-100 kg, but feed conversion will normally be reduced with the lower-energy diets.

[b]Approximate protein levels required to meet the essential amino acid needs. If cereal grains other than corn are used, an increase of 1 or 2 percent of protein may be required.

[c]Carotene and vitamin A values are based on 1 mg of beta-carotene equaling 500 IU of biologically active vitamin A. Vitamin A requirements can be met by carotene or vitamin A or both.

[d]It is assumed that all the niacin in the cereal grains and their by-products is in a bound form and thus is largely unavailable.

[e]Methionine can fulfill the total requirement; cystine can meet at least 50 percent of the total requirement.

[f]Phenylalanine can fulfill the total requirement; tyrosine can fulfill 30 percent of the requirement.

TABLE 34
NUTRIENT REQUIREMENTS OF BREEDING SWINE: PERCENTAGE OR AMOUNT PER KILOGRAM OF AIR-DRY DIET

NUTRIENTS	BRED GILTS AND SOWS	LACTATING GILTS AND SOWS	YOUNG AND ADULT BOARS
LIVEWEIGHT (kg):	110-250	140-250	110-250
(lb):	243-551	309-551	243-551
		REQUIREMENTS	
Energy and Protein			
TDN (%)	74.8	74.8	74.8
Digestible energy (kcal)	3,300	3,300	3,300
Metabolizable energy (kcal)	3,170	3,170	3,170
Crude protein (%)	14	15	14
Inorganic Nutrients (%)			
Calcium	0.75	0.75	0.75
Phosphorus	0.50	0.50	0.50
NaCl (salt)	0.5	0.5	0.5
Vitamins			
Beta-carotene (mg)	8.2	6.6	8.2
Vitamin A (IU)	4,100	3,300	4,100
Vitamin D (IU)	275	220	275
Vitamin E (mg)	11.0	11.0	11.0
Thiamine (mg)	1.5	1.0	1.5
Riboflavin (mg)	4.0	3.5	4.0
Niacin (mg)	22.0	17.5	22.0
Pantothenic acid (mg)	16.5	13.0	16.5
Vitamin B_{12} (μg)	14.0	11.0	14.0
Amino Acids (%)			
Arginine	—	0.34[b]	[c]
Histidine	0.20[a]	0.26[b]	[c]
Isoleucine	0.37	0.67[b]	[c]
Leucine	0.66[a]	0.99[b]	[c]
Lysine	0.42	0.60[b]	[c]
Methionine + cystine	0.28	0.36[b]	[c]
Phenylalanine + tyrosine	0.52[a]	1.00[b]	[c]
Threonine	0.34	0.51[b]	[c]
Tryptophan	0.07	0.13[b]	[c]
Valine	0.46	0.68[b]	[c]

[a]This level is adequate; the minimum requirement has not been established.

[b]All suggested requirements for lactation are based on the requirement for maintenance + amino acids produced in milk by sows fed 5-5.5 kg of feed per day from which amino acids are 80 percent available.

[c]No data available; it is suggested that the requirement will not exceed that of bred gilts and sows.

Table 35

DAILY NUTRIENT REQUIREMENTS OF BREEDING SWINE

	BRED GILTS	BRED SOWS	LACTATING GILTS	LACTATING SOWS	YOUNG BOARS	ADULT BOARS
Liveweight (kg)[a]:	110-160	160-250	140-200	200-250	110-180	180-250
(lb):	243-353	353-551	309-441	441-551	243-400	400-551
Feed Intake (air-dry) (kg):	2.000	2.000	5.000	5.500	2.500	2.000
(lb):	4.4	4.4	11.0	12.1	5.5	4.4
Nutrients	*Requirements*					
Dry Matter (kg)	1.8(4.0 lb)	1.8(4.0 lb)	4.5(9.9 lb)	5.0(11.0 lb)	2.3(5.1 lb)	1.8(4.0 lb)
Energy and Protein						
TDN (kg)	1.5(3.3 lb)	1.5(3.3 lb)	3.7(8.2 lb)	4.1(9.0 lb)	1.9(4.2 lb)	1.5(3.3 lb)
Digestible energy (kcal)	6,600	6,600	16,500	18,150	8,250	6,600
Metabolizable energy (kcal)	6,340	6,340	15,840	17,420	7,920	6,340
Crude protein (kg)	0.280(0.62 lb)	0.280(0.62 lb)	0.750(1.65 lb)	0.825(1.82 lb)	0.350(0.77 lb)	0.280(0.62 lb)
Inorganic Nutrients						
Calcium (kg)	0.0150(0.033 lb)	0.0150(0.033 lb)	0.0375(0.083 lb)	0.0412(0.091 lb)	0.0188(0.041 lb)	0.0150(0.033 lb)
Phosphorus (kg)	0.0100(0.022 lb)	0.0100(0.022 lb)	0.0250(0.055 lb)	0.0275(0.061 lb)	0.0125(0.028 lb)	0.0100(0.022 lb)
NaCl (salt) (kg)	0.0100(0.022 lb)	0.0100(0.022 lb)	0.0250(0.055 lb)	0.0275(0.061 lb)	0.0125(0.028 lb)	0.0100(0.022 lb)
Vitamins						
Beta-carotene (mg)	16.4	16.4	33.0	36.3	20.5	16.4
Vitamin A (IU)	8,200	8,200	16,500	18,150	10,250	8,200
Vitamin D (IU)	550	550	1,100	1,210	690	550
Vitamin E (mg)	22.0	22.0	55.0	60.5	27.5	22.0
Thiamine (mg)	3.0	3.0	5.0	5.5	3.8	3.0
Riboflavin (mg)	8.0	8.0	17.5	19.3	10.0	8.0
Niacin (mg)	44.0	44.0	87.5	96.3	55.0	44.0
Pantothenic acid (mg)	33.0	33.0	65.0	71.5	41.3	33.0
Vitamin B$_{12}$ (µg)	28.0	28.0	55.0	60.5	35.0	28.0
Amino Acids (g)						
Arginine			17.0	18.7	*b*	*c*
Histidine	4.0	4.0	13.0	14.3	*b*	*c*
Isoleucine	7.4	7.4	33.5	36.9	*b*	*c*
Leucine	13.2	13.2	46.4	51.0	*b*	*c*
Lysine	8.4	8.4	30.0	33.0	*b*	*c*
Methionine + cystine	5.6	5.6	18.0	19.8	*b*	*c*
Phenylalanine + tyrosine	10.4	10.4	46.9	51.6	*b*	*c*
Threonine	6.8	6.8	25.5	28.1	*b*	*c*
Tryptophan	1.4	1.4	6.5	7.2	*b*	*c*
Valine	9.2	9.2	34.0	37.4	*b*	*c*

[a]Expected daily gain for bred gilts is 0.35-0.45 kg; for bred sows, 0.15-0.30 kg; and for young boars, 0.25-0.45 kg.

[b]Data unavailable; intakes 25 percent greater than those of bred gilts are suggested as adequate.

[c]Data unavailable; intakes equal to those of bred sows are suggested as adequate.

Table 35a
MICROMINERAL REQUIREMENTS OF SWINE

MINERAL ELEMENT	REQUIREMENT (*mg/kg diet*)
Copper	6[a]
Iodine	0.2
Iron	80[a]
Manganese	20
Selenium	0.1
Zinc	50[b]

[a]Baby pig requirement.

[b]Higher levels may be needed if excessive calcium is fed.

57 Tables on Feed Composition

In the following table is provided information on the composition of 135 different feedstuffs. For the most part, these are the more commonly used feeds, although a few less commonly used materials have been included for instructional purposes. Information such as is normally used in balancing rations for the different classes of livestock has been included in this table.

In subsequent tables may be found information on the mineral (other than calcium and phosphorus, which have been included in the Table 37) content of these feeds and on the amino acid and vitamin content of the more commonly used swine feeds.

For the most part the information in these tables has come directly from publications of the National Research Council of the National

Academy of Sciences and is presented here with the Academy's permission. The specific NRC publications from which composition information has been obtained include the following:

- *ATLAS OF NUTRITIONAL DATA ON UNITED STATES AND CANADIAN FEEDS*, Publication ISBN 0-309-01919-2, Committee on Animal Nutrition, National Academy of Sciences — National Research Council, Washington, D.C. 1972.

- *UNITED STATES-CANADIAN TABLES OF FEED COMPOSITION*, Second Revised Edition, Publication 1684, Committee on Animal Nutrition, National Academy of Sciences — National Research Council, Washington, D.C. 1969.

- *NUTRIENT REQUIREMENTS OF BEEF CATTLE*, Fourth Revised Edition, Publication ISBN 0-309-01754-8, Committee on Animal Nutrition, National Academy of Sciences — National Research Council, Washington, D.C. 1970.

- *NUTRIENT REQUIREMENTS OF DAIRY CATTLE*, Fourth Revised Edition, Publication ISBN 0-309-01916-8, Committee on Animal Nutrition, National Academy of Sciences — National Research Council, Washington, D.C. 1971.

- *NUTRIENT REQUIREMENTS OF SWINE*, Seventh Revised Edition, Publication ISBN 0-309-02140-5, Committee on Animal Nutrition, National Academy of Sciences — National Research Council, Washington, D.C. 1973.

In listing those feeds on which composition information has been included, the standard NRC nomenclature as outlined in the above listed *Atlas of Nutritional Data on United States and Canadian Feeds* has been followed. According to the NRC nomenclature scheme the name of a feed should embrace information about the feed on the following points insofar as such information is available and applicable:

1. Origin (or parent material).
2. Species, variety, or kind.
3. Part eaten.
4. Process(es) and treatment(s) to which the parent material or the part eaten has been subjected.
5. Stage of maturity (applicable only to forages).

6.   Cutting or crop (applicable only to forages).

7.   Grade, quality designations, and guarantees.

8.   Classification.

The order of listing of the various feeds is alphabetically based on each feed's origin or parent material.

According to the NRC system, feeds are grouped into eight classes.

The particular class for each feed is indicated in the NRC name by a number in parentheses. The numbers and the classes they designate are as follows:

(1)   Dry forages or dry roughages.

(2)   Pasture, range plants, and fed green.

(3)   Silages.

(4)   Energy feeds.

(5)   Protein supplements.

(6)   Minerals.

(7)   Vitamins.

(8)   Additives.

Within the various NRC feed composition tables is to be found a considerable number of abbreviations. Most of these same abbreviations have been used in the feed composition tables of this book and are given below as a guide to the user of these tables:

blm	bloom
by-prod	by-product
Ca	calcium
comm	commercial
cond	condensed
DE	digestible energy
dehy	dehydrated
dig	digestible
dry-mil	dry-milled
dry-rend	dry-rendered
equiv	equivalent
extd	extracted
extn	extraction
F	fluorine
fbr	fiber
fm	foreign material
g	gram
gr	grade

grnd	ground
hydro	hydrolyzed
insol	insoluble
IU	international units
kcal	kilocalories
kg	kilogram
m	maintenance
mech-extd	mechanically extracted
Mcal	megacalories
ME	metabolizable energy
μg	microgram
mg	milligram
mil-rn	mill-run
mn	minimum
mx	maximum
N	nitrogen
P	phosphorus
precip	precipitated
proc	processed, processing
prot	protein
pt	part(s)
res	residue
s-c	sun-cured
shred	shredded
skim	skimmed
sol	solubles
solv-extd	solvent-extracted
w	with
wet-rend	wet-rendered
wo	without
wt	weight

Where species differences in nutritive values prevail, only figures for cattle and hogs are included in the feed composition table of this text. It was considered to be too overly cumbersome to try to include separate figures for all the different livestock species in this table, and certainly based on the total amount of feed consumed, cattle and hogs are by far the more important. When balancing rations for horses and sheep, it is proposed that cattle feed values be used. Such a practice should not introduce any great amount of error in the balancing of rations for these species.

In some instances, instead of numerical values, asterisks (***) or dashes (---) appear in the feed composition table. Where asterisks appear, this simply indicates that no specific composition figure is available, but it is known that there is very little if any of that nutrient in that feed and for all practical purposes the value is 0.0. Where dashes appear, this indicates that composition values simply are not available.

Finally, rather than include the botanical names of the different feed crops as a part of the feed composition table, it was decided that it might be simpler and just as effective to list these separately, and this has been done in Table 36.

TABLE 36
BOTANICAL NAMES OF COMMON FEED CROPS

COMMON NAME OF CROP*	BOTANICAL NAME
Alfalfa	*Medicago sativa*
Bahiagrass hay	*Paspalum notatum*
Barley	*Hordeum vulgare*
Sugar beet	*Beta saccharifera*
Bermudagrass	*Cynodon dactylon*
Coastal bermudagrass	*Cynodon dactylon*
Kentucky bluegrass	*Poa pratensis*
Bromegrass	*Bromus inermis*
Cabbage	*Brassica oleracea capitata*
Reed canarygrass	*Phalaris arundinacea*
Carrot	*Daucus spp*
Citrus	*Citrus spp*
Alsike clover	*Trifolium hybridum*
Crimson clover	*Trifolium incarnutum*
Ladino clover	*Trifolium repens*
Red clover	*Trifolium pratense*
White clover	*Trifolium repens*
Copra	*Cocos nucifera*
Corn	*Zea mays*
Cotton	*Gossypium spp*
Cowpea	*Vigna spp*
Dallisgrass	*Paspalum dilatatum*
Fescue	*Festuca arundinacea*
Flax	*Linum usitatissimum*
Kudzu	*Pueraria spp*
Lespedeza	*Lespedeza spp*
Sericea	*Lespedeza cuneata*
Millet	*Setaria italica*
Oats	*Avena sativa*
Orchardgrass	*Dactylis glomerata*

TABLE 36 (Continued)

COMMON NAME OF CROP*	BOTANICAL NAME
Peanut	*Arachis hypogaea*
Potato	*Solanum tuberosum*
Redtop	*Agrostis alba*
Rice	*Oryza sativa*
Rye	*Secale cereale*
Ryegrass	*Lolium spp*
Safflower	*Carthamus tinctorius*
Sesame	*Sesamum indicum*
Grain sorghum	*Sorghum vulgare*
Hegari	*Sorghum vulgare*
Johnsongrass	*Sorghum halepense*
Kafir	*Sorghum vulgare*
Milo	*Sorghum vulgare*
Sorgo	*Sorghum vulgare saccharatum*
Sudangrass	*Sorghum vulgare sudanense*
Soybean	*Glycine max*
Sugarcane	*Saccharum officinarum*
Sunflower	*Helianthas spp*
Sweet clover	*Melilotus spp*
Sweet potato	*Ipomoea batatas*
Timothy	*Phleum pratense*
Turnips	*Brassica spp*
Wheat	*Triticum spp*

*Crops are listed in the same order as they appear in Table 37.

TABLE 37
COMPOSITION OF FEEDS

SCIENTIFIC NAME FOLLOWED BY COMMON NAME IN PARENTHESES	BASIS	DRY MATTER %	ASH %	CRUDE FIBER %	ETHER EXTRACT %	N-FREE EXTRACT %	CRUDE PROTEIN %	DIG PROT CATTLE %	DIG PROT SWINE %
1. Alfalfa, aerial part, dehy grnd, (1) (Dehydrated alfalfa meal)	As fed	92.2	9.7	22.9	2.9	38.8	17.9	14.0	8.4
	Dry	100.0	10.5	24.8	3.2	42.1	19.4	15.1	9.1
2. Alfalfa, hay, s-c, (1) (Alfalfa hay)	As fed	91.4	9.0	28.0	1.7	37.1	15.5	10.2	7.3
	Dry	100.0	9.9	30.6	1.9	40.6	17.0	11.1	8.0
3. Alfalfa, leaves, dehy grnd, (1) (Dehydrated alfalfa leaf meal)	As fed	92.3	11.2	18.5	3.1	38.6	20.8	15.2	13.5
	Dry	100.0	12.2	20.1	3.4	41.8	22.6	16.5	14.6
4. Alfalfa, stems, dehy grnd, (1) (Dehydrated alfalfa stem meal)	As fed	90.9	7.4	33.9	1.7	35.5	12.5	8.0	---
	Dry	100.0	8.2	37.2	1.8	39.0	13.7	8.8	---
5. Alfalfa, aerial part, fresh, (2) (Fresh alfalfa forage)	As fed	25.9	2.4	5.5	1.1	11.3	5.7	4.4	---
	Dry	100.0	9.1	21.1	4.1	43.8	21.9	17.0	---
6. Animal, fat, hydrolyzed, feed gr, mn 85% fatty acids, mx 6% unsaponifiable matter, mx 1% insoluble matter, (4) (Animal fat, feed grade)	As fed	95.0	***	***	95.0	***	***	***	***
	Dry	100.0	***	***	100.0	***	***	***	***
7. Animal, blood, dehy grnd, (5) (Blood meal)	As fed	89.3	4.4	0.6	1.4	2.7	80.2	---	62.5
	Dry	100.0	4.9	0.7	1.5	3.0	89.8	---	70.0
8. Animal, carcass residue, dry rendered dehy grnd, mn 9% indigestible material, mx 4.4% phosphorus, (5) (Meat scrap or meat meal)	As fed	94.2	24.9	2.5	9.4	2.5	54.9	50.0	47.9
	Dry	100.0	26.4	2.7	10.0	2.7	58.3	53.0	50.8
9. Animal, carcass residue w blood, dry or wet rendered dehy grnd, mn 9% indigestible material, mx 4.4% phosphorus, (5) (Digester tankage)	As fed	93.7	21.2	1.8	9.0	2.4	59.2	53.9	37.8
	Dry	100.0	22.7	1.9	9.6	2.6	63.2	57.6	40.3

TABLE 37 (Continued)

	DE CATTLE Mcal/kg	DE SWINE kcal/kg	ME CATTLE Mcal/kg	ME SWINE kcal/kg	NE M CATTLE Mcal/kg	NE GAIN CATTLE Mcal/kg	NE LACTATING COWS Mcal/kg	TDN CATTLE %	TDN SWINE %	CALCIUM %	PHOSPHORUS %	CAROTENE mg/kg	CAROTENE mg/lb
1.	2.50	1472	2.05	1356	1.21	0.64	1.30	56.7	33.4	1.36	0.28	88.0	39.9
	2.71	1597	2.22	1471	1.31	0.69	1.41	61.5	36.2	1.48	0.31	95.4	43.3
2.	2.17	1419	1.78	1313	1.23	0.45	1.14	49.1	32.2	1.29	0.19	65.2	29.6
	2.37	1553	1.94	1438	1.35	0.49	1.25	53.8	35.2	1.41	0.24	71.4	32.4
3.	2.63	2279	2.16	2079	---	---	---	59.6	51.7	1.64	0.23	149.2	67.7
	2.85	2469	2.34	2252	---	---	---	64.6	56.0	1.78	0.25	161.6	73.3
4.	2.12	---	1.74	---	---	---	---	48.2	---	1.00	0.02	6.6	3.0
	2.34	---	1.92	---	---	---	---	53.0	---	1.10	0.02	7.2	3.3
5.	0.74	---	0.61	---	0.34	0.18	0.36	16.8	---	0.44	0.07	62.6	28.3
	2.87	---	2.35	---	1.32	0.71	1.39	65.1	---	1.68	0.28	241.6	109.6
6.	---	---	---	---	4.34	2.49	---	175.2	---	0.00	0.00	***	***
	---	---	---	---	4.57	2.62	---	184.4	---	0.00	0.00	***	***
7.	2.60	2475	2.13	1927	---	---	---	58.8	56.1	0.30	0.23	***	***
	2.91	2772	2.39	2159	---	---	---	65.9	62.9	0.33	0.26	***	***
8.	2.73	2957	2.24	2490	1.63	1.07	1.79	62.0	66.9	8.49	4.18	***	***
	2.90	3139	2.38	2644	1.73	1.14	1.90	65.8	71.2	9.01	4.44	***	***
9.	2.89	3004	2.37	2145	---	---	---	65.6	68.1	6.43	3.39	***	***
	3.08	3206	2.53	2290	---	---	---	69.9	72.7	6.86	3.62	***	***

417

TABLE 37 (Continued)

SCIENTIFIC NAME FOLLOWED BY COMMON NAME IN PARENTHESES	BASIS	DRY MATTER %	ASH %	CRUDE FIBER %	ETHER EXTRACT %	N-FREE EXTRACT %	CRUDE PROTEIN %	DIG PROT CATTLE %	DIG PROT SWINE %
10. Animal, carcass residue w blood w bone, dry- or wet-rendered dehy grnd, mn 9% indigestible material, mn 4.4% phosphorus, (5) (Tankage with bone)	As fed	94.1	28.0	2.4	12.4	4.2	47.1	---	---
	Dry	100.0	29.8	2.6	13.2	4.4	50.0	---	---
11. Animal, carcass residue w bone, dry-rendered dehy grnd, mn 9% indigestible material, mn 4.4% phosphorus, (5) (Meat and bone scrap or meal)	As fed	93.6	28.6	1.8	11.1	2.7	49.5	45.0	44.0
	Dry	100.0	30.5	1.9	11.9	2.9	52.9	48.1	47.1
12. Animal, bone, steamed dehy grnd, mn 12% phosphorus, (6) (Steamed bone meal)	As fed	95.7	80.4	1.9	1.9	4.4	7.1	---	---
	Dry	100.0	84.0	2.0	2.0	4.6	7.4	---	---
13. Bahiagrass, hay, s-c, mature, (1) (Bahiagrass hay)	As fed	90.8	5.4	30.5	1.5	49.2	4.3	0.9	---
	Dry	100.0	5.9	33.6	1.6	54.2	4.7	1.0	---
14. Bakery, refuse, dehy grnd, mx salt declared above 3.5%, (4) (Dried bakery product)	As fed	91.6	3.5	0.7	13.7	62.8	10.9	6.3	9.6
	Dry	100.0	3.8	0.8	15.0	68.6	11.9	6.9	10.5
15. Barley, straw, (1) (Barley straw)	As fed	86.9	6.0	36.2	1.7	39.5	3.6	0.4	---
	Dry	100.0	6.9	41.6	1.9	45.4	4.1	0.5	---
16. Barley, grain, (4) (Barley grain)	As fed	89.0	3.0	5.3	1.7	67.4	11.6	8.5	8.7
	Dry	100.0	3.4	6.0	1.9	75.7	13.0	9.6	9.8
17. Beet, sugar, molasses, mn 48% invert sugar, mn 79.5° Brix, (4) (Beet molasses)	As fed	77.5	8.9	0.0	0.1	61.9	6.6	3.8	---
	Dry	100.0	11.4	0.0	0.1	79.9	8.5	4.9	---
18. Beet, sugar, pulp, dehy, (4) (Dried beet pulp)	As fed	90.6	4.8	18.2	0.5	58.4	8.7	3.9	3.6
	Dry	100.0	5.3	20.1	0.6	64.5	9.6	4.3	3.9

TABLE 37 (Continued)

	DE CATTLE Mcal/kg	DE SWINE kcal/kg	ME CATTLE Mcal/kg	ME SWINE kcal/kg	NE M CATTLE Mcal/kg	NE GAIN CATTLE Mcal/kg	NE LACTATING COWS Mcal/kg	TDN CATTLE %	TDN SWINE %	CAL-CIUM %	PHOS-PHORUS %	CAROTENE mg/kg	CAROTENE mg/lb
10.	2.76	3198	2.27	2748	--	--	--	62.7	72.6	11.47	5.25	***	***
	2.94	3399	2.41	2920	--	--	--	66.6	77.1	12.19	5.58	***	***
11.	2.67	1974	2.19	1647	1.51	0.96	1.66	60.6	81.6	11.42	5.69	***	***
	2.86	2110	2.34	1760	1.61	1.03	1.77	64.8	87.2	12.20	6.08	***	***
12.	--	--	--	--	--	--	--	--	--	30.92	14.01	***	***
	--	--	--	--	--	--	--	--	--	32.30	14.63	***	***
13.	2.24	--	1.84	--	--	--	--	50.8	--	0.45	0.20	--	--
	2.47	--	2.02	--	--	--	--	55.9	--	0.50	0.22	--	--
14.	3.65	3800	2.99	3557	--	--	--	82.7	86.1	0.06	0.47	4.6	2.1
	3.98	4148	3.27	3883	--	--	--	90.3	94.1	0.06	0.51	5.0	2.3
15.	1.68	--	1.38	--	0.88	0.12	0.84	38.2	--	0.31	0.09	***	***
	1.94	--	1.59	--	1.01	0.14	0.97	43.9	--	0.35	0.10	***	***
16.	3.17	3120	2.60	2914	1.90	1.25	1.90	71.9	70.8	0.07	0.40	***	***
	3.56	3506	2.92	3274	2.13	1.40	2.14	80.8	79.5	0.08	0.45	***	***
17.	2.69	2463	2.20	2322	1.58	1.05	1.47	61.1	55.9	0.12	0.03	***	***
	3.47	3178	2.84	2996	2.04	1.36	1.90	78.8	72.1	0.16	0.03	***	***
18.	2.87	3000	2.36	2820	1.45	0.93	1.60	65.2	68.0	0.68	0.09	***	***
	3.17	3310	2.60	3113	1.60	1.03	1.77	71.9	75.1	0.75	0.10	***	***

TABLE 37 (Continued)

SCIENTIFIC NAME FOLLOWED BY COMMON NAME IN PARENTHESES	BASIS	DRY MATTER %	ASH %	CRUDE FIBER %	ETHER EXTRACT %	N-FREE EXTRACT %	CRUDE PROTEIN %	DIG PROT CATTLE %	SWINE %
19. Beet, sugar, pulp, wet, (4) (Wet beet pulp)	As fed	11.3	0.5	3.4	0.2	5.8	1.3	0.6	---
	Dry	100.0	4.7	30.1	2.1	51.4	11.7	5.3	---
20. Beet, sugar, pulp w molasses, dehy, (4) (Dried beet pulp with molasses)	As fed	92.3	5.7	13.0	0.7	63.8	9.0	6.0	2.3
	Dry	100.0	6.2	14.1	0.7	69.2	9.8	6.5	2.4
21. Bermudagrass, hay, s-c, (1) (Bermudagrass hay)	As fed	91.2	7.5	26.8	1.7	48.1	7.2	3.4	---
	Dry	100.0	8.2	29.4	1.8	52.7	7.9	3.8	---
22. Bermudagrass, coastal, aerial part, dehy grnd, cut 1, (1) (Dehydrated coastal bermudagrass meal)	As fed	87.0	6.1	24.3	3.3	38.2	15.1	10.9	---
	Dry	100.0	7.0	27.9	3.8	43.9	17.4	12.5	---
23. Bermudagrass, coastal, hay, s-c (1) (Coastal bermudagrass hay)	As fed	91.0	4.5	29.6	1.5	46.5	9.0	4.9	---
	Dry	100.0	4.9	32.5	1.7	51.1	9.9	5.4	---
24. Bermudagrass, coastal, aerial part, fresh, (2) (Fresh coastal bermudagrass forage)	As fed	28.8	1.8	8.2	1.1	13.4	4.3	3.1	---
	Dry	100.0	6.3	28.4	3.8	46.6	15.0	10.7	---
25. Bluegrass, Kentucky, hay, s-c, (1) (Kentucky bluegrass hay)	As fed	88.9	5.9	26.7	3.0	44.2	9.1	5.1	---
	Dry	100.0	6.6	30.0	3.4	49.7	10.2	5.8	---
26. Bluegrass, Kentucky, aerial part, fresh, (2) (Fresh Kentucky bluegrass forage)	As fed	29.0	3.1	7.5	1.4	12.4	4.5	3.0	---
	Dry	100.0	10.8	25.9	4.8	42.9	15.6	10.4	---
27. Brome, smooth; hay, s-c, (1) (Bromegrass hay)	As fed	89.7	8.3	28.5	2.1	40.3	10.5	6.4	---
	Dry	100.0	9.2	31.8	2.4	44.9	11.7	7.1	---
28. Brome, smooth, aerial part, fresh, immature, (2) (Fresh bromegrass forage)	As fed	30.0	2.9	7.0	1.2	12.8	6.1	4.6	---
	Dry	100.0	9.6	23.2	4.0	42.8	20.4	15.2	---

420

TABLE 37 (Continued)

	DE CATTLE Mcal/kg	DE SWINE kcal/kg	ME CATTLE Mcal/kg	ME SWINE kcal/kg	NE M CATTLE Mcal/kg	NE GAIN CATTLE Mcal/kg	NE LACTATING COWS Mcal/kg	TDN CATTLE %	TDN SWINE %	CAL-CIUM %	PHOS-PHORUS %	CAROTENE mg/kg	CAROTENE mg/lb
19.	0.32	---	0.26	---	0.17	0.11	0.18	7.2	---	0.10	0.01	***	***
	2.81	---	2.31	---	1.52	0.95	1.63	63.8	---	0.86	0.10	***	***
20.	2.99	3074	2.45	2891	1.87	1.24	1.69	67.7	69.7	0.56	0.08	***	***
	3.24	3331	2.65	3132	2.03	1.34	1.83	73.4	75.6	0.61	0.08	***	***
21.	2.01	---	1.64	---	0.97	0.23	---	44.7	---	0.37	0.19	88.3	40.0
	2.20	---	1.80	---	1.06	0.25	---	49.9	---	0.41	0.21	96.9	43.9
22.	2.44	---	2.00	---	---	---	---	54.6	---	0.29	0.23	179.3	81.3
	2.80	---	2.30	---	---	---	---	62.8	---	0.33	0.26	206.1	93.5
23.	2.15	---	1.76	---	1.04	0.38	1.03	49.1	---	0.31	0.14	74.4	33.7
	2.36	---	1.94	---	1.14	0.42	1.13	53.9	---	0.34	0.16	81.7	37.1
24.	0.85	---	0.70	---	---	---	---	19.4	---	0.14	0.08	95.2	43.2
	2.96	---	2.43	---	---	---	---	67.2	---	0.49	0.27	330.5	149.9
25.	2.47	---	2.02	---	---	---	---	56.0	---	0.40	0.27	300.3	136.2
	2.78	---	2.28	---	---	---	---	63.0	---	0.45	0.30	337.7	153.2
26.	0.82	---	0.67	---	0.46	0.30	0.51	18.6	---	0.15	0.12	58.0	26.3
	2.83	---	2.32	---	1.59	1.02	1.77	64.2	---	0.53	0.43	200.2	90.8
27.	2.20	---	1.80	---	0.97	0.23	1.02	49.8	---	0.32	0.17	36.8	16.9
	2.45	---	2.01	---	1.08	0.26	1.14	55.5	---	0.36	0.19	41.0	18.6
28.	0.94	---	0.77	---	0.46	0.28	0.49	21.4	---	0.17	0.13	174.9	79.3
	3.15	---	2.58	---	1.52	0.95	1.63	71.4	---	0.55	0.45	582.9	264.3

TABLE 37 (Continued)

SCIENTIFIC NAME FOLLOWED BY COMMON NAME IN PARENTHESES	BASIS	DRY MATTER %	ASH %	CRUDE FIBER %	ETHER EXTRACT %	N-FREE EXTRACT %	CRUDE PROTEIN %	DIG PROT CATTLE %	SWINE %
29. Cabbage, aerial part, fresh, (2) (Cabbage heads)	As fed	9.6	1.0	1.0	0.2	5.3	2.0	1.5	---
	Dry	100.0	10.3	10.5	2.6	55.8	20.8	15.5	---
30. Canarygrass, reed, hay, s-c, (1) (Reed canarygrass hay)	As fed	88.8	7.3	30.0	2.8	38.5	10.2	6.1	---
	Dry	100.0	8.2	33.8	3.1	43.4	11.5	6.9	---
31. Carrot, root, fresh, (4) (Carrots)	As fed	12.9	1.2	1.2	0.2	9.0	1.3	0.7	0.9
	Dry	100.0	9.7	9.1	1.4	69.6	10.3	5.1	7.3
32. Cattle, whey, dehy, mn 65 lactose, (4) (Dried whey)	As fed	93.2	8.4	0.2	1.1	68.6	14.9	10.0	13.6
	Dry	100.0	9.0	0.2	1.2	73.6	16.0	10.7	14.6
33. Cattle, milk, dehy, feed gr, mx 8% moisture, mn 26% fat, (5) (Dried whole milk)	As fed	96.2	5.6	0.1	26.7	38.3	25.5	23.7	24.3
	Dry	100.0	5.8	0.1	27.8	39.8	26.5	24.6	25.3
34. Cattle, milk, fresh, (5) (Fresh cow's milk)	As fed	12.6	0.7	0.0	3.6	4.8	3.5	3.3	3.4
	Dry	100.0	5.8	0.0	28.5	37.9	27.8	26.7	26.9
35. Cattle, milk, skimmed, centrifuged, (5) (Fresh skimmed milk)	As fed	9.3	0.7	0.0	0.1	5.3	3.3	3.1	3.2
	Dry	100.0	7.5	0.0	0.8	56.5	35.1	33.7	34.4
36. Cattle, milk, skimmed dehy, mx 8% moisture, (5) (Dried skimmed milk)	As fed	94.3	8.0	0.3	1.0	51.1	34.0	30.6	33.3
	Dry	100.0	8.5	0.3	1.1	54.2	36.0	32.4	35.3
37. Citrus, pulp, fresh, (4) (Fresh citrus pulp)	As fed	18.3	1.4	2.3	0.6	12.8	1.2	0.4	---
	Dry	100.0	7.7	12.6	3.3	69.9	6.6	2.0	---
38. Citrus, pulp w molasses, dehy grnd (4) (Citrus pulp with molasses, dried)	As fed	90.3	6.5	10.3	3.6	63.6	6.2	2.1	---
	Dry	100.0	7.2	11.5	4.0	70.5	6.9	2.3	---
39. Citrus, pulp wo fines, shredded dehy, (4) (Dried citrus pulp)	As fed	90.2	6.3	11.6	3.4	62.5	6.4	2.3	2.6
	Dry	100.0	7.0	12.9	3.8	69.3	7.1	2.5	2.9

TABLE 37 (Continued)

	DE CATTLE Mcal/kg	DE SWINE kcal/kg	ME CATTLE Mcal/kg	ME SWINE kcal/kg	NE M CATTLE Mcal/kg	NE GAIN CATTLE Mcal/kg	NE LACTATING COWS Mcal/kg	TDN CATTLE %	TDN SWINE %	CALCIUM %	PHOSPHORUS %	CAROTENE mg/kg	CAROTENE mg/lb
29.	0.36	---	0.30	---	---	---	---	8.3	---	0.06	0.03	---	---
	3.80	---	3.12	---	---	---	---	86.2	---	0.64	0.35	---	---
30.	2.07	---	1.70	---	1.06	0.46	---	46.9	---	0.37	0.23	6.2	2.8
	2.33	---	1.91	---	1.19	0.52	---	52.8	---	0.41	0.25	7.0	3.2
31.	0.47	469	0.38	440	0.27	0.18	0.27	10.6	10.6	0.04	0.04	114.9	52.1
	3.62	3638	2.97	3417	2.06	1.37	2.11	82.1	82.5	0.37	0.32	890.8	404.0
32.	3.43	3085	2.81	2973	---	---	1.27	77.8	75.0	0.91	0.76	***	***
	3.68	3310	3.02	3190	---	---	1.36	83.5	80.4	0.98	0.81	***	***
33.	4.85	5884	3.98	5336	4.42	1.93	---	110.1	133.5	0.90	0.72	7.2	3.3
	5.05	6119	4.14	5547	4.59	2.01	---	114.5	138.8	0.94	0.74	7.5	3.4
34.	0.71	690	0.58	624	0.58	0.25	0.47	16.1	15.6	0.12	0.09	0.9	0.4
	5.66	5476	4.64	4950	4.59	2.01	3.76	128.3	124.2	0.93	0.75	7.5	3.4
35.	0.37	403	0.31	358	0.22	0.14	0.23	8.5	9.2	0.13	0.10	***	***
	4.01	4330	3.29	3849	2.32	1.50	2.48	91.0	98.2	1.37	1.05	***	***
36.	3.41	3750	2.80	3327	1.95	1.29	---	77.4	85.0	1.27	1.03	***	***
	3.62	3977	2.97	3528	2.07	1.37	---	82.1	90.2	1.35	1.09	***	***
37.	0.67	414	0.55	---	---	---	---	15.2	---	0.40	0.02	---	---
	3.67	2261	3.01	---	---	---	---	83.2	---	2.22	0.15	---	---
38.	3.40	---	2.78	---	---	---	---	77.0	---	1.66	0.11	***	***
	3.76	---	3.08	---	---	---	---	85.3	---	1.84	0.12	***	***
39.	3.37	1988	2.77	1880	1.78	1.19	1.75	76.6	47.4	2.00	0.14	***	***
	3.74	2205	3.07	2084	1.97	1.32	1.94	84.9	52.5	2.22	0.15	***	***

TABLE 37 (Continued)

SCIENTIFIC NAME FOLLOWED BY COMMON NAME IN PARENTHESES	BASIS	DRY MATTER %	ASH %	CRUDE FIBER %	ETHER EXTRACT %	N-FREE EXTRACT %	CRUDE PROTEIN %	DIG PROT CATTLE %	SWINE %
40. Citrus, syrup, mn 45% invert sugar, mn 71° Brix, (4) (Citrus molasses)	As fed	67.7	5.4	0.0	0.2	56.4	5.7	2.5	---
	Dry	100.0	8.0	0.0	0.3	83.4	8.4	3.7	---
41. Clover, alsike, hay, s-c, (1) (Alsike clover hay)	As fed	87.4	7.6	26.3	2.4	38.8	12.4	7.8	---
	Dry	100.0	8.7	30.1	2.7	44.3	14.2	8.9	---
42. Clover, crimson, aerial part, fresh, (2) (Fresh crimson clover forage)	As fed	17.6	1.7	4.9	0.6	7.5	3.0	2.1	---
	Dry	100.0	9.5	27.7	3.3	42.6	17.0	11.9	---
43. Clover, ladino, aerial part, fresh, (2) (Fresh ladino clover forage)	As fed	18.0	1.9	2.5	0.9	8.2	4.5	3.5	---
	Dry	100.0	10.8	14.1	4.8	45.5	24.7	19.5	---
44. Clover, red, hay, s-c, (1) (red clover hay)	As fed	90.0	7.1	29.1	2.3	38.3	13.2	7.9	---
	Dry	100.0	7.9	32.3	2.6	42.6	14.7	8.8	---
45. Clover, red, aerial part, fresh, (2) (Fresh red clover forage)	As fed	22.7	2.0	5.0	1.0	10.5	4.2	2.7	---
	Dry	100.0	8.7	22.3	4.2	46.5	18.3	11.9	---
46. Clover, white, aerial part, fresh, (2) (Fresh white clover forage)	As fed	17.7	2.1	2.8	0.6	7.2	5.0	3.5	---
	Dry	100.0	11.9	15.7	3.3	40.9	28.2	19.8	---
47. Coconut, meat, extn unspecified grnd, (5) (Copra meal)	As fed	88.2	5.9	12.3	7.4	39.9	22.7	18.4	16.6
	Dry	100.0	6.7	14.0	8.4	45.2	25.7	20.8	18.8
48. Corn, aerial part wo ears wo husks, s-c, mature, (1) (Corn stover)	As fed	84.4	6.2	28.2	1.1	43.2	5.7	2.3	---
	Dry	100.0	7.3	33.4	1.3	51.2	6.8	2.7	---
49. Corn, cobs, grnd, (1) (Ground corn cob)	As fed	89.8	1.6	31.1	0.6	53.7	2.8	-.02	0.0
	Dry	100.0	1.8	34.6	0.7	59.8	3.1	-.03	0.0
50. Corn, aerial part, ensiled, mature (3) (Corn fodder silage or corn silage)	As fed	27.8	1.9	7.6	1.2	14.9	2.2	1.4	---
	Dry	100.0	7.0	27.2	4.3	53.5	8.0	4.9	---

TABLE 37 (Continued)

	DE CATTLE Mcal/kg	DE SWINE kcal/kg	ME CATTLE Mcal/kg	ME SWINE kcal/kg	NE M CATTLE Mcal/kg	NE GAIN CATTLE Mcal/kg	NE LACTATING COWS Mcal/kg	TDN CATTLE %	TDN SWINE %	CAL-CIUM %	PHOS-PHORUS %	CAROTENE mg/kg	CAROTENE mg/lb
40.	2.32	2400	1.90	2264	1.33	0.89	1.31	52.5	54.4	1.20	0.12	***	***
	3.42	3546	2.80	3344	1.97	1.32	1.94	77.6	80.4	1.77	0.18	***	***
41.	2.31	- -	1.89	- -	1.13	0.58	1.19	52.4	- -	1.13	0.23	163.4	74.1
	2.64	- -	2.17	- -	1.29	0.66	1.36	59.9	- -	1.29	0.26	187.0	84.8
42.	0.52	- -	0.43	- -	- -	- -	- -	11.8	- -	0.24	0.05	- -	- -
	2.96	- -	2.43	- -	- -	- -	- -	67.1	- -	1.38	0.29	- -	- -
43.	0.60	- -	0.49	- -	- -	- -	- -	13.5	- -	0.23	0.08	57.5	26.1
	3.31	- -	2.71	- -	- -	- -	- -	75.0	- -	1.27	0.42	319.4	144.9
44.	2.32	- -	1.91	- -	1.13	0.56	1.18	52.7	- -	1.30	0.20	33.1	15.0
	2.58	- -	2.12	- -	1.26	0.62	1.31	58.5	- -	1.45	0.23	36.8	16.7
45.	0.65	- -	0.53	- -	0.35	0.22	0.39	14.6	- -	0.41	0.06	41.8	19.0
	2.84	- -	2.33	- -	1.56	0.99	1.70	64.5	- -	1.80	0.26	184.3	83.6
46.	0.52	- -	0.43	- -	- -	- -	- -	11.8	- -	0.25	0.09	26.3	11.9
	2.94	- -	2.41	- -	- -	- -	- -	66.6	- -	1.40	0.51	149.0	67.6
47.	3.25	3232	2.66	2937	1.66	1.10	1.83	73.7	73.4	0.17	0.59	***	***
	3.68	3666	3.02	3329	1.88	1.25	2.07	83.5	83.1	0.20	0.67	***	***
				- -									
48.	2.26	- -	1.86	- -	1.02	0.46	1.11	51.4	- -	0.50	0.08	- -	- -
	2.68	- -	2.20	- -	1.21	0.55	1.31	60.9	- -	0.60	0.10	- -	- -
49.	1.98	317	1.62	303	0.95	0.22	0.81	44.8	7.2	0.11	0.04	0.6	0.3
	2.20	353	1.80	337	1.06	0.25	0.90	49.9	8.0	0.12	0.04	0.7	0.3
50.	0.87	- -	0.71	- -	0.43	0.28	0.47	19.7	- -	0.08	0.06	4.4	2.0
	3.12	- -	2.56	- -	1.56	0.99	1.70	70.7	- -	0.28	0.21	15.7	7.1

425

TABLE 37 (Continued)

SCIENTIFIC NAME FOLLOWED BY COMMON NAME IN PARENTHESES	BASIS	DRY MATTER %	ASH %	CRUDE FIBER %	ETHER EXTRACT %	N-FREE EXTRACT %	CRUDE PROTEIN %	DIG PROT CATTLE %	SWINE %
51. Corn, aerial part wo ears wo husks, ensiled, (3) (Corn stover silage)	As fed	35.1	3.0	11.8	0.8	16.8	2.6	1.1	...
	Dry	100.0	8.6	33.6	2.4	47.9	7.5	3.0	...
52. Corn, ears w husks, ensiled, (3) (Corn ear silage)	As fed	43.4	2.1	5.1	1.6	30.7	3.8	2.1	...
	Dry	100.0	4.9	11.8	3.8	70.7	8.8	4.8	...
53. Corn, bran, wet or dry milled dehy, (4) (Corn bran)	As fed	88.7	1.9	9.6	4.5	64.6	8.0	3.8	...
	Dry	100.0	2.2	10.9	5.1	72.8	9.1	4.3	...
54. Corn, ears, grnd, (4) (Ground ear corn or corn and cob meal)	As fed	85.4	1.5	7.1	3.4	65.4	8.0	3.9	5.7
	Dry	100.0	1.7	8.3	4.0	76.6	9.3	4.6	6.7
55. Corn, ears w husks, grmd, (4) (Ground snapped corn)	As fed	88.5	2.7	10.4	3.0	64.7	7.7	3.6	...
	Dry	100.0	3.0	11.8	3.4	73.1	8.7	4.0	...
56. Corn, grits by-products, mn 5% fat, (4) (Hominy feed)	As fed	89.2	2.7	4.4	6.8	64.6	10.6	7.1	...
	Dry	100.0	3.0	5.0	7.7	72.5	11.9	8.0	...
57. Corn oil (Corn oil)	As fed	100.0	***	***	100.0	***	***	***	***
	Dry	100.0	***	***	100.0	***	***	***	***
58. Corn, starch, dehy grnd, (4) (Corn starch)	As fed	90.4	0.2	0.2	0.2	89.3	0.6	0.0	0.0
	Dry	100.0	0.2	0.2	0.2	98.8	0.7	0.0	0.0
59. Corn, gluten, wet milled dehy, (5) (Corn gluten meal)	As fed	91.0	3.3	4.6	2.3	41.7	39.0	33.5	...
	Dry	100.0	3.6	5.1	2.6	45.9	42.9	36.9	...
60. Corn, dent, yellow grain, (4) (Yellow shelled corn)	As fed	86.0	1.1	2.0	3.8	70.3	8.8	6.5	7.0
	Dry	100.0	1.3	2.3	4.4	81.8	10.2	7.6	8.2
61. Cotton, seed hulls, (1) (Cottonseed hulls)	As fed	90.2	2.6	40.6	1.4	34.6	4.0	-0.6	...
	Dry	100.0	2.9	45.0	1.6	38.4	4.4	-0.7	...
62. Cotton, seeds, grnd, (5) (Ground cottonseed)	As fed	92.7	3.5	16.9	22.9	26.3	23.1	14.6	...
	Dry	100.0	3.8	18.2	24.7	28.4	24.9	15.7	...

TABLE 37 (Continued)

	DE CATTLE Mcal/kg	DE SWINE kcal/kg	ME CATTLE Mcal/kg	ME SWINE kcal/kg	NE_M CATTLE Mcal/kg	NE GAIN CATTLE Mcal/kg	NF LACTATING COWS Mcal/kg	TDN CATTLE %	TDN SWINE %	CALCIUM %	PHOSPHORUS %	CAROTENE mg/kg	CAROTENE mg/lb
51.	0.89	---	0.73	---	0.44	0.21	0.45	20.2	---	0.13	0.07	---	---
	2.54	---	2.08	---	1.24	0.59	1.28	57.5	---	0.38	0.19	---	---
52.	1.38	---	1.13	---	0.69	0.45	0.77	31.0	---	0.03	0.12	3.3	1.5
	3.18	---	2.60	---	1.60	1.03	1.77	72.0	---	0.06	0.27	7.7	3.5
53.	3.01	2739	2.47	2579	---	---	---	68.1	62.1	0.03	0.16	---	---
	3.39	3087	2.78	2907	---	---	---	76.8	70.0	0.04	0.18	---	---
54.	3.20	3043	2.63	2865	1.90	1.19	1.89	72.6	69.0	0.04	0.23	3.5	1.6
	3.75	3563	3.08	3354	2.23	1.39	2.21	85.0	80.8	0.04	0.26	4.0	1.8
55.	2.94	2541	2.41	2394	---	---	---	66.7	57.6	0.06	0.34	4.1	1.9
	3.32	2871	2.73	2705	---	---	---	75.4	65.1	0.07	0.38	4.6	2.1
56.	3.74	3298	3.07	3085	2.19	1.38	2.28	84.9	74.8	0.05	0.54	9.1	4.1
	4.20	3697	3.44	3460	2.45	1.55	2.56	95.2	83.9	0.06	0.61	10.2	4.6
57.	7.62	7618	---	---	---	---	---	172.8	172.8	0.00	0.00	---	---
	7.62	7618	---	---	---	---	---	172.8	172.8	0.00	0.00	---	---
58.	3.78	3671	3.10	3680	---	---	---	85.8	86.7	***	***	***	***
	4.18	4060	3.43	4070	---	---	---	94.9	95.8	***	***	***	***
59.	3.32	3514	2.72	3069	1.81	1.21	1.98	79.7	0.14	0.46	16.4	7.4	
	3.65	3862	3.00	3373	1.99	1.33	2.18	82.8	87.6	0.16	0.51	18.0	8.2
60.	3.45	3488	2.83	3275	1.96	1.27	2.08	78.0	79.0	0.03	0.27	4.1	1.8
	4.01	4056	3.29	3808	2.28	1.48	2.42	91.0	92.0	0.03	0.31	4.8	2.2
61.	2.06	---	1.69	---	0.93	0.17	0.69	46.5	---	0.13	0.06	***	***
	2.28	---	1.87	---	1.03	0.19	0.77	51.6	---	0.14	0.07	***	***
62.	3.82	4839	3.13	4402	1.86	1.11	2.47	86.6	109.7	0.14	0.68	***	***
	4.12	5220	3.38	4749	2.01	1.20	2.66	93.5	118.4	0.15	0.73	***	***

427

TABLE 37 (Continued)

SCIENTIFIC NAME FOLLOWED BY COMMON NAME IN PARENTHESES	BASIS	DRY MATTER %	ASH %	CRUDE FIBER %	ETHER EXTRACT %	N-FREE EXTRACT %	CRUDE PROTEIN %	DIG PROT CATTLE %	SWINE %
63. Cotton, seeds w some hulls, mech-extd grnd, mn 36% protein, mx 17% fiber, mn 2% fat, (5) (36% cottonseed meal)	As fed	91.8	6.0	14.0	6.4	29.5	35.9	28.4	30.9
	Dry	100.0	6.5	15.3	7.0	32.1	39.1	30.9	33.7
64. Cotton, seeds w some hulls, mech-extd grnd, mn 41% protein, mx 14% fiber, mn 2% fat, (5) (41% cottonseed meal)	As fed	92.7	6.1	10.9	5.6	28.6	41.4	33.5	35.3
	Dry	100.0	6.6	11.8	6.0	30.9	44.7	36.2	38.1
65. Cottonseeds w some hulls, solv-extd grnd, mn 41% protein, mx 14% fiber, mn 0.5% fat, (5) (41% solv-extd cottonseed meal)	As fed	91.1	6.2	11.4	2.1	31.8	41.9	33.9	---
	Dry	100.0	6.8	12.5	2.3	34.9	46.0	37.3	---
66. Cowpea, hay, s-c, (1) (Cowpea hay)	As fed	90.5	10.6	24.3	2.6	37.0	16.0	11.1	---
	Dry	100.0	11.7	26.9	2.9	40.9	17.7	12.2	---
67. Dallisgrass, hay, s-c, (1) (Dallisgrass hay)	As fed	90.7	8.8	29.1	2.2	44.1	6.5	2.9	---
	Dry	100.0	9.7	32.1	2.4	48.6	7.2	3.2	---
68. Dallisgrass, aerial part, fresh, (2) (Fresh Dallisgrass forage)	As fed	25.0	3.2	7.2	0.6	11.0	3.0	2.0	---
	Dry	100.0	12.8	28.8	2.4	44.0	12.0	8.1	---
69. Fescue, meadow, hay, s-c, (1) (Tall fescue hay)	As fed	86.5	8.0	28.1	2.2	40.0	8.2	4.5	---
	Dry	100.0	9.2	32.5	2.6	46.2	9.5	5.2	---
70. Fescue, meadow, aerial part, fresh, (2) (Fresh fescue forage)	As fed	28.5	2.5	8.2	1.2	13.1	3.5	2.4	---
	Dry	100.0	8.6	28.6	4.2	46.1	12.4	8.4	---
71. Fish, menhaden, whole or cuttings, cooked mech-extd dehy grnd, (5) (Fish meal)	As fed	91.4	19.0	0.6	9.8	1.6	60.4	49.3	56.0
	Dry	100.0	20.8	0.6	10.7	1.8	66.1	53.9	61.3

TABLE 37 (Continued)

	DE CATTLE Mcal/kg	DE SWINE kcal/kg	ME CATTLE Mcal/kg	ME SWINE kcal/kg	NE M CATTLE Mcal/kg	NE GAIN CATTLE Mcal/kg	NE LACTATING COWS Mcal/kg	TDN CATTLE %	TDN SWINE %	CAL-CIUM %	PHOS-PHORUS %	CAROTENE mg/kg	CAROTENE mg/lb
63.	2.96	3157	2.42	2768	1.52	0.99	2.26	67.1	71.6	0.23	0.92	***	***
	3.22	3439	2.64	3016	1.66	1.08	2.46	73.1	78.0	0.25	1.00	***	***
64.	3.21	3359	2.37	2927	1.68	1.11	1.83	73.6	62.1	0.19	1.09	***	***
	3.47	3623	2.56	3158	1.81	1.20	1.97	79.4	67.0	0.20	1.18	***	***
65.	3.28	---	2.50	---	1.54	1.01	1.70	68.5	---	0.16	1.06	***	***
	3.60	---	2.74	---	1.69	1.11	1.87	75.2	---	0.18	1.16	***	***
66.	2.35	---	1.93	---	1.23	0.69	1.32	53.3	---	1.37	0.34	---	---
	2.60	---	2.13	---	1.36	0.76	1.46	58.9	---	1.52	0.37	---	---
67.	2.33	---	1.91	---	---	---	---	52.9	---	0.39	0.15	---	---
	2.57	---	2.11	---	---	---	---	58.3	---	0.43	0.17	---	---
68.	0.67	---	0.55	---	---	---	---	15.2	---	0.14	0.05	75.6	34.4
	2.68	---	2.19	---	---	---	---	60.7	---	0.56	0.20	302.2	137.1
69.	2.32	---	1.90	---	1.15	0.62	1.22	52.6	---	0.43	0.31	61.8	28.0
	2.68	---	2.20	---	1.33	0.72	1.41	60.8	---	0.50	0.36	71.4	32.4
70.	0.81	---	0.67	---	---	---	---	18.4	---	0.12	0.08	96.1	43.6
	2.85	---	2.34	---	---	---	---	64.6	---	0.43	0.30	337.5	153.1
71.	2.99	3282	2.45	2843	1.52	0.99	---	67.8	69.7	5.14	2.91	***	***
	3.27	3590	2.68	3110	1.66	1.08	---	74.2	76.2	5.62	3.18	***	***

TABLE 37 (Continued)

SCIENTIFIC NAME FOLLOWED BY COMMON NAME IN PARENTHESES	BASIS	DRY MATTER %	ASH %	CRUDE FIBER %	ETHER EXTRACT %	N-FREE EXTRACT %	CRUDE PROTEIN %	DIG PROT CATTLE %	SWINE %
72. Flax, seeds, mech-extd grnd, mn 10% fiber (5) (Linseed meal)	As fed	91.1	5.6	8.9	5.1	35.6	35.9	31.6	32.3
	Dry	100.0	6.2	9.7	5.6	39.1	39.4	34.6	35.4
73. Garbage, hotel and restaurant, cooked wet grnd, (4) (Restaurant garbage)	As fed	26.3	1.4	0.7	5.9	14.0	4.2	2.9	3.8
	Dry	100.0	5.3	2.7	22.4	53.2	16.3	11.0	14.4
74. Grains, brewers grains, dehy, mx 3% dried spent hops, (5) (Brewers dried grains)	As fed	91.0	3.8	14.7	6.6	40.2	25.8	19.1	20.3
	Dry	100.0	4.2	16.1	7.2	44.2	28.3	21.0	22.4
75. Grains, brewers grains, wet, (5) (Wet brewers grains)	As fed	23.8	1.1	3.8	1.5	11.8	5.5	4.0	4.3
	Dry	100.0	4.8	16.1	6.5	49.6	23.0	16.8	18.2
76. Grains, distillers grains, dehy, (5) (Distillers dried grains)	As fed	92.5	1.6	12.8	7.4	43.4	27.4	21.6	21.6
	Dry	100.0	1.7	13.8	8.0	46.9	29.6	23.4	23.4
77. Kudzu, hay, s-c, (1) (Kudzu hay)	As fed	91.5	6.7	29.9	2.3	37.0	15.5	10.6	---
	Dry	100.0	7.3	32.7	2.6	40.4	16.9	11.6	---
78. Lespedeza, hay, s-c, (1) (Lespedeza hay)	As fed	91.7	6.6	28.0	2.8	41.7	12.7	8.2	---
	Dry	100.0	7.2	30.5	3.0	45.5	13.8	8.9	---
79. Lespedeza, sericea, hay, s-c, (1) (Sericea hay)	As fed	90.8	4.9	28.4	1.8	43.1	12.6	8.2	---
	Dry	100.0	5.4	31.3	2.0	47.4	13.9	9.0	---
80. Limestone, grnd, mn 33% calcium, (6) (Ground limestone)	As fed	99.9	96.8	***	***	***	***	***	***
	Dry	100.0	96.9	***	***	***	***	***	***
81. Millet, foxtail, hay, s-c, (1) (Millet hay)	As fed	85.8	7.0	25.0	2.8	42.5	8.5	4.7	---
	Dry	100.0	8.1	29.2	3.3	49.5	9.9	5.5	---
82. Native plants, intermountain, hay, s-c, (1) (Meadow hay)	As fed	93.5	8.0	30.5	2.3	44.8	7.8	3.9	---
	Dry	100.0	8.6	32.7	2.5	47.9	8.3	4.1	---

TABLE 37 (Continued)

	DE CATTLE Mcal/kg	DE SWINE kcal/kg	ME CATTLE Mcal/kg	ME SWINE kcal/kg	NE M CATTLE Mcal/kg	NE GAIN CATTLE Mcal/kg	NE LACTATING COWS Mcal/kg	TDN CATTLE %	TDN SWINE %	CAL-CIUM %	PHOS-PHORUS %	CAROTENE mg/kg	CAROTENE mg/lb
72.	3.24	3074	2.66	2706	1.73	1.16	1.89	73.6	69.7	0.39	0.87	***	***
	3.56	3374	2.92	2970	1.90	1.27	2.07	80.7	76.5	0.43	0.95	***	***
73.	0.99	1041	0.81	965	...	...	...	22.4	23.6	0.11	0.07	...	...
	3.75	3957	3.07	3668	...	...	...	85.0	89.8	0.42	0.27	...	...
74.	2.66	1779	2.18	1607	1.29	0.76	1.41	60.3	40.4	0.27	0.48	***	***
	2.92	1956	2.40	1766	1.42	0.83	1.55	66.3	44.4	0.30	0.53	***	***
75.	0.70	958	0.58	875	...	...	...	15.9	21.7	0.07	0.12	***	***
	2.96	4026	2.42	3678	...	...	...	67.1	91.3	0.30	0.51	***	***
76.	3.37	3914	2.76	3523	1.84	1.23	2.02	76.4	88.7	0.16	1.06	7.8	3.5
	3.64	4231	2.99	3809	1.99	1.33	2.18	82.6	95.9	0.17	1.15	8.4	3.8
77.	2.25	...	1.84	...	...	...	...	51.0	...	2.15	0.32	40.3	18.3
	2.46	...	2.02	...	...	...	...	55.8	...	2.35	0.35	44.1	20.0
78.	2.34	...	1.91	...	1.12	0.50	0.92	53.0	...	1.09	0.24	10.7	4.9
	2.55	...	2.09	...	1.22	0.55	1.00	57.8	...	1.19	0.26	11.7	5.3
79.	1.79	...	1.47	...	...	...	...	40.5	...	0.94	0.22	35.8	16.2
	1.97	...	1.62	...	...	...	...	44.6	...	1·03	0.25	39.5	17.9
80.	***	***	***	***	***	***	***	***	***	35.85	0.02	***	***
	***	***	***	***	***	***	***	***	***	35.89	0.02	***	***
81.	2.16	...	1.78	...	...	...	...	49.0	...	0.28	0.16	...	...
	2.52	...	2.07	...	...	...	...	57.1	...	0.33	0.18	...	...
82.	2.47	...	2.03	...	0.94	0.10	...	56.1	...	0.55	0.15	40.0	18.1
	2.65	...	2.17	...	1.00	0.11	...	60.0	...	0.58	0.16	42.8	19.4

431

TABLE 37 (Continued)

SCIENTIFIC NAME FOLLOWED BY COMMON NAME IN PARENTHESES	BASIS	DRY MATTER %	ASH %	CRUDE FIBER %	ETHER EXTRACT %	N-FREE EXTRACT %	CRUDE PROTEIN %	DIG PROT CATTLE %	SWINE %
83. Native plants, midwest, hay, s-c, (1) (Prairie hay)	As fed	91.0	7.2	30.7	2.1	45.2	5.8	1.6	---
	Dry	100.0	8.0	33.7	2.3	49.6	6.4	1.8	---
84. Oats, hay, s-c, (1) (Oat hay)	As fed	90.7	7.5	27.9	1.9	45.7	7.7	3.7	---
	Dry	100.0	8.2	30.8	2.1	50.4	8.5	4.1	---
85. Oats, hulls, (1) (Oat hulls)	As fed	92.7	6.0	29.8	1.4	51.8	3.6	1.2	---
	Dry	100.0	6.5	32.2	1.5	55.9	3.8	1.3	---
86. Oats, straw, (1) (Oat straw)	As fed	88.6	6.8	36.3	2.1	39.6	3.8	1.2	---
	Dry	100.0	7.6	41.0	2.3	44.7	4.3	1.4	---
87. Oats, grain, (4) (Oats grain)	As fed	88.9	3.4	10.6	4.5	58.7	11.7	8.7	9.5
	Dry	100.0	3.8	11.9	5.1	66.0	13.2	9.8	10.7
88. Oats, groats, (4) (Oat groats)	As fed	90.7	2.2	2.6	6.0	63.5	16.5	11.5	13.8
	Dry	100.0	2.4	2.8	6.6	70.0	18.2	12.7	15.2
89. Orchardgrass, hay, s-c, (1) (Orchardgrass hay)	As fed	88.6	6.6	30.3	2.9	39.0	9.8	5.8	---
	Dry	100.0	7.4	34.2	3.3	44.0	11.1	6.6	---
90. Orchardgrass, aerial part, fresh, (2) (Fresh orchardgrass forage)	As fed	24.9	---	6.2	1.6	11.2	3.8	2.8	---
	Dry	100.0	---	25.0	6.5	45.0	15.4	11.2	---
91. Oysters, shells, fine grnd, mn 33% calcium, (6) (Oyster shell flour)	As fed	99.6	90.2	***	***	***	1.0	***	***
	Dry	100.0	90.6	***	***	***	1.0	***	***
92. Peanut, hay, s-c, (1) (Peanut hay)	As fed	91.2	9.7	23.7	5.1	42.1	10.6	6.4	---
	Dry	100.0	10.6	26.0	5.6	46.2	11.6	7.0	---
93. Peanut, hulls, (1) (Peanut hulls)	As fed	91.5	4.4	59.8	1.2	19.4	6.6	3.0	---
	Dry	100.0	4.8	65.4	1.3	21.2	7.3	3.2	---
94. Peanut, kernels, solv-extd grnd, mx 7% fiber, (5) (Peanut oil meal)	As fed	91.5	4.5	13.1	1.2	25.3	47.4	42.7	44.6
	Dry	100.0	4.9	14.3	1.3	27.7	51.8	46.6	48.7

TABLE 37 (Continued)

	DE CATTLE Mcal/kg	DE SWINE kcal/kg	ME CATTLE Mcal/kg	ME SWINE kcal/kg	NE M CATTLE Mcal/kg	NE GAIN CATTLE Mcal/kg	NE LACTATING COWS Mcal/kg	TDN CATTLE %	TDN SWINE %	CAL-CIUM %	PHOS-PHORUS %	CAROTENE mg/kg	CAROTENE mg/lb
83.	2.12	---	1.74	---	0.97	0.25	---	48.1	---	0.32	0.12	9.5	4.3
	2.33	---	1.91	---	1.07	0.28	---	52.9	---	0.35	0.14	10.4	4.7
84.	2.40	---	1.97	---	1.19	0.63	1.26	54.4	---	0.22	0.20	47.5	21.5
	2.64	---	2.17	---	1.31	0.70	1.39	60.0	---	0.24	0.22	52.4	23.8
85.	1.42	---	1.17	---	---	---	---	32.2	---	0.15	0.10	***	***
	1.53	---	1.26	---	---	---	---	34.8	---	0.16	0.11	***	***
86.	2.05	---	1.68	---	0.98	0.31	0.95	46.5	---	0.24	0.09	***	***
	2.32	---	1.90	---	1.11	0.35	1.07	52.5	---	0.27	0.10	***	***
87.	2.92	2731	2.40	2549	1.54	1.01	1.69	66.3	61.9	0.09	0.33	***	***
	3.29	3072	2.70	2868	1.73	1.14	1.90	74.5	69.7	0.10	0.37	***	***
88.	3.99	3690	3.27	3407	2.14	1.38	2.25	90.5	83.7	0.08	0.45	***	***
	4.40	4068	3.61	3756	2.36	1.52	2.48	99.8	92.3	0.08	0.49	***	***
89.	2.33	---	1.91	---	1.08	0.49	1.11	52.9	---	0.28	0.23	17.2	7.8
	2.63	---	2.16	---	1.22	0.55	1.25	59.7	---	0.31	0.26	19.4	8.9
90.	0.78	---	0.64	---	0.35	0.20	0.40	17.6	---	0.13	0.12	79.4	36.0
	3.12	---	2.56	---	1.41	0.82	1.59	70.7	---	0.54	0.50	319.0	44.7
91.	***	***	***	***	***	***	***	***	***	37.95	0.07	***	***
	***	***	***	***	***	***	***	***	***	38.10	0.07	***	***
92.	2.57	---	2.11	---	---	---	---	58.3	---	1.16	0.21	45.8	20.8
	2.82	---	2.31	---	---	---	---	63.9	---	1.28	0.23	50.3	22.8
93.	0.74	---	0.60	---	---	---	---	16.7	---	0.25	0.06	0.8	0.4
	0.81	---	0.66	---	---	---	---	18.3	---	0.27	0.07	0.9	0.4
94.	3.10	2845	2.54	2433	1.61	1.06	1.78	70.3	64.5	0.20	0.65	***	***
	3.39	3109	2.78	2659	1.76	1.16	1.94	76.8	70.5	0.22	0.71	***	***

433

TABLE 37 (Continued)

SCIENTIFIC NAME FOLLOWED BY COMMON NAME IN PARENTHESES	BASIS	DRY MATTER %	ASH %	CRUDE FIBER %	ETHER EXTRACT %	N-FREE EXTRACT %	CRUDE PROTEIN %	DIG PROT CATTLE %	SWINE %
95. Peanut, kernels wo shells, (5) (Peanut kernels)	As fed	94.8	2.4	2.8	47.7	13.6	28.4	23.6	---
	Dry	100.0	2.5	3.0	50.3	14.3	29.9	24.9	---
96. Phosphate, defluorinated grnd, mn 1 pt F per 100 pt P, (6) (Defluorinated phosphate)	As fed	99.8	---	***	***	***	***	***	***
	Dry	100.0	---	***	***	***	***	***	***
97. Potato, tubers, dehy grnd, (4) (Potato meal)	As fed	91.4	4.3	2.1	0.3	75.0	9.7	5.3	8.3
	Dry	100.0	4.7	2.3	0.3	82.1	10.6	5.8	9.1
98. Potato, tubers, fresh, (4) (White potatoes)	As fed	23.1	1.1	0.6	0.1	19.1	2.2	1.2	0.7
	Dry	100.0	4.8	2.4	0.3	82.9	9.6	5.3	3.2
99. Poultry, feathers, hydrolyzed dehy grnd, mn 75% of protein digestible, (5) (Feather meal)	As fed	94.6	3.7	0.6	2.9	0.0	87.4	---	61.6
	Dry	100.0	3.9	0.6	3.1	0.0	92.4	---	65.1
100. Poultry, viscera w feet w heads, dry- or wet-rend dehy grnd, mx 16% ash, 4% acid insoluble ash, (5) (Poultry by-product meal)	As fed	93.4	18.7	1.6	13.1	4.6	55.4	---	47.1
	Dry	100.0	20.0	1.7	14.0	4.9	59.3	---	50.4
101. Redtop, hay, s-c, (1) (Redtop hay)	As fed	92.3	6.0	28.5	2.8	47.5	7.4	3.6	---
	Dry	100.0	6.6	30.9	3.1	51.5	8.1	3.9	---
102. Rice, hulls, (1) (Rice hulls)	As fed	92.4	18.4	41.1	0.8	29.2	2.8	-0.3	---
	Dry	100.0	19.9	44.5	0.9	31.6	3.1	-0.3	---
103. Rice, bran w germ, dry milled, mx 13% fiber calcium carbonate declared above 3% mn, (4) (Rice bran)	As fed	90.7	12.1	11.2	14.4	40.1	13.0	8.4	9.8
	Dry	100.0	13.3	12.4	15.8	44.2	14.3	9.3	10.8
104. Rye, aerial part, fresh, (2) (Fresh rye forage)	As fed	20.6	2.1	5.2	0.9	8.2	4.3	3.4	---
	Dry	100.0	10.3	25.2	4.1	39.7	20.7	16.5	---

TABLE 37 (Continued)

	DE CATTLE Mcal/kg	DE SWINE kcal/kg	ME CATTLE Mcal/kg	ME SWINE kcal/kg	NE M CATTLE Mcal/kg	NE GAIN CATTLE Mcal/kg	NE LACTATING COWS Mcal/kg	TDN CATTLE %	TDN SWINE %	CAL-CIUM %	PHOS-PHORUS %	CAROTENE mg/kg	CAROTENE mg/lb
95.	5.78	7254	4.74	6525	--	--	--	131.1	164.5	0.06	0.43	***	***
	6.10	7652	5.00	6883	--	--	--	138.3	173.6	0.06	0.45	***	***
96.	***	***	***	***	***	***	***	***	***	33.00	18.00	***	***
	***	***	***	***	***	***	***	***	***	33.07	18.04	***	***
97.	3.16	3366	2.59	3159	--	--	--	71.7	76.3	0.07	0.20	***	***
	3.46	3682	2.84	3456	--	--	--	78.4	83.5	0.08	0.22	***	***
98.	0.32	858	0.67	807	0.45	0.30	0.46	18.5	19.4	0.01	0.05	***	***
	3.53	3715	2.90	3494	1.95	1.30	2.00	80.2	84.2	0.05	0.24	***	***
99.	2.75	2839	2.26	2196	--	--	--	62.2	64.4	0.42	0.51	***	***
	2.90	3001	2.39	2321	--	--	--	65.7	68.1	0.45	0.54	***	***
100.	3.22	3483	2.64	2926	--	--	--	73.0	79.0	3.00	1.70	***	***
	3.45	3729	2.83	3133	--	--	--	78.2	84.6	3.21	1.82	***	***
101.	2.51	--	2.06	--	--	--	--	57.0	--	0.39	0.20	3.7	1.7
	2.72	--	2.23	--	--	--	--	61.7	--	0.43	0.22	4.0	1.8
102.	0.44	--	0.36	--	--	--	--	10.0	--	0.08	0.07	***	***
	0.48	--	0.39	--	--	--	--	10.8	--	0.09	0.08	***	***
103.	2.60	2907	2.13	2706	1.30	0.77	1.41	58.9	65.9	0.07	1.59	***	***
	2.86	3205	2.35	2984	1.43	0.85	1.55	64.9	72.7	0.08	1.75	***	***
104.	0.65	--	0.53	--	--	--	--	14.8	--	0.11	0.09	70.6	32.0
	3.16	--	2.59	--	--	--	--	71.8	--	0.51	0.41	342.6	155.4

TABLE 37 (Continued)

SCIENTIFIC NAME FOLLOWED BY COMMON NAME IN PARENTHESES	BASIS	DRY MATTER %	ASH %	CRUDE FIBER %	ETHER EXTRACT %	N-FREE EXTRACT %	CRUDE PROTEIN %	DIG PROT CATTLE %	SWINE %
105. Rye, grain, (4) (Rye grain)	As fed	88.2	1.7	2.0	1.5	71.6	11.3	6.8	9.1
	Dry	100.0	2.0	2.3	1.7	81.2	12.8	7.7	10.3
106. Ryegrass, annual, aerial part, fresh, (2) (Fresh ryegrass forage)	As fed	24.1	3.2	5.6	0.9	10.5	3.9	2.8	---
	Dry	100.0	13.2	23.2	3.9	43.5	16.2	11.7	---
107. Safflower, seeds wo hulls, mech-extd grnd, (5) (Safflower meal without hulls)	As fed	88.5	6.4	8.5	6.7	26.4	40.4	34.0	---
	Dry	100.0	7.2	9.6	7.6	29.9	45.7	38.4	---
108. Sesame, seeds, mech-extd grnd, (5) (Sesame oil meal)	As fed	92.2	10.3	5.4	8.6	23.6	44.3	35.4	41.5
	Dry	100.0	11.2	5.8	9.4	25.6	48.0	38.4	45.1
109. Sorghum, grain variety, aerial part, ensiled, mature, (3) (Grain sorghum silage)	As fed	30.7	2.3	7.3	0.9	17.3	3.0	1.3	---
	Dry	100.0	7.3	23.8	2.8	56.4	9.6	4.3	---
110. Sorghum, grain variety, grain, (4) (Grain sorghum grain)	As fed	88.5	2.1	2.3	3.1	72.0	8.9	5.1	6.3
	Dry	100.0	2.4	2.6	3.5	81.5	10.0	5.7	7.1
111. Sorghum, hegari, grain, (4) (Hegari grain)	As fed	89.4	1.6	2.0	2.6	73.7	9.6	5.5	6.8
	Dry	100.0	1.8	2.2	2.9	82.4	10.7	6.1	7.6
112. Sorghum, Johnsongrass, hay, s-c, (1) (Johnsongrass hay)	As fed	90.6	8.1	30.3	1.8	43.8	6.9	3.2	---
	Dry	100.0	8.9	33.1	2.0	48.4	7.6	3.6	---
113. Sorghum, kafir, grain, (4) (Kafir grain)	As fed	90.0	1.5	2.0	2.9	71.8	11.8	6.8	7.8
	Dry	100.0	1.7	2.2	3.2	79.8	13.1	7.6	8.7
114. Sorghum, milo, grain, (4) (Milo grain)	As fed	88.9	1.8	2.3	2.9	71.2	10.9	6.2	7.7
	Dry	100.0	2.0	2.5	3.2	80.1	12.2	7.0	8.7
115. Sorghum, sorgo, aerial part, ensiled, (3) (Sorgo silage)	As fed	28.2	2.2	6.6	0.7	16.9	1.9	0.5	---
	Dry	100.0	7.7	23.3	2.5	59.7	6.8	1.8	---

TABLE 37 (Continued)

	DE CATTLE Mcal/kg	DE SWINE kcal/kg	ME CATTLE Mcal/kg	ME SWINE kcal/kg	NE M CATTLE Mcal/kg	NE GAIN Mcal/kg	NE LACTATING COWS Mcal/kg	TDN CATTLE %	TDN SWINE %	CALCIUM %	PHOSPHORUS %	CAROTENE mg/kg	CAROTENE mg/lb
105.	3.15	3434	2.58	3208	1.80	1.20	---	71.4	77.9	0.07	0.34	***	***
	3.57	3894	2.93	3638	2.04	1.36	---	80.9	88.3	0.08	0.39	***	***
106.	0.65	---	0.54	---	0.32	0.17	0.34	14.8	---	0.15	0.10	96.6	43.8
	2.71	---	2.22	---	1.33	0.72	1.41	61.5	---	0.64	0.41	401.0	181.9
107.	3.07	3354	2.51	2910	1.38	0.88	1.68	69.5	76.0	0.32	0.59	---	---
	3.47	3790	2.84	3288	1.56	0.99	1.90	78.6	85.9	0.36	0.67	---	---
108.	3.20	3557	2.62	3069	1.56	1.02	1.72	72.6	80.6	1.99	1.33	---	---
	3.47	3858	2.85	3329	1.69	1.11	1.87	78.8	87.5	2.16	1.44	---	---
109.	0.76	---	0.62	---	0.37	0.17	0.38	17.6	---	0.08	0.06	---	---
	2.46	---	2.02	---	1.22	0.57	1.24	57.4	---	0.25	0.18	---	---
110.	3.14	3467	2.58	3259	1.73	1.16	1.89	71.3	78.6	0.03	0.29	***	***
	3.55	3918	2.91	3682	1.96	1.31	2.14	80.6	88.9	0.03	0.32	***	***
111.	3.18	3605	2.60	3383	---	---	---	72.0	81.8	0.10	0.27	***	***
	3.55	4033	2.91	3784	---	---	---	80.6	91.5	0.11	0.31	***	***
112.	2.37	---	1.94	---	1.08	0.46	1.10	53.8	---	0.80	0.27	38.6	17.5
	2.62	---	2.15	---	1.19	0.51	1.21	59.3	---	0.89	0.30	42.7	19.4
113.	2.86	3532	2.34	2896	1.42	0.91	---	65.0	80.0	0.04	0.33	***	***
	3.18	3924	2.60	3218	1.60	1.03	---	72.0	89.0	0.04	0.37	***	***
114.	3.14	3400	2.58	3180	1.64	1.09	---	71.3	77.1	0.03	0.28	***	***
	3.54	3825	2.90	3577	1.85	1.23	---	80.2	86.7	0.03	0.31	***	***
115.	0.71	---	0.58	---	0.35	0.17	0.36	16.1	---	0.09	0.06	7.3	3.3
	2.51	---	2.06	---	1.25	0.61	1.28	57.0	---	0.33	0.20	25.9	11.7

TABLE 37 (Continued)

SCIENTIFIC NAME FOLLOWED BY COMMON NAME IN PARENTHESES	BASIS	DRY MATTER %	ASH %	CRUDE FIBER %	ETHER EXTRACT %	N-FREE EXTRACT %	CRUDE PROTEIN %	DIG PROT CATTLE %	DIG PROT SWINE %
116. Sorghum, sudangrass, hay, s-c, (1) (Sudangrass hay)	As fed	89.6	8.6	27.5	1.6	43.1	8.7	3.8	...
	Dry	100.0	9.6	30.7	1.8	48.1	9.7	4.2	...
117. Sorghum, sudangrass, aerial part, fresh, (2) (Fresh sudangrass forage)	As fed	20.8	1.6	5.7	0.7	9.8	2.9	2.1	...
	Dry	100.0	7.8	27.5	3.3	47.2	14.1	9.9	...
118. Soybean, hay, s-c, (1) (Soybean hay)	As fed	88.9	7.1	33.2	2.0	33.6	13.1	8.1	...
	Dry	100.0	8.0	37.4	2.2	37.7	14.7	9.1	...
119. Soybean, hulls, (1) (Soybean hulls)	As fed	91.6	3.8	33.1	2.5	40.9	11.3	7.0	...
	Dry	100.0	4.2	36.1	2.8	44.6	12.4	7.6	...
120. Soybean, seeds, (5) (Soybean seed)	As fed	90.9	4.9	5.3	17.4	25.4	37.9	34.1	31.1
	Dry	100.0	5.4	5.8	19.2	27.9	41.7	37.5	34.2
121. Soybean, seeds, solv-extd grnd, mx 7% fiber, (5) (44% soybean oil meal)	As fed	89.0	5.8	6.0	0.9	30.5	45.8	39.0	41.7
	Dry	100.0	6.5	6.7	1.0	34.3	51.5	43.8	46.9
122. Soybean, seeds wo hulls, solv-extd grnd, mx 3% fiber, (5) (49% soybean oil meal)	As fed	91.3	5.9	2.9	1.1	30.5	50.8	46.6	47.1
	Dry	100.0	6.5	3.2	1.2	33.5	55.6	51.0	51.6
123. Sugarcane, molasses, mn 48% invert sugar, mn 79.5° Brix, (4) (Cane molasses)	As fed	77.0	7.8	0.0	0.0	64.7	4.5	2.6	...
	Dry	100.0	10.1	0.0	0.0	84.0	5.9	3.4	...
124. Sunflower, seeds wo hulls, solv-extd grnd, (5) (Sunflower meal)	As fed	93.0	7.1	10.8	2.9	24.8	46.8	41.7	42.1
	Dry	100.0	8.3	11.6	3.1	26.7	50.3	44.8	45.3
125. Sweetclover, hay, s-c, (1) (Sweetclover hay)	As fed	91.3	8.0	27.4	2.2	38.7	15.0	10.2	...
	Dry	100.0	8.8	30.0	2.4	42.4	16.4	11.2	...

TABLE 37 (Continued)

	DE CATTLE Mcal/kg	DE SWINE kcal/kg	ME CATTLE Mcal/kg	ME SWINE kcal/kg	NE M CATTLE Mcal/kg	NE GAIN Mcal/kg	NE LACTATING COWS Mcal/kg	TDN CATTLE %	TDN SWINE %	CALCIUM %	PHOSPHORUS %	CAROTENE mg/kg	CAROTENE mg/lb
116.	2.34	--	1.92	--	1.13	0.56	1.17	53.1	--	0.36	0.27	4.7	2.1
	2.61	--	2.14	--	1.26	0.63	1.31	59.3	--	0.40	0.30	5.3	2.4
117.	0.63	--	0.52	--	0.28	0.16	0.30	14.3	--	0.10	0.09	38.0	17.2
	3.04	--	2.49	--	1.37	0.77	1.46	68.9	--	0.49	0.44	182.8	82.9
118.	2.02	--	1.66	--	0.99	0.32	0.95	45.8	--	1.08	0.29	31.8	14.4
	2.27	--	1.86	--	1.11	0.36	1.07	51.5	--	1.22	0.33	35.7	16.2
119.	2.59	--	2.13	--	0.90	0.06	1.80	58.8	--	0.54	0.16	***	***
	2.83	--	2.32	--	0.98	0.07	1.97	64.2	--	0.59	-.17	***	***
120.	3.66	4032	3.00	3531	2.19	1.39	2.29	83.1	91.4	0.24	0.58	***	***
	4.03	4436	3.30	3885	2.41	1.53	2.52	91.4	100.6	0.27	0.63	***	***
121.	3.18	3300	2.61	2825	1.72	1.15	1.84	72.0	75.0	0.32	0.67	***	***
	3.57	3708	2.93	3174	1.93	1.29	2.07	81.0	84.0	0.36	0.75	***	***
122.	3.37	4008	2.77	3542	--	--	--	76.3	71.2	0.29	0.65	***	***
	3.69	4390	3.03	3880	--	--	--	83.6	78.0	0.32	0.71	***	***
123.	3.26	2524	2.67	2393	1.75	1.14	1.86	73.9	57.2	0.81	0.08	***	***
	4.23	3277	3.47	3107	2.27	1.48	2.42	96.0	74.3	1.05	0.11	***	***
124.	2.67	3012	2.19	2586	1.21	0.77	1.42	60.6	68.3	0.53	0.50	***	***
	2.88	3239	2.36	2780	1.41	0.83	1.53	65.2	73.5	0.57	0.54	***	***
125.	2.36	--	1.94	--	1.11	0.50	1.14	53.5	--	1.31	0.24	90.2	40.9
	2.59	--	2.12	--	1.22	0.55	1.25	58.7	--	1.44	0.27	98.8	44.8

TABLE 37 (Continued)

SCIENTIFIC NAME FOLLOWED BY COMMON NAME IN PARENTHESES	BASIS	DRY MATTER %	ASH %	CRUDE FIBER %	ETHER EXTRACT %	N-FREE EXTRACT %	CRUDE PROTEIN %	DIG PROT CATTLE %	SWINE %
126. Sweet potato, roots, dehy grnd, (4) (Sweet potato meal)	As fed	90.2	4.1	3.3	0.9	77.0	4.9	0.9	---
	Dry	100.0	4.5	3.7	1.0	85.4	5.4	1.0	---
127. Sweet potato, roots, fresh, (4) (Sweet potatoes)	As fed	30.6	1.1	1.3	0.4	26.2	1.7	0.3	---
	Dry	100.0	3.6	4.2	1.3	85.5	5.4	1.0	---
128. Timothy, hay, s-c, (1) (Timothy hay)	As fed	88.6	4.5	30.2	2.3	45.4	6.3	2.6	---
	Dry	100.0	5.1	34.0	2.5	51.2	7.1	2.9	---
129. Turnips, roots, fresh, (4) (Turnips)	As fed	9.7	0.8	1.1	0.1	6.5	1.1	06.	0.4
	Dry	100.0	8.6	11.1	1.5	67.4	11.4	6.4	3.9
130. Wheat, straw, (1) (Wheat straw)	As fed	87.8	6.3	38.3	1.4	38.6	3.2	0.4	---
	Dry	100.0	7.2	43.6	1.5	44.0	3.7	0.5	---
131. Wheat, bran, dry milled, (4) (Wheat bran)	As fed	88.7	5.9	8.6	4.4	54.0	15.7	12.3	12.0
	Dry	100.0	6.7	9.7	5.0	60.9	17.8	13.8	13.5
132. Wheat, flour by-products, coarse sift, mx 7% fiber, (4) (Wheat shorts)	As fed	87.1	4.0	6.3	4.8	56.2	15.8	11.0	14.9
	Dry	100.0	4.5	7.3	5.5	64.6	18.1	12.6	17.1
133. Wheat, flour by-products, fine sift, mx 4% fiber, (4) (Wheat middlings)	As fed	87.4	2.5	2.5	3.4	63.5	15.5	10.7	13.8
	Dry	100.0	2.8	2.8	3.9	72.7	17.7	12.3	15.8
134. Wheat, grain, (4) (Wheat grain)	As fed	88.9	1.9	2.5	1.9	70.8	11.9	9.3	10.5
	Dry	100.0	2.1	2.8	2.1	79.6	13.4	10.5	11.8
135. Wood, molasses, (4) (Wood molasses)	As fed	62.4	3.1	0.0	0.1	58.5	0.6	-1.8	---
	Dry	100.0	5.0	0.0	0.2	93.8	1.0	-3.0	---

***Very little, if any.
- -Figures not available.

TABLE 37 (Continued)

	DE CATTLE Mcal/kg	DE SWINE kcal/kg	ME CATTLE Mcal/kg	ME SWINE kcal/kg	NE M CATTLE Mcal/kg	NE GAIN CATTLE Mcal/kg	NE LACTATING COWS Mcal/kg	TDN CATTLE %	TDN SWINE %	CAL-CIUM %	PHOS-PHORUS %	CAROTENE mg/kg	CAROTENE mg/lb
126.	3.20	3141	2.62	2981	---	---	---	72.7	71.2	0.15	0.14	---	---
	3.55	3483	2.91	3305	---	---	---	80.6	79.0	0.17	0.16	---	---
127.	1.12	1081	0.92	1026	---	---	---	25.4	24.5	0.03	0.05	54.9	24.9
	3.66	3533	3.00	3353	---	---	---	83.1	80.1	0.10	0.15	179.5	81.4
128.	2.15	---	1.76	---	1.16	0.61	1.13	48.8	---	0.36	0.15	47.3	21.5
	2.43	---	1.99	---	1.31	0.69	1.28	55.1	---	0.40	0.16	53.4	24.2
129.	0.37	324	0.30	304	0.19	0.13	0.21	8.3	7.3	0.03	***	***	***
	3.78	3347	3.10	3136	1.97	1.31	2.18	85.7	75.9	0.56	0.28	***	***
130.	1.89	---	1.55	---	0.90	0.17	0.77	42.8	---	0.14	0.07	1.9	0.9
	2.15	---	1.76	---	1.03	0.19	0.88	48.8	---	0.16	0.08	2.2	1.0
131.	2.76	2511	2.27	2321	1.36	0.85	1.51	62.7	56.9	0.14	1.16	2.6	1.2
	3.12	2832	2.56	2617	1.53	0.96	1.70	70.7	64.2	0.16	1.31	3.0	1.4
132.	3.20	2924	2.63	2700	1.82	1.21	1.96	72.7	66.3	0.09	0.81	***	***
	3.68	3358	3.02	3101	2.08	1.38	2.24	83.5	76.2	0.10	0.93	***	***
133.	3.45	3210	2.83	2966	1.71	1.14	1.86	78.3	72.8	0.08	0.50	***	***
	3.95	3673	3.24	3394	1.96	1.31	2.14	89.6	83.3	0.09	0.57	***	***
134.	3.45	3607	2.83	3376	1.91	1.26	2.06	78.3	77.8	0.08	0.34	***	***
	3.88	4056	3.18	3798	2.15	1.42	2.32	88.0	87.5	0.09	0.39	***	***
135.	2.35	2331	1.93	2233	---	---	---	53.4	52.8	1.45	0.03	***	***
	3.77	3735	3.09	3579	---	---	---	85.6	84.7	2.33	0.05	***	***

Table 38
MINERAL CONTENT OF FEEDS–DRY BASIS
(See Table 37 for Ca and P)

Name of feed*	Chlorine %	Cobalt mg/kg	Copper mg/kg	Iodine mg/kg	Iron %	Magnesium %	Manganese mg/kg	Potassium %	Sodium %	Sulfur %	Zinc mg/kg
Dehydrated alfalfa meal	0.72	0.119	8.0	—	0.023	0.22	21.7	1.46	0.14	—	—
Alfalfa hay	0.31	0.225	20.2	—	-.019	0.34	61.9	2.18	0.17	0.32	33.7
Dehy alfalfa leaf meal	0.34	0.216	—	—	—	—	—	—	—	—	—
Fresh alfalfa forage	—	—	—	—	—	—	—	—	0.07	—	—
Blood meal	0.45	—	9.5	—	0.031	0.30	50.1	2.10	0.16	0.37	—
Meat scrap or meat meal	0.29	—	10.8	—	0.411	0.24	5.8	0.10	0.35	0.36	—
Digester tankage	1.39	0.165	10.3	—	0.047	0.29	10.1	0.58	1.78	0.53	—
Meat and bone meal	1.89	0.195	35.1	—	0.207	0.46	16.3	0.55	1.81	0.76	—
Steamed bone mea[l]	0.80	0.065	1.6	—	0.053	1.20	13.2	1.56	0.78	—	—
Dried bakery product	—	0.000	13.6	—	3.148	0.73	30.8	0.19	21.85	0.30	405.2
Barley straw	0.68	—	0.0	—	0.007	0.35	0.0	0.91	—	0.02	—
Barley grain	0.17	—	6.5	—	0.033	0.13	16.6	1.88	0.14	0.17	—
Beet molasses	1.62	0.492	22.5	—	0.009	0.15	8.9	0.55	0.07	0.17	—
Dried beet pulp	0.04	0.112	13.8	—	0.006	0.29	5.9	6.07	1.49	0.61	—
Dried beet pulp/ molasses	—	—	—	—	0.033	0.30	38.5	0.21	0.19	0.22	0.7
Bermudagrass hay	—	—	—	0.115	—	0.16	—	1.78	—	0.42	—
Coastal bermuda-grass hay	—	—	—	—	—	0.16	—	1.57	—	—	—

442

TABLE 38 (Continued)

Name of feed*	Chlorine %	Cobalt mg/kg	Copper mg/kg	Iodine mg/kg	Iron %	Magnesium %	Manganese mg/kg	Potassium %	Sodium %	Sulfur %	Zinc mg/kg
Kentucky bluegrass hay	0.62	–	9.9	–	0.028	0.21	85.6	1.87	0.11	0.13	–
Fresh Kentucky bluegrass forage	–	–	13.9	–	0.017	0.23	80.3	1.95	0.23	0.66	–
Bromegrass hay	0.54	–	8.6	–	0.012	0.15	58.0	2.07	0.63	0.19	–
Fresh bromegrass forage	–	–	–	–	–	0.32	–	3.16	–	–	–
Cabbage heads	0.53	–	14.1	–	0.008	0.21	30.5	2.81	0.18	1.17	–
Reed canarygrass hay	–	–	–	–	–	0.24	–	3.15	–	–	–
Carrots	0.50	–	11.1	–	0.011	0.17	31.5	2.50	1.00	0.17	–
Dried whey	1.55	0.116	51.4	–	0.014	0.14	4.6	–	–	1.12	–
Dried whole milk	–	–	0.9	–	0.009	–	0.5	1.18	0.40	–	–
Fresh skimmed milk	–	–	11.6	–	0.005	0.11	2.3	1.58	0.32	–	–
Fresh cow's milk	1.56	–	0.3	–	–	–	–	1.11	0.39	–	–
Dried skimmed milk	–	0.117	12.2	–	0.004	0.12	2.3	1.71	0.55	0.34	–
Dried citrus pulp	–	–	6.4	–	0.018	0.18	7.6	0.69	–	–	–
Citrus molasses	0.10	0.159	108.0	–	0.050	0.21	38.5	0.13	0.40	–	137.0
Alsike clover hay	0.78	–	6.0	0.183	0.045	0.32	117.0	2.74	0.46	0.21	–
Fresh crimson clover forage	0.61	–	–	–	–	0.29	245.8	31.0	0.40	0.28	–
Fresh ladino clover forage	–	–	–	–	0.036	0.48	71.7	1.87	0.12	0.12	–
Red clover hay	0.37	–	10.7	–	-.011	0.42	76.9	1.87	0.20	0.16	–
Fresh red clover forage	0.77	0.141	9.0	–	0.033	0.43	123.3	2.46	0.20	0.17	–

443

TABLE 38 (Continued)

Name of feed*	Chlorine %	Cobalt mg/kg	Copper mg/kg	Iodine mg/kg	Iron %	Magnesium %	Manganese mg/kg	Potassium %	Sodium %	Sulfur %	Zinc mg/kg
Fresh white clover forage	0.61	—	—	—	0.034	0.45	307.2	2.13	0.39	0.33	—
Copra meal	—	—	10.4	—	0.075	0.39	82.9	2.09	—	0.37	—
Corn stover	0.31	—	5.1	—	0.022	0.45	135.5	1.64	0.07	0.17	—
Ground corn cob	—	—	—	—	0.07	—	—	0.91	—	—	—
Corn (fodder) silage	0.06	0.060	—	—	—	—	—	—	—	—	—
Corn bran	—	—	—	—	—	0.29	17.9	0.72	—	0.08	—
Ground ear corn	—	0.340	7.7	—	0.008	0.15	13.0	0.50	—	—	—
Ground snapped corn	—	—	—	—	—	0.15	12.1	0.97	—	—	—
Hominy feed	0.008	0.066	14.8	—	0.008	0.26	16.5	0.71	0.14	0.03	—
Corn gluten meal	—	0.078	31.1	—	0.044	0.05	11.6	0.03	0.11	—	—
Yellow shelled corn	0.03	0.10	3.4	—	0.003	0.15	4.1	0.33	0.01	0.12	10.4
Cottonseed hulls	—	—	—	—	—	0.14	—	0.96	—	—	—
Ground cottonseed	—	—	54.0	—	0.015	0.35	13.1	1.20	0.31	0.26	—
36% cottonseed meal	0.02	—	—	—	—	0.53	—	1.53	0.07	0.28	—
41% cottonseed meal	0.06	2.0	19.6	—	0.01	0.53	24.8	1.35	0.05	0.43	84.1
41% solv-extd cottonseed meal	0.04	2.10	19.9	—	0.01	0.52	23.9	1.38	0.07	0.23	66.0
Cowpea hay	0.17	—	—	—	0.091	0.41	—	2.39	0.22	0.35	—
Fresh Dallisgrass forage	—	0.073	—	—	0.016	0.40	—	1.72	0.34	—	—

444

TABLE 38 (Continued)

Name of feed*	Chlorine %	Cobalt mg/kg	Copper mg/kg	Iodine mg/kg	Iron %	Magnesium %	Manganese mg/kg	Potassium %	Sodium %	Sulfur %	Zinc mg/kg
Tall fescue hay	—	0.14	—	—	—	0.50	24.5	1.87	—	—	—
Fish meal	0.34	0.215	12.0	2.193	0.050	0.16	37.3	0.80	0.37	—	164.7
Linseed meal	0.04	0.471	46.2	—	0.019	0.62	35.7	1.34	0.11	0.43	—
Brewers dried grains	0.19	0.067	23.1	—	0.027	0.15	40.7	0.10	0.28	0.33	—
Wet brewers grains	—	—	—	—	—	—	—	0.08	—	—	—
Distillers dried grains	0.05	0.099	48.9	—	0.035	0.15	57.9	1.28	0.04	0.49	210.0
Kudzu hay	—	—	—	—	—	0.80	—	—	—	—	—
Lespedeza hay	0.05	0.201	—	0.159	0.029	0.22	100.8	1.10	0.07	0.19	29.3
Sericea hay	—	—	—	—	—	2.06	269.6	0.12	0.06	0.04	—
Ground limestone	—	—	—	—	—	0.23	138.2	1.94	0.10	0.16	—
Millet hay	0.13	—	—	—	—	0.24	—	1.08	—	—	—
Prairie hay	—	—	—	—	0.009	0.18	91.6	0.94	0.17	—	—
Oat hay	0.52	—	—	—	0.056	—	20.1	0.52	—	—	—
Oat hulls	—	—	3.3	—	0.011	0.20	33.1	2.23	0.45	—	—
Oat straw	0.78	—	11.0	—	0.020	0.18	48.6	0.48	0.08	0.22	—
Oat grain	0.12	0.064	9.3	—	0.008	0.13	31.6	0.40	0.06	0.23	—
Oat groats	0.10	—	7.1	—	0.008	0.17	78.3	3.28	—	0.22	—
Orchardgrass hay	0.42	0.022	16.7	—	0.014	0.30	—	—	—	0.26	18.1
Oyster shell flour	0.01	—	—	—	0.287	—	134.1	0.10	0.21	—	—
Peanut hay	—	0.079	—	—	—	0.04	—	—	—	—	—
Peanut oil meal	—	—	—	—	—	—	—	—	—	—	—
Peanut kernels	0.02	—	—	—	0.002	0.19	—	0.64	0.30	0.26	—

TABLE 38 (Continued)

Name of feed*	Chlorine %	Cobalt mg/kg	Copper mg/kg	Iodine mg/kg	Iron %	Magnesium %	Manganese mg/kg	Potassium %	Sodium %	Sulfur %	Zinc mg/kg
Potato meal	0.39	—	—	—	—	—	3.1	2.16	—	—	—
White potatoes	0.28	—	17.7	—	0.009	0.14	41.6	2.26	0.09	0.09	—
Redtop hay	0.07	0.146	3.9	0.099	0.015	0.22	225.5	1.89	0.07	0.25	17.9
Rice hulls	—	—	—	—	—	—	333.3	0.34	—	—	—
Rice bran	0.08	—	14.3	—	0.021	1.05	461.6	1.92	0.00	0.20	33.1
Fresh rye forage	—	—	—	—	—	0.36	—	—	—	—	—
Rye grain	0.03	—	8.6	—	0.007	0.13	83.3	0.52	0.02	0.17	34.4
Sesame oil meal	0.08	—	—	—	—	—	51.8	—	—	—	—
Grain sorghum grain	0.10	0.304	—	—	—	—	—	—	0.05	0.18	15.4
Johnsongrass hay	—	—	—	—	0.058	0.35	—	1.35	—	—	—
Kafir grain	0.11	0.434	3.3	—	0.001	0.17	8.2	0.38	0.07	0.18	—
Milo grain	0.09	0.062	18.1	—	0.005	0.15	14.9	0.39	0.01	—	—
Sorgo silage	0.06	—	31.3	—	0.020	0.27	67.3	1.12	0.03	—	—
Sudangrass hay	—	0.124	—	—	0.019	0.35	91.5	2.10	0.02	0.06	—
Fresh sudangrass forage	—	—	35.9	—	0.021	0.35	81.3	2.14	—	—	—
Soybean hay	1.5	0.132	9.0	—	0.030	0.79	92.6	0.97	0.12	0.11	24.0
Soybean hulls	—	0.090	—	—	—	—	13.9	—	—	0.26	—
Soybean seed	0.03	—	17.4	—	0.009	0.31	32.8	1.77	0.13	0.24	—
44% soybean oil meal	—	0.100	40.8	—	0.013	0.30	30.9	2.21	0.38	—	—

TABLE 38 (Continued)

Name of feed*	Chlorine %	Cobalt mg/kg	Copper mg/kg	Iodine mg/kg	Iron %	Magnesium %	Manganese mg/kg	Potassium %	Sodium %	Sulfur %	Zinc mg/kg
49% soybean oil meal	0.08	—	—	—	—	—	50.8	2.25	0.56	—	—
Cane molasses	3.72	1.216	80.2	—	0.024	0.47	57.2	4.02	0.20	0.46	—
Sweetclover hay	0.37	—	10.0	—	0.013	0.25	117.9	1.84	0.09	0.45	—
Sweet potatoes	0.06	—	4.2	—	0.005	0.16	11.1	1.01	0.05	0.13	—
Timothy hay	0.62	—	5.0	—	0.014	0.18	47.1	1.59	0.18	0.13	—
Turnips	0.65	—	21.3	—	0.011	0.22	42.7	2.99	1.05	0.43	—
Wheat straw	0.33	—	3.1	—	0.017	0.12	55.0	0.67	0.14	0.18	—
Wheat bran	0.04	0.109	13.9	—	0.019	0.62	130.1	1.39	0.07	0.25	—
Wheat middlings	0.12	0.128	7.2	—	0.005	0.19	60.8	0.62	0.14	0.27	72.9
Wheat shorts	—	0.120	13.3	—	0.008	0.29	133.0	1.07	0.02	—	123.3
Wheat grain	0.09	—	9.1	—	0.007	0.16	49.0	0.47	0.07	0.22	—
Wood molasses	0.20	—	—	—	—	0.11	20.3	0.06	0.05	0.05	—

*Only the common names are given here. The comparable scientific name may be obtained from Table 37. The different feeds are listed in the same order as they appear in Table 37.

Table 39
AMINO ACID CONTENT OF SOME COMMON SWINE FEEDS—AS FED BASIS

Name of feed*	Arginine %	Cystine %	Histidine %	Isoleucine %	Leucine %	Lysine %	Methionine %	Phenylalanine %	Threonine %	Tryptophan %	Tyrosine %	Valine %
Dehydrated alfalfa meal	0.70	0.32	0.40	0.70	1.30	0.80	0.20	0.80	0.80	0.40	0.50	0.90
Blood meal	3.50	1.40	4.20	1.00	10.30	6.90	0.90	6.10	3.70	1.10	1.80	6.50
Meat scrap	3.70	0.60	1.10	1.90	3.50	3.80	0.80	1.90	1.80	0.30	0.90	2.60
Digester tankage	3.60	—	1.90	1.90	5.10	4.00	0.80	2.70	2.40	0.70	—	4.20
Meat and bone meal	4.00	0.60	0.90	1.70	3.10	3.50	0.70	1.80	1.80	0.20	0.80	2.40
Barley grain	0.53	0.18	0.27	0.53	0.80	0.53	0.18	0.62	0.36	0.18	0.36	0.62
Dried beet pulp	0.30	—	0.20	0.30	0.60	0.60	—	0.30	0.40	0.10	0.40	0.40
Dried whey	0.40	0.30	0.20	0.90	1.40	1.10	0.20	0.40	0.80	0.20	0.30	0.70
Dried whole milk	0.90	—	0.70	1.30	2.50	2.20	0.60	1.30	1.00	0.40	1.30	1.70
Dried skimmed milk	1.20	0.50	0.90	2.30	3.30	2.80	0.80	1.50	1.40	0.40	1.30	2.20
Copra meal	2.70	0.30	0.56	0.66	1.49	0.64	0.29	0.90	0.65	0.20	0.56	0.98
Ground ear corn	0.45	0.16	0.18	0.36	1.00	0.18	0.16	0.45	0.36	0.09	—	0.36
Hominy feed	0.50	0.18	0.20	0.40	0.80	0.40	0.18	0.30	0.40	0.10	0.50	0.50
Corn gluten meal	1.40	0.60	1.00	2.30	7.60	0.80	1.00	2.90	1.40	0.20	1.00	2.20

TABLE 39 (Continued)

Name of feed*	Arginine %	Cystine %	Histidine %	Isoleucine %	Leucine %	Lysine %	Methionine %	Phenylalanine %	Threonine %	Tryptophan %	Tyrosine %	Valine %
Yellow shelled corn	0.45	0.09	0.18	0.45	0.99	0.18	0.09	0.45	0.36	0.09	–	0.36
41% cottonseed meal	4.25	0.85	1.10	1.60	2.50	1.70	0.65	2.35	1.45	0.65.	0.70	2.05
41% solv-extd cottonseed meal	4.25	0.85	1.10	1.60	2.50	1.70	0.65	2.35	1.45	0.65	–	2.05
Fish meal	4.00	0.94	1.60	4.10	5.00	5.30	1.80	2.70	2.90	0.60	1.60	3.60
Linseed meal	–	–	–	–	–	–	0.70	–	–	–	–	–
Brewers dried grains	1.30	–	0.50	1.50	2.30	0.90	0.40	1.30	0.90	0.40	1.20	1.60
Oats grain	0.71	0.18	0.18	0.53	0.89	0.36	0.18	0.62	0.36	0.18	0.53	0.62
Peanut oil meal	5.90	0.60	1.20	2.00	3.70	2.30	0.40	2.70	1.50	0.50	1.80	2.80
Hydrolyzed poultry feathers	5.90	3.00	–	–	–	2.00	0.60	–	–	0.50	–	–
Rice bran	0.50	0.10	0.20	0.40	0.60	0.50	0.29	0.40	0.40	0.10	0.68	0.60
Rye grain	0.53	0.18	0.27	0.53	0.71	0.45	0.18	0.62	0.36	0.09	0.27	0.62
Sesame oil meal	4.80	0.60	1.10	2.10	3.40	1.30	1.40	2.20	1.60	0.78	2.00	2.40

TABLE 39 (Continued)

Name of feed*	Arginine %	Cystine %	Histidine %	Isoleucine %	Leucine %	Lysine %	Methionine %	Phenylalanine %	Threonine %	Tryptophan %	Tyrosine %	Valine %
Grain sorghum grain	0.36	0.18	0.27	0.53	1.42	0.27	—	0.45	0.27	0.09	0.36	0.53
Milo grain	0.36	0.18	0.27	0.53	1.42	0.27	0.09	0.45	0.27	0.09	0.36	0.53
44% soybean oil meal	3.20	0.67	1.10	2.50	3.40	2.90	0.60	2.20	1.70	0.60	1.40	2.40
49% soybean oil meal	3.80	0.80	1.20	2.60	3.80	3.20	0.73	2.70	2.00	0.65	2.00	2.70
Cane molasses	—	—	—	—	—	—	—	—	—	—	—	—
Sunflower meal	3.50	0.70	1.00	2.10	2.60	1.70	1.50	2.20	1.50	0.50	—	2.30
Wheat bran	1.00	0.30	0.30	0.60	0.90	0.60	0.10	0.50	0.40	0.30	0.40	0.70
Wheat shorts	0.95	0.20	0.32	0.70	1.20	0.70	0.18	0.70	0.50	0.20	0.40	0.77
Wheat middlings	1.00	0.20	0.40	0.70	1.20	0.60	0.10	0.50	0.50	0.20	0.50	0.80
Wheat grain	0.71	0.18	0.27	0.53	0.89	0.45	0.18	0.62	0.36	0.18	0.45	0.53

*Only the common names are given here. The comparable scientific names may be obtained from Table 37. The different feeds are listed in the same order as they appear in Table 37.

Table 40
VITAMIN CONTENT OF SOME COMMON SWINE FEEDS—AS FED BASIS

Name of feed*	a-Tocopherol mg/kg	Biotin mg/kg	Carotene mg/kg	Choline mg/kg	Folic acid mg/kg	Niacin mg/kg	Pantothenic acid mg/kg	Pyridoxine mg/kg	Riboflavin mg/kg	Thiamine mg/kg	Vitamin B_{12} µg/kg	Vitamin K mg/kg
Dehydrated alfalfa meal	128.0	0.33	88.0	1518	2.10	45.8	30.0	6.30	12.3	3.5	—	8.70
Blood meal	0.0	—	—	757	—	31.5	1.1	—	1.5	—	—	—
Meat scrap	1.0	0.09	—	1955	0.05	56.9	4.8	3.00	5.3	0.2	51.1	—
Digester tankage	—	—	—	2169	1.50	39.2	2.4	—	2.4	—	—	—
Meat and bone meal	1.0	0.14	—	2189	0.05	47.8	3.7	2.50	4.4	1.1	44.8	—
Barley grain	—	—	—	988	—	52.9	8.1	—	1.8	—	—	—
Dried beet pulp	—	—	—	829	—	16.3	1.5	—	0.7	0.4	—	—
Dried whey	—	0.40	—	20	0.90	11.2	47.7	2.50	29.9	3.7	30.0	—
Dried whole milk	—	0.37	7.2	—	—	8.4	22.7	4.63	19.6	3.7	—	—
Dried skimmed milk	9.2	0.33	—	1426	0.62	11.5	33.7	3.97	20.1	3.5	60.0	—
Copra meal	—	—	—	920	1.30	24.9	6.6	—	3.1	0.7	—	—
Ground ear corn	20.0	0.05	3.5	550	0.30	20.0	5.0	5.00	1.1	—	—	—
Hominy feed	—	0.13	9.1	1000	0.28	51.1	7.5	11.00	2.0	7.9	—	—

TABLE 40 (Continued)

Name of feed*	a-Tocopherol mg/kg	Biotin mg/kg	Carotene mg/kg	Choline mg/kg	Folic acid mg/kg	Niacin mg/kg	Pantothenic acid mg/kg	Pyridoxine mg/kg	Riboflavin mg/kg	Thiamine mg/kg	Vitamin B$_{12}$ μg/kg	Vitamin K mg/kg
Corn gluten meal	42.0	0.15	16.4	330	0.20	49.9	10.3	8.00	1.5	0.2	–	–
Yellow shelled corn	–	–	4.1	–	–	26.3	3.9	–	1.3	3.6	–	–
41% cottonseed meal	40.0	–	–	2780	2.30	39.5	14.0	5.30	5.0	6.5	–	–
41% solv-extd cottonseed meal	–	–	–	2860	2.30	39.5	14.0	–	5.0	6.5	–	–
Fish meal	9.0	0.26	–	3080	0.20	55.9	8.8	–	4.8	0.7	88.0	–
Linseed meal	–	–	–	1225	–	30.1	–	–	2.9	9.5	–	–
Brewers dried grains	–	–	–	1587	0.22	43.4	8.6	0.66	1.5	0.7	–	–
Oats grain	5.9	0.30	–	1073	0.40	15.8	12.9	1.20	1.6	6.2	–	–
Peanut oil meal	3.0	0.39	–	2000	0.36	170.1	53.0	10.00	11.0	7.3	–	–
Hydrolyzed poultry feathers	–	–	–	882	–	30.9	11.0	–	2.2	–	–	–
Rice bran	60.0	4.20	–	1254	–	303.2	23.5	–	2.6	22.4	–	–
Rye grain	15.0	0.06	–	–	0.60	1.2	6.9	–	1.6	3.9	–	–

452

TABLE 40 (Continued)

Name of feed*	a-Toco-pherol mg/kg	Bio-tin mg/kg	Caro-tene mg/kg	Cho-line mg/kg	Folic acid mg/kg	Nia-cin mg/kg	Panto-thenic acid mg/kg	Pyri-doxine mg/kg	Ribo-flavin mg/kg	Thia-mine mg/kg	Vita-min B$_{12}$ µg/kg	Vita-min K mg/kg
Sesame oil meal	—	—	—	1533	—	30.0	6.4	12.50	3.7	2.9	—	—
Grain sorghum grain	—	2.60	—	678	0.20	43.1	11.1	5.30	1.3	4.1	—	—
Milo grain	12.0	0.18	—	678	0.24	42.7	11.4	4.10	1.2	3.9	—	—
44% soybean oil meal	3.0	0.32	—	2743	0.70	26.8	14.5	8.00	3.3	6.6	—	—
49% soybean oil meal	3.3	0.32	—	2761	3.60	21.6	14.5	8.00	3.1	2.40	—	—
Cane molasses	—	—	—	876	—	34.3	38.3	—	3.3	0.9	—	—
Sunflower meal	11.0	—	—	2900	—	220.0	10.0	16.00	3.1	—	—	—
Wheat bran	10.8	0.48	2.6	988	1.80	209.2	29.0	10.00	3.1	7.9	—	—
Wheat shorts	29.9	0.37	—	928	1.10	94.6	17.6	11.00	2.0	15.8	—	—
Wheat middlings	57.6	0.37	—	1100	1.10	52.6	13.6	11.00	1.5	18.9	—	—
Wheat grain	15.5	0.10	—	830	0.40	56.6	12.1	—	1.2	4.9	—	—

*Only the common names are given here. The comparable scientific names may be obtained from Table 37. The different feeds are listed in the same order as they appear in Table 37.

Table 41
TABLE OF EQUIVALENTS

Weight

1 ounce (avdp.)	=	28.50	grams
1 pound (avdp.)	=	16.0	ounces
	=	453.6	grams
1 kilogram	=	1,000	grams
	=	2.205	pounds
1 ton	=	2,000	pounds
	=	907.0	kilograms
1 metric ton	=	1,000	kilograms
	=	2,205	pounds
	=	1.102	tons

Length

1 inch	=	2.54	centimeters
	=	25.4	millimeters
1 foot	=	12.0	inches
	=	30.48	centimeters
1 yard	=	3.0	feet
	=	0.9144	meters
1 rod	=	16.5	feet
1 mile	=	5,280	feet
	=	1,760	yards
	=	320	rods
	=	1.615	kilometers
1 centimeter	=	10.0	millimeters
	=	0.3937	inch
1 meter	=	100	centimeters
	=	1,000	millimeters
1 kilometer	=	1,000.0	meters
	=	0.6412	miles

Volume or capacity

1 cubic inch	=	16.387	cubic centimeters
1 cubic foot	=	1,728.0	cubic inches
	=	0.0283	cubic meters
1 cubic yard	=	27.0	cubic feet
	=	0.7646	cubic meters
1 cubic centimeter	=	1.0	milliliter
	=	0.061	cubic inch
1 cubic meter	=	35.315	cubic feet
	=	1.308	cubic yards

TABLE 41 (Continued)

1 liquid pint (U.S.)	=	28.875	cubic inches
	=	0.5	liquid quart (U.S.)
	=	0.47316	liter
1 liquid quart (U.S.)	=	57.75	cubic inches
	=	2.0	liquid pints (U.S.)
	=	0.9463	liter
1 liquid gallon (U.S.)	=	231	cubic inches
	=	8.0	liquid pints (U.S.)
	=	4.0	liquid quarts (U.S.)
	=	3.7853	liters
1 liter	=	2.1134	liquid pints (U.S.)
	=	1.057	liquid quarts (U.S.)
	=	0.2642	liquid gallons (U.S.)
1 bushel	=	2,150.42	cubic inches
	=	1.244	cubic feet
	=	9.309	liquid gallons (U.S.)
	=	4	pecks

Area

1 square inch	=	6.452	square centimeters
1 square foot	=	0.0929	square meter
1 square yard	=	9.0	square feet
	=	0.8361	square meters
1 acre	=	43,560	square feet
	=	4,840.0	square yards
	=	160.0	square rods
	=	0.4047	hectares
1 square mile	=	640.0	acres
	=	259.0	hectares
1 square centimeter	=	0.155	square inch
1 square meter	=	1.196	square yards
	=	10.764	square feet
1 hectare	=	10,000	square meters
	=	2.471	acres
1 square kilometer	=	0.386	square mile
	=	247.1	acres

Temperature

| Degrees Centigrade | = | 5/9 (Degrees Fahrenheit − 32) |
| Degrees Fahrenheit | = | (9/5 × Degrees Centigrade) + 32 |

58 Glossary of Terms Frequently Used in Discussing Matters Related to Feeds and Feeding

ABOMASUM   The fourth compartment of a ruminant's stomach. Sometimes called the *true stomach*.

ABORTION   The expulsion of a nonviable, immature fetus.

ABSCESS   A collection of pus in any part of the body.

ABSORPTION   The movement of nutrients or other substances from the digestive tract or through the skin into the blood and/or lymph system.

ACETIC ACID   One of the volatile fatty acids with the formula CH_3COOH. Commonly found in silage, rumen contents, and vinegar.

ADDITIVE   An ingredient or a combination of ingredients added, usually in small quantities, to a basic feed mix for the purpose of fortifying the basic mix with certain essential nutrients and/or medicines.

ADIPOSE   Of a fatty nature.

AD LIBITUM   As desired by the animal.

ADRENAL   Near the kidney.

AERIAL PART   The above-ground part of a plant.

AEROBIC   Living or functioning in the presence of air or molecular oxygen.

AFTERBIRTH   The membranes expelled from the uterus following delivery of a fetus.

ALANINE   One of the nonessential amino acids.

ALIMENTARY   Having to do with feed or food.

ALIMENTARY TRACT   Same as *digestive tract.*

AMINO ACID   Any one of a class of organic compounds which contain both the amino (NH_2) group and the carboxyl (COOH) group.

AMMONIATED   Combined or impregnated with ammonia or an ammonium compound.

AMYLASE   Any one of several enzymes which effect a hydrolysis of starch to maltose. Examples are pancreatic amylase (amylopsin) and salivary amylase (ptyalin).

ANABOLISM   The conversion of simple substances into more complex substances by living cells. Constructive metabolism.

ANAEROBIC   Living or functioning in the absence of air or molecular oxygen.

ANEMIC   Lacking in size and/or number of red blood cells.

ANIMAL PROTEIN FACTOR   What was once an unidentified growth factor essential for poultry and swine and present in protein feeds of animal origin. It is now known to be the same as vitamin B_{12}.

ANTACID   A substance that counteracts acidity.

ANTIBIOTIC   A substance produced by one microorganism which has an inhibiting effect on the growth of another.

ANTIBODY   Substance produced in the body that acts against disease.

ANTIOXIDANT   A material capable of chemically protecting other substances from oxidation.

ANUS   The posterior end and opening of the digestive tract.

ARACHIDONIC ACID   A 20-carbon unsaturated fatty acid having four double bonds.

ARGININE   One of the essential amino acids.

ARTIFICIALLY DRIED   Dried by other than natural means. Dehydrated.

ASCORBIC ACID   Same as *vitamin C*, the antiscorbutic vitamin.

AS FED   As consumed by the animal.

ASH   The incombustible residue remaining after incineration at 600°C for several hours.

ASPARTIC ACID   One of the nonessential amino acids.

ASPHYXIA   Suffocation or the suspension of animation as the result of suffocation.

ASPIRATED   Removal of light materials from heavier material by use of air.

ATROPHY   A wasting away of a part of the body.

AVIDIN   A protein in egg albumen which can combine with biotin to render the latter unavailable to the animal.

BACTERIA   Very small, unicellular plant organisms.

BALANCED   Containing essential nutrients in the proper proportions.

BALANCED DAILY RATION   Such a combination of feeds as will provide the essential nutrients in such amounts as will properly nourish a given animal for a 24-hour period.

BALANCED RATION   Such a combination of feeds as will provide the essential nutrients in the proper proportions.

BASAL METABOLISM   The heat production of an animal during physical, digestive, and emotional rest.

BILE   A greenish-yellow fluid formed in the liver, stored in the gall bladder (except in the horse which has no gall bladder), and secreted via bile duct into the upper small intestine. It functions in digestion.

BIOCHEMISTRY   The chemistry of living things.

BIOLOGICAL   Pertaining to the science of life.

BIOLOGICAL FUNCTION   The role played by a chemical compound in living organisms.

BIOLOGICAL VALUE   The efficiency with which a protein furnishes the proper proportions and amounts of the essential amino acids. A protein which has a high biological value is said to be of *good quality*.

BIOSYNTHESIS   The formation of chemical substances from other chemical substances in a living organism.

BIOTIN   One of the B vitamins.

BOILING POINT   The temperature at which the vapor pressure of a liquid equals the atmospheric pressure.

BOLTED   Separated from parent material by means of a bolting cloth.

BOMB CALORIMETER   An instrument used for determining the gross energy content of a material.

BRAN   The pericarp or seed coat of grain removed during processing.

BUFFER   Any substance that can counteract changes in free acid or alkali concentration.

BUSHEL   A certain volume equal to 2150.42 cubic inches (approximately 1.25 cubic feet).

BUTYRIC ACID   One of the volatile fatty acids with the formula $CH_3CH_2CH_2COOH$. Commonly found in rumen contents and poor quality silage.

CAECUM   Same as *cecum.*

CALCIFICATION   Process by which organic tissue becomes hardened by a deposit of calcium salts.

CALORIC   Pertaining to heat or energy.

CALORIE   The amount of energy as heat required to raise one gram of water $1°C$ (precisely from $14.5°$ to $15.5°C$).

CALORIMETER   An instrument for measuring energy.

CALORIMETRY   The science of measuring heat.

CARBOHYDRATE   Organic substances that contain carbon, hydrogen, and oxygen, with the hydrogen and oxygen present in the same proportions as in water.

CARCASS   The body of an animal less the viscera and usually the head, skin and lower leg.

CARCINOGEN   Any cancer-producing substance.

CARCINOGENIC   Cancer producing.

CARDIOVASCULAR   Pertaining to the heart and blood vessels.

CARIES   Areas of tooth decay.

CAROTENE   A yellow organic compound that is a precursor of vitamin A.

CARRIER   An edible material which is used to facilitate the addition of micronutrients to a ration.

CARTILAGE   The gristle or connective tissue attached to the ends of bones.

CASEIN   The protein precipitated from milk by acid and/or rennin.

CATABOLISM   The conversion of complex substances into more simple compounds by living cells. Distructive metabolism.

CATALYST   A substance that speeds up the rate of a chemical reaction but is not itself used up in the reaction.

CECUM   An intestinal pouch located at the junction of the large and small intestine. Also *caecum.*

CELL   The structural and functional microscopic unit of plant and animal organisms.

CELL PLATELET   A small, colorless, disk-shaped cell in the blood concerned with blood coagulation.

CELLULOSE   A polysaccharide having the formula $(C_6H_{10}O_5)_n$. Found in the fibrous portion of plants. Low in digestibility.

CELSIUS   Same as *Centigrade.*

CENTIGRADE   A thermometer scale in which water freezes at $0°$ and boils at $100°$. Same as *Celsius.*

CHLOROPHYLL   The green coloring matter present in growing plants.

CHOLESTEROL   The most common member of the sterol group.

CHOLINE   One of the B vitamins.

CHOPPED   Reduced in particle size by cutting.

CHROMATOGRAPHY   A technique for separating complex mixtures of chemical substances.

CITRULLINE   One of the nonessential amino acids.

CLIPPED   With oat grain, the more fibrous end has been removed.

COAGULATED   Curdled, clotted, or congealed.

COAGULATION   The change from a fluid state to a thickened jelly, curd, or clot.

COENZYME   A partner required by some enzymes to produce enzymatic activity.

COLLAGEN   The main supportive protein of connective tissue.

COLOSTRUM MILK   The milk secreted during the first few days of lactation.

COMBUSTION   The combination of substances with oxygen accompanied by the liberation of heat.

COMMERCIAL FEED   Any material produced by a commercial company and distributed for use as a feed or feed component.

COMPLETE RATION   A single feed mixture into which has been included all of the dietary essentials, except water, of a given class of livestock.

CONCENTRATE   Any feed low (under about 20%) in crude fiber and high (over about 60%) in TDN on an air-dry basis. Opposite of roughage. Also, a concentrated source of one or more nutrients used to enhance the nutritional adequacy of a supplement mix.

CONGENITAL   Existing at birth.

CONGESTION   Excessive accumulation of blood in a part of the body.

CONVULSION   A violent involuntary contraction or series of contractions of the voluntary muscles.

CORONARY   Refers to the arteries that supply the heart.

CREATININE   A nitrogenous compound arising from protein metabolism and secreted in the urine.

CRIMPED   Having been passed between rollers with corrugated surfaces.

CRUDE FAT   That part of a feed which is soluble in ether. Also referred to as *ether extract.*

CRUDE FIBER   The more fibrous, less digestible portion of a feed. Consists primarily of cellulose and lignin.

CRUDE PROTEIN   Total ammoniacal nitrogen $\times$ 6.25, based on the fact that feed protein on the average contains 16.0% nitrogen.

CUD   A bolus of previously eaten food which has been regurgitated by a ruminant animal for further chewing.

CURD   The semi-solid mass that is formed when milk comes in contact with an acid or the enzyme rennin. It consists mainly of the protein casein.

CYANOCOBALAMIN   Same as *vitamin B_{12}.*

CYSTINE   One of the nonessential amino acids. It is sulfur containing and may be used to meet in part the need for methionine.

CYSTITIS   Inflammation of the bladder.

DEFICIENCY DISEASE   A disease resulting from an inadequate dietary intake of some nutrient.

DEFLUORINATED   Having had the fluorine content reduced to a level which is nontoxic under normal use.

DEHYDRATED   Having had most of the moisture removed through artificial drying.

DERMATITIS   Inflammation of the skin.

DESICCATE   To dry completely.

DEXTRIN   An intermediate polysaccharide product obtained during starch hydrolysis.

DIGESTIBLE ENERGY   That part of the gross energy of a feed which does not appear in the feces.

DIGESTION   The processes involved in the conversion of feed into absorbable forms.

DIGESTIVE TRACT   The passage from the mouth to the anus through which feed passes following consumption as it is subjected to various digestive processes. Primarily the stomach and intestines.

DISACCHARIDE   Any one of several so-called compound sugars which yield two monosaccharide molecules upon hydrolysis. Sucrose, maltose, and lactose are the most common.

DISPENSABLE AMINO ACID   Basically the same as *nonessential amino acid.*

DRY MATTER   That part of feed which is not water. Sometimes referred to as *dry substance* or *total solids.* Is the sum of the crude protein, crude fat, crude fiber, nitrogen-free extract, and ash.

DRY-RENDERED   Having been heat processed for the removal of fat without the addition of water or steam.

DUODENUM   The upper portion of the small intestine which extends from the stomach to the jejunum.

DYSTOCIA   Difficult parturition.

EDEMA   Swelling of a part of or of the entire body due to an accumulation of an excess of water.

ELEMENT   Any one of the fundamental atoms of which all matter is composed.

EMACIATED   An excessively thin condition of the body.

EMULSIFY   To disperse small drops of one liquid into another liquid.

ENDEMIC   Occurring in low incidence but more or less constantly in a given population.

ENDOCRINE   Pertaining to internal secretions.

ENDOGENOUS   Originating from within the organism.

ENDOMETRIUM   The mucous membrane that lines the uterus.

ENERGY   The capacity to perform work.

ENSILAGE   The same as *silage.*

ENSILED   Having been subjected to anaerobic fermentation to form silage.

ENTERITIS   Inflammation of the intestines.

ENVIRONMENTAL   Pertaining to external influences.

ENZYMATIC   Related to an enxyme.

ENZYME   One of a class of organic compounds, formed by living cells, capable of producing or accelerating specific organic reactions. An organic catalyst.

EPIDEMIC   When many people in a given region are attacked by some disease at the same time.

EPITHELIAL   Refers to those cells that form the outer layer of the skin and other membranes.

ERGOSTEROL   One of the sterols which upon exposure to ultraviolet light is converted to vitamin D_2.

ESOPHAGUS   The passageway leading from the mouth to the stomach. Sometimes called the *gullet.*

ESSENTIAL AMINO ACID   Any one of several amino acids that are needed by animals and cannot be synthesized by them in the amount needed and so must be present in the protein of the feed as such.

ESTROGENS   Estrus-producing hormones secreted by the ovaries.

ESTRUS   The recurring periods of sexual receptivity in female mammals. The period of *heat.*

ETIOLOGY   The causes of a disease or disorder.

EXCRETA   The products of excretion—primarily feces and urine.

EXOGENOUS   Originating from outside of the organism.

EXPANDED   As applied to feed—having been increased in volume as the result of a sudden reduction in surrounding pressure.

EXPELLER PROCESS   A process for the mechanical extraction of oil from seeds, involving the use of a screw press.

EXTRINSIC FACTOR   A factor coming from or originating from outside an organism.

EXTRUDED   As applied to feed—having been forced through a die under pressure.

FACTOR   In nutrition, any chemical substance found in feed.

FAHRENHEIT   A thermometer scale in which water freezes at $32°$ and boils at $212°$.

FAT   The product formed when a fatty acid reacts with glycerol. The glyceryl ester of a fatty acid. Stearin, palmitin, and olein are examples.

FAT SOLUBLE   Soluble in fats and fat solvents but generally not soluble in water.

FATTENING   This is the deposition of unused energy in the form of fat within the body tissues.

FATTY ACID   Any one of several organic compounds containing carbon, hydrogen, and oxygen which combine with glycerol to form fat.

FAUNA   The animal life present. Frequently used to refer to the overall protozoal population present.

FECES   The excreta discharged from the digestive tract through the anus.

FEED   Any material eaten by an animal as a part of its daily ration.

FEED GRADE   Suitable for animal but not for human consumption.

FERMENTATION   Chemical changes brought about by enzymes produced by various microorganisms.

FETUS   The unborn young of animals.

FIBROUS   High in content of cellulose and/or lignin.

FINISH   To fatten a slaughter animal. Also, the degree of fatness of such an animal.

FISTULA   An abnormal tube-like passage from some part of the body to another part or to the exterior—sometimes surgically inserted.

FLAKED   Rolled or cut into flat pieces.

FLORA   The plant life present. In nutrition it generally refers to the bacteria present in the digestive tract.

FODDER   The entire above-ground part of nearly mature corn or sorghum in the fresh or cured form.

FOLACIN   Same as *folic acid*. One of the B vitamins.

FOLIC ACID   Same as *folacin*, which is one of the B vitamins.

FORAGE   Crops used in the whole plant form (except roots) as pasture, hay, silage, or green chop for feeding purposes.

FORMULA FEED   A feed consisting of two or more ingredients mixed in specified proportions.

FORTIFY   Nutritionally, to add one or more nutrients to a feed.

FRACTIONATION   The laboratory separation of natural materials into their component parts.

FREE CHOICE   Free to eat of two or more feeds at will.

FRESH   Usually denotes the green or wet form of a feed material.

FRUCTOSE   A hexose monosaccharide found especially in ripe fruits and honey. Obtained along with glucose from sucrose hydrolysis. Commonly known as *fruit sugar.*

GALACTOSE   A hexose monosaccharide obtained along with glucose from lactose hydrolysis.

GALL BLADDER   A membranous sac lying next to the liver of all farm livestock except the horse in which bile is stored.

GASTRIC   Pertaining to the stomach.

GASTRIC JUICE   A clear liquid secreted by the wall of the stomach. It contains hydrochloric acid and the enzymes rennin, pepsin, and gastric lipase.

GASTRITIS   Inflammation of the stomach.

GASTROENTERITIS   Inflammation of the stomach and intestines.

GASTROINTESTINAL   Pertaining to the stomach and intestines.

GENETIC   Pertaining to heredity.

GENITOURINARY   Refers to the organs of reproduction and urine excretion.

GERM   Embryo of a seed.

GESTATION   The condition of bearing an unborn fetus. Pregnancy.

GINGIVITIS   Inflammation of the gums.

GLAND   An organ that produces and secretes a chemical substance in the body.

GLUCOSE   A hexose monosaccharide obtained upon the hydrolysis of starch and certain other carbohydrates. Also called *dextrose*.

GLUTAMIC ACID   One of the nonessential amino acids.

GLYCEROL   An alcohol containing three carbons and three hydroxy groups.

GYLCINE   One of the nonessential amino acids.

GLYCOGEN   A polysaccharide with the formula $(C_6H_{10}O_5)_n$ which is formed in the liver and depolymerized to glucose to serve as a ready source of energy when needed by the animal. Known also as *animal starch*.

GOITER   An enlargement of the thyroid gland located in the neck. Sometimes caused by an iodine deficiency.

GOSSYPOL   A substance present in cottonseed and cottonseed meal which is toxic to swine and certain other nonruminant animals.

GRAVID   Pregnant.

GREEN CHOP   Forage harvested and fed in the green, chopped form.

GROAT   Grain from which the hull has been removed.

GROSS ENERGY   The total heat of combustion of a material as determined by the use of a bomb calorimeter.

GROUND   Reduced in particle size by impact, shearing, or attrition.

GROWTH   An increase in muscle, bone, vital organs, and connective tissue as contrasted to fattening or fat deposition.

HAY   The aerial part of finer-stemmed forage crops stored in the dry form for animal feeding.

HEAT INCREMENT   The heat which is unavoidably produced by an animal incidental with nutrient digestion and utilization. Was originally called *work of digestion*.

HEAT LABILE   Unstable to heat.

HEMOGLOBIN   The oxygen-carrying, red-pigmented protein of the red corpuscles.

HEMORRHAGE   Copious loss of blood through bleeding.

HEPATITIS   Inflammation of the liver.

HEXOSAN   A hexose-based polysaccharide having the general formula $(C_6H_{10}O_5)_n$. Cellulose, starch, and glycogen are the most common.

HEXOSE   A 6-carbon monosaccharide having the formula $C_6H_{12}O_6$. Glucose, frustose, and galactose are common examples.

HISTIDINE   One of the essential amino acids.

HOMOGENIZED   The fat within a fluid having been reduced to globules so small they remain in suspension for an extended period of time.

HORMONE   A chemical substance secreted into the body fluids by an endocrine gland.

HULLS   The outer protective covering of seeds.

HUSKS   Usually refers to the fibrous covering of an ear of corn.

HYDRAULIC PROCESS   A process for the mechanical extraction of oil from seeds, involving the use of a hydraulic press. Sometimes referred to as the *old process.*

HYDROGENATION   The chemical addition of hydrogen to any unsaturated compound.

HYDROLYSIS   The splitting of a substance into the smaller units by its chemical reaction with water.

HYDROXYPROLINE   One of the nonessential amino acids.

HYPER   A prefix meaning in excess of the normal.

HYPEREMIA   An excess of blood in any part of the body.

HYPERTENSION   An abnormally high tension—usually associated with high blood pressure.

HYPERTHYROIDISM   Overactivity of the thyroid gland.

HYPERTROPHIED   Having increased in size independent of natural growth.

HYPERVITAMINOSIS   An abnormal condition resulting from the intake of an excess of one or more vitamins.

HYPO   A prefix denoting less than the normal amount.

HYPOMAGNESEMIA   An abnormally low level of magnesium in the blood.

HYSTERITIS   Inflammation of the uterus.

ILEUM   The lower portion of the small intestine extending from the jejunum to the cecum.

IMPERMEABLE   Not capable of being penetrated.

INACTIVATE   To render a substance inactive.

INCIDENCE   The frequency of occurrence of a situation or a condition.

INDISPENSABLE AMINO ACID   Basically the same as *essential amino acid.*

INERT   Relatively inactive.

INGEST   To eat or take in through the mouth.

INORGANIC   Denotes chemical compounds that do not contain carbon in chain structure.

INOSITOL   One of the B vitamins.

INSULIN   A hormone secreted by the pancreas into the blood. It regulates sugar metabolism.

INTESTINAL JUICE   A clear liquid secreted by glands in the wall of the small intestine. It contains the enzymes intestinal lactase, maltase, and sucrase, and several peptidases.

INTESTINAL TRACT   The small and large intestine.

INTESTINE, LARGE   The tube-like part of the digestive tract lying between the small intestine and the anus. Larger in diameter but shorter in length than the small intestine.

INTESTINE, SMALL   The long, tortuous, tube-like part of the digestive tract leading from the stomach to the cecum and large intestine. Smaller in diameter but longer than the large intestine.

INTRINSIC FACTOR   A chemical substance in normal stomach juice necessary for the absorption of vitamin B_{12}.

INULIN   A polysaccharide found especially in Jerusalem artichokes which yields fructose upon hydrolysis.

IODINE NUMBER   A number which denotes the degree of unsaturation of a fat or fatty acid. It is the amount of iodine in grams which can be taken up by 100 g of fat.

IRRADIATION   The act of treating with ultraviolet light.

ISOLEUCINE   One of the essential amino acids.

JEJUNUM   The middle portion of the small intestine which extends from the duodenum to the ileum.

KERATIN   A sulfur-containing protein which is the primary component of epidermis, hair, wool, hoof, horn, and the organic matrix of the teeth.

KERNEL   A dehulled seed.

KILOCALORIE   1,000 calories.

LABILE   Unstable. Easily destroyed.

LACTASE   An enzyme present in intestinal juice which acts on lactose to produce glucose and galactose.

LACTATION   The secretion of milk.

LACTIC ACID   An organic acid, one form ($CHOH \cdot CH_2 \cdot COOH$) of which is commonly found in sour milk, sauerkraut, and silage. Other forms enter into body metabolism.

LACTOSE   A disaccharide found in milk having the formula $C_{12}H_{22}O_{11}$. It hydrolyzes to glucose and galactose. Commonly known as *milk sugar.*

LD$_{50}$ A dose which is lethal for 50% of the test animals.

LESION   Any unhealthy change in the structure of a part of the body.

LEUCINE   One of the essential amino acids.

LIGNIN   An indigestible compound which along with cellulose is a major component of the cell wall of certain plant materials such as wood, hulls, straws, and overripe hays.

LINOLEIC ACID   An 18-carbon unsaturated fatty acid having two double bonds. It reacts with glycerol to form linolein.

LINOLEIN   An unsaturated fat formed from the reaction of linoleic acid with glycerol.

LINOLENIC ACID   An 18-carbon unsaturated fatty acid having three double bonds.

LIPASE   A fat-splitting enzyme. Gastric lipase is present in gastric juice and pancreatic lipase is present in pancreatic juice. Both act on fats to produce fatty acids and glycerol.

LIPIDS   A broad term for fats and fat-like substances.

LYMPH   The slightly yellow, transparent fluid occupying the lymphatic channels of the body.

LYSINE   One of the essential amino acids.

MALFORMATION   Any abnormal development of a part of the body.

MALIGNANT   Virulent or destructive as applied to cancer.

MALTASE   An enzyme which acts on maltose to produce glucose. Salivary amylase is present in saliva, and intestinal maltase is present in intestinal juice.

MALTOSE   A disaccharide having the formula $C_{12}H_{22}O_{11}$. Obtained from the partial hydrolysis of starch. It hydrolyzes to glucose.

MAMMARY GLANDS   The milk-secreting glands.

MANURE　　The refuse from animal quarters consisting of excreta with or without litter or bedding.

MATRIX　　The intercellular framework of a tissue.

MEAL　　A feed ingredient having a particle size somewhat larger than flour.

MECHANICALLY EXTRACTED　　Having had its fat content removed by the application of heat and mechanical pressure. The hydraulic and expeller processes are both methods of mechanical extraction.

MEDIUM　　A nutrient substrate used for supporting the growth of micro-organisms.

MEGACALORIE　　1,000 kilocalories or 1,000,000 calories.

METABOLISM　　The sum of all the physical and chemical processes taking place in a living organism.

METABOLITE　　Any substance produced by metabolism.

METABOLIZABLE ENERGY　　Digestible energy minus the energy of the urine and fermentation gases.

METHIONINE　　One of the essential amino acids. It is sulfur containing and may be replaced in part by cystine.

METRITIS　　Inflammation of the uterus.

MICROBE　　Same as *microorganism.*

MICROBIOLOGICAL　　Pertaining to microorganisms.

MICROFLORA　　The gross overall bacterial population present. Is sometimes used to include the protozoa as well as the bacteria.

MICROGRAM　　One millionth of a gram or one thousandth of a milligram.

MICRO-INGREDIENT　　Any ration component normally measured in milligrams or micrograms per kilogram or in parts per million.

MICROORGANISM　　A very small living organism—usually microscopic in size.

MIDDLINGS　　A by-product of flour milling consisting of varying proportions of small particles of bran, endosperm, and germ.

MILLIGRAM　　One-thousandth of a gram.

MILL RUN　　A product as it comes from the mill, having no definite specifications.

MISCIBLE　　Capable of being mixed easily with another substance.

MOLASSES　　A thick, viscous, usually dark colored, liquid product containing a high concentration of soluble carbohydrates, minerals, and certain other materials.

MOLECULE A chemical combination of two or more atoms.

MONOSACCHARIDE Any one of several simple, nonhydrolyzable sugars. Glucose, fructose, galactose, arabinose, xylose, and ribose are examples.

MORBIDITY A state of sickness.

MORIBUND In a dying state—near death.

MUCOSA The membrane that lines the passages and cavities of the body.

MUCOUS MEMBRANE A membrane lining the cavities and canals of the body, kept moist by mucus.

MUCUS A slimy liquid secreted by the mucous glands and membranes.

MYCOTOXIN A fungous or bacterial toxin. Sometimes present in feed material.

NECROSIS Death of a part of the cells making up a living tissue.

NEONATE A newly born animal.

NEPHRITIS Inflammation of the kidneys.

NET ENERGY This is that part of metabolizable energy over the use of which the animal has complete control. It is metabolizable energy minus the heat increment.

NEURITIC Pertaining to the nerves.

NEW PROCESS Pertains to the extraction of oil from seeds. Same as *expeller process.*

NIACIN Same as *nicotinic acid.* Is one of the B vitamins. Nicotinamide also has niacin activity.

NICOTINAMIDE The amid of nicotinic acid. It has niacin activity.

NICOTINIC ACID Same as *niacin.* One of the B vitamins.

NITROGEN-FREE EXTRACT That part of feed dry matter which is not crude protein, crude fat, crude fiber, or ash. It consists mostly of sugars and starches. Sometimes referred to as *NFE.*

NONESSENTIAL AMINO ACID Any one of several amino acids that are required by animals but which can be synthesized in adequate amounts by an animal in its tissues from other amino acids.

NONPROTEIN Any one of a group of ammoniacal nitrogen containing compounds which are not true proteins. Urea is a common example.

NONRUMINANT A simple-stomached animal that does not ruminate. Examples are swine, horses, dogs, and humans.

NUTRIENT Any chemical compound having specific functions in the nutritive support of animal life.

NUTRITURE   Nutritional status.

OBESE   Being overweight due to a surplus of body fat.

OIL   Usually a mixture of pure fats which is liquid at room temperature.

OLD PROCESS   Pertains to the extraction of oil from seeds. Same as *hydraulic process.*

OLEIC ACID   An 18-carbon unsaturated fatty acid (one double bond) which reacts with glycerol to form olein.

OLEIN   The fat formed from the reaction of oleic acid with glycerol.

OMASUM   The third compartment of a ruminant's stomach. Sometimes called the *manyplies.*

ORGANIC   Refers to chemical compounds that contain carbon in chain structure.

ORGANIC ACID   Any organic compound that contains a carboxyl group (COOH).

ORTS   That portion of an animal's feed which it refuses to eat.

OSMOSIS   The passage of a solute or a solution through a semi-permeable membrane toward effecting an equalization of the concentration of the fluids on opposite sides of the membrane.

OSMOTIC PRESSURE   The pressure exerted by the movement of a solvent through a semi-permeable membrane toward equalizing solution concentration on opposite sides of the membrane.

OSSIFICATION   The process of bone formation.

OSTEITIS   Inflammation of a bone.

OSTEOMALACIA   A weakening of the bones due to a calcium, phosphorus, and/or vitamin D deficiency.

OSTEOPOROSIS   An abnormal porousness of bone as the result of a calcium, phosphorus, and/or vitamin D deficiency.

OVULATION   The discharge of the ovum or egg from the graafian follicle of the ovary.

PABA   Para-aminobenzoic acid.

PALMITIC ACID   A 16-carbon saturated fatty acid.

PALMITIN   The fat formed from the reaction of palmitic acid with glycerol.

PANCREAS   A large, elongated gland located near the stomach. It produces pancreatic juice which is secreted into the upper small intestine via pancreatic duct.

PANCREATIC JUICE   A thick, transparent liquid secreted by the pancreas into the upper small intestine. It contains the enzymes pancreatic amylase, pancreatic lipase, and trypsin; also the hormone insulin.

PANDEMIC   Widely spread throughout several countries.

PANTOTHENIC ACID   One of the B vitamins.

PAPILLAE   Small nipple-shaped projections located on the interior of the rumen wall.

PARA-AMINOBENZOIC ACID   One of the B vitamins. Often abbreviated *PABA*.

PARAKERATOSIS   Any abnormality of the outermost or horny layer of the skin.

PARALYSIS   Loss of power of voluntary motion.

PARATHYROID   Any one of four small glands situated beside the thyroid gland, concerned chiefly with calcium and phosphorus metabolism.

PARTURITION   The act of giving birth to young.

PASTURE   Forages which are harvested by grazing animals.

PATHOGEN   Any disease-producing microorganism or material.

PATHOLOGY   The branch of medicine that deals with the special nature of disease.

PELLETS   Compacted particles of feed formed by forcing ground material through die openings.

PENTOSAN   A pentose-based polysaccharide having the general formula $(C_5H_8O_4)_n$. Araban and xylan are examples. Not nearly as abundant as the hexosans.

PENTOSE   A 5-carbon monosaccharide having the formula $C_5H_{10}O_5$. Arabinose, xylose, and ribose are examples. Not abundant in the free form in nature.

PEPSIN   The proteolytic enzyme present in the gastric juice. It acts on protein to form proteoses, peptones, and peptides.

PERMEABLE   Capable of being penetrated.

PERSPIRATION   Sweat or the act of sweating.

pH   A measure of hydrogen ion concentration or the degree of acidity.

PHAGOCYTE   Any cell that can ingest particles or cells that are foreign or harmful to the body.

PHENYLALANINE   One of the essential amino acids.

PHOSPHOLIPIDS   Fat-like substances containing phosphorus and nitrogen, along with fatty acids and cholesterol.

PHYSIOLOGICAL   Pertaining to the science which deals with the functions of living organisms or their parts.

PITUITARY   A gland in the lower part of the brain which produces a number of hormones.

PLASMA   The colorless fluid portion of the blood in which the corpuscles are suspended.

POLYSACCHARIDE   Any one of a group of carbohydrates consisting of a combination of a large but undetermined number of monosaccharide molecules, such as starch, dextrin, glycogen, cellulose, inulin, etc.

POSTPARTUM   Following the birth of young.

POTENT   Strong, powerful, concentrated.

POULTRY LITTER   The fibrous material used on the floor of poultry houses along with the excreta which accumulates therein.

PRECONCEPTIONAL   Before pregnancy.

PRECURSOR   A compound that can be used by the body to form another compound.

PREGNANT   The state of having a developing embryo in the body. Gravid.

PRE-MIX   A uniform mixture of one or more microingredients and a carrier, used in the introduction of microingredients into a larger mixture.

PRESSURE COOKER   An airtight container for the cooking of feed at high temperature under steam pressure.

PROGESTERONE   A sex hormone produced by the corpus lutea of the ovary.

PROLINE   One of the nonessential amino acids.

PROPIONIC ACID   One of the volatile fatty acids with the formula CH_3CH_2COOH, commonly found in rumen contents but not in silage.

PROTEIN   Any one of many complex organic nitrogenous compounds formed from various combinations of different amino acids.

PROTOPLASM   The essential protein substance of living cells.

PROTOZOA   Very small, unicellular animal organisms.

PROVITAMIN A   Carotene.

PUBERTY   The age at which the reproductive organs become functionally active.

PULP   The solid residue which remains following the removal of the juices from plant materials.

PUTREFACTION   The decomposition of proteins by microorganisms under anaerobic conditions.

PYREXIA   A feverish condition.

PYRIDOXINE   The same as *vitamin B_6*.

RADIOACTIVE   Giving off atomic energy in the form of alpha, beta, or gamma rays.

RADIOISOTOPE   A radioactive form of an element.

RANCID   A term used to describe fats that have undergone partial decomposition.

RANGE CUBES   Large pellets produced for feeding in the pasture on the ground.

RENNIN   The milk-curdling enzyme present in the gastric juice of milk-consuming animals.

RESIDUE   That which remains of any particular substance.

RESORPTION   A return of the nutritive components of a partially formed fetus and fetal membrane to the system of the mother.

RESPIRATION   The act of breathing.

RETICULUM   The second compartment of a ruminant's stomach. Also called the *honeycomb* or *waterbag*.

RIBOFLAVIN   Same as *vitamin B_2*. Formerly known as vitamin G.

ROLLED   Compressed into flat particles by having been passed between rollers.

ROUGHAGE   Any feed high (over about 20%) in crude fiber and low (under about 60%) in TDN, on an air-dry basis. Opposite of concentrate.

RUMEN   The first compartment of a ruminant's stomach. Also called the *paunch*.

RUMINANT   Any of a group of hoofed mammals that have a four-compartment stomach and that ruminate or chew a cud. Examples are cattle, sheep, goats, and deer.

RUMINATE   To regurgitate previously eaten feed for further chewing. To chew a cud.

SALIVA   A clear, somewhat viscid solution secreted by glands into the mouth. It contains the enzymes salivary amylase and salivary maltase.

SALMONELLA   A pathogenic, diarrhea-producing organism sometimes present in contaminated feeds.

SAPONIFIABLE   Having the capacity to react with alkali to form soap.

SAPONIFICATION   The formation of soap and glycerol from the reaction of fat with alkali.

SARCOMA   A tumor of fleshy consistency—often highly malignant.

SATURATED FAT   A fat formed from the reaction of glycerol with any one of several saturated fatty acids. Stearin and palmitin are examples.

SATURATED FATTY ACID   Any one of several fatty acids containing no double bonds. Stearic and palmitic acids are examples.

SEDENTARY   Sitting most of the time.

SELF-FED   Provided with a part or all of the ration on a continuous basis, thus permitting the animal to eat at will.

SEMI-DISPENSABLE AMINO ACID   An amino acid which is essential only under certain circumstances or which may replace in part one of the essential amino acids. Arginine, cystine, and tyrosine fall into this group.

SEPTICEMIA   A diseased condition resulting from the presence of pathogenic bacteria and their associated poisons in the blood.

SERINE   One of the nonessential amino acids.

SERUM   The colorless fluid portion of blood remaining after clotting and removal of corpuscles. It differs from plasma in that the fibrinogen has been removed.

SHORTS   A by-product of flour milling consisting of a mixture of small particles of bran and germ, the aleurone layer, and coarse flour.

SILAGE   The feed resulting from the storage and fermentation of green or wet crops under anaerobic conditions.

SILO   A semi-airtight to airtight structure designed for use in the production and storage of silage.

SOAP   A compound formed along with glycerol from the reaction of fat with alkali.

SOLID   A substance that does not perceptibly flow.

SOLUTION   A uniform liquid mixture of two or more substances molecularly dispersed within one another.

SOLVENT PROCESS   A process for the chemical extraction of oil from seeds involving the use of an organic solvent.

SPECIFIC GRAVITY   The ratio of the weight of a body to the weight of an equal volume of water.

SPECIFIC HEAT   The heat-absorbing capacity of a substance in relation to that of water.

SPORE   An inactive reproductive form of certain microorganisms.

STABILIZED   Made more resistant to chemical change by the addition of a particular substance.

STARCH   A polysaccharide having the formula $(C_6H_{10}O_5)_n$. An important source of energy for livestock. Yields glucose upon complete hydrolysis.

STEARIC ACID   An 18-carbon saturated fatty acid which reacts with glycerol to form stearin.

STEARIN   The fat formed from the reaction of stearic acid with glycerol.

STERILE   Free from living microorganisms. Also, not capable of producing young.

STERILITY   An inability to produce young.

STEROL   One of a class of complex, fat-like substances widely distributed in nature.

STOMACH   That part of the digestive tract lying between the esophagus and the small intestine. A 4-compartment organ in ruminants; a single compartment organ in nonruminants.

STRAW   That part of the mature plant remaining after the removal of the seed by threshing or combining.

STRESS   Any circumstance which tends to disrupt the normal, steady functioning of the body and its parts.

SUBSTRATE   A substance upon which an enzyme acts. Same as *zymolyte*.

SUCRASE   An enzyme present in intestinal juice which acts on sucrose to produce glucose and fructose.

SUCROSE   A disaccharide having the formula $C_{12}H_{22}O_{11}$. It hydrolyzes to glucose and fructose. Commonly known as *cane, beet,* or *table sugar.*

SUN-CURED   Dried by exposure to the sun.

SUPPLEMENT   A semi-concentrated source of one or more nutrients used to enhance the nutritional adequacy of a daily ration or a complete ration mixture.

SYNDROME   A medical term meaning a set of symptoms that occur together.

SYNTHESIS   The bringing together of two or more substances to form a new material.

TDN   Total digestible nutrients.

TETANY   A syndrome involving sharp flexion of the wrist and ankle joints, muscle twitching, cramps, and convulsions.

THERAPEUTIC   Pertaining to the medical treatment of disease.

THERAPY   The medical treatment of disease.

THERMAL   Refers to heat.

THIAMINE   The same as *thiamin, thiamine hydrochloride,* or *vitamin B₁*.

THREONINE   One of the essential amino acids.

THROMBOSIS   The obstruction of a blood vessel by the formation of a blood clot.

THYROID   The gland in the neck that secretes the hormone thyroxin.

TOCOPHEROL   Any of four different forms of an alcohol which is also known as *vitamin E.*

TOTAL DIGESTIBLE NUTRIENTS   A figure which indicates the relative energy value of a feed to an animal. It is the sum of the digestible protein, digestible nitrogen-free extract, digestible crude fiber, and (2.25 X the digestible fat).

TOXIC   Of a poisonous nature.

TRACE MINERAL   Any one of several mineral elements that are required by animals in very minute amounts. Same as *micromineral.*

TRACER ELEMENT   A radioactive element used in biological and other research to trace the fate of a substance.

TRAUMA   A wound or injury.

TRUE PROTEIN   A nitrogenous compound which will hydrolyze completely to amino acids.

TRYPTOPHAN   One of the essential amino acids.

TYROSINE   One of the nonessential amino acids.

UNSATURATED FAT   A fat formed from the reaction of glycerol with any one of several unsaturated fatty acids. Olein and linolein are examples.

UNSATURATED FATTY ACID   Any one of several fatty acids containing one or more double bonds. Oleic, linoleic, linolenic, and arachidonic acids are examples.

UREA   A white, crystalline, water-soluble substance with the formula $CO(NH_2)_2$. It is the most extensively used source of nonprotein nitrogen for animal feeding.

UREASE   An enzyme which acts on urea to produce carbon dioxide and ammonia. It is found in the jackbean and the soybean, and is produced by certain microorganisms in the rumen.

UREMIA   A toxic accumulation of urinary constituents in the blood.

VALINE   One of the essential amino acids.

VASCULAR   Pertaining to the blood vessels of the body.

VERTEBRATES   Animals with backbones.

VFA   Volatile fatty acid(s).

VILLI   Small thread-like projections attached to the interior side of the wall of the small intestine.

VISCERA   The organs of the great cavities of the body which are normally removed at slaughter.

VISCOSITY   The freedom of flow of liquids.

VITAMIN   One of a group of organic substances which in relatively small amounts are essential for life.

VOLATILE FATTY ACID   Any one of several volatile organic acids found especially in rumen contents and/or silage. Acetic, propionic, and butyric acids are ordinarily the most prevalent.

WET-RENDERED   Cooked with steam under pressure in closed tanks.

WHEY   The watery portion of milk remaining after the removal of the fat and curd.

WORK   Movement of matter through space.

Index

Abdominal deposition, 134-35
Absorption of nutrients, 36-40
Acetic acid-buffered potassium permanganate solution, 31
Acetic acids, 38
Acid-base balance, 98
Acid detergent fiber (ADF), 30
Acid detergent lignin (ADL), 30
Acid insoluble lignin, 30
Acids, fatty, 7, 133
ADF (acid detergent fiber), 30
ADL (acid detergent lignin), 30
Air-dry, 20
    concentrates, 211-12
    energy feeds, 162-74
    feeds, 8, 211
    materials, 14, 91
    roughages, 188-202, 211-12
Alanine, 72
Alfalfa, 344-45, 354 *ff.*
    forage, fresh, 23, 24, 26, 50, 80, 88, 93, 263

    hay, 21, 24, 25, 49, 79, 88, 93, 188-89, 263
    leaf meal, dehydrated, 21, 24, 26, 49, 79, 88, 92, 193, 215, 263
    meal, 160, 215
        dehydrated, 21, 24, 26, 49, 79, 88, 92, 192, 263
    stem meal, dehydrated, 22, 24, 25, 49, 88, 93, 193, 215, 263
Alpha, 125
Alpha tocopherol, 125
Alsike clover, 345, 354 *ff.*
    hay, 22, 24, 26, 49, 79, 88, 92, 190, 264
Amids, 15
Amino acid content of feeds (table), 448-50
    limiting, 72
Amino acids, 8, 15, 36, 38, 40, 71, 99, 133, 163
    essential, 71-73
    nonessential, 71-72
Ammonium hydroxide, 14
Ammonium sulfate, 14

Analysis of feed stuffs, 12-27
Anemia, 105, 127
Animal protein factor (APF), 77
Annual lespedeza (*see* Lespedeza)
Annuals, 343
Antineuritic factor, 126
Anti-rachitic factor, 125
Antiscorbutic factor, 129
APF (animal protein factor), 77
Arachidonic, 7, 133
Arginine, 72, 73
Arrowleaf clover, 352-53, 354 *ff.*
Artifact lignin, 31
Ascorbic acid, 128
Ash, 12, 17-18, 137
Aspartic acid, 72

Bahiagrass hay, 22, 24, 25, 49, 80, 88, 93, 196, 263, 347, 354 *ff.*
Bakery product, dried, 21, 23, 27, 48, 79, 89, 92, 171, 263
Balanced rations, 255-59, 267-305
    formulation of, 310-334
Barley, 79, 159, 349
    grain, 22, 24, 26, 48, 79, 89, 92, 165, 259, 263
    straw, 22, 24, 25, 49, 80, 88, 93, 200-201, 263
    winter, 354 *ff.*
Basal metabolism, 131
Beet molasses, 25, 27, 49, 79, 89, 93, 203-204, 205, 263
Beet pulp:
    dried, 22, 25, 26, 49, 79, 88, 93, 170, 263
        with molasses, 21, 25, 26, 49, 79, 88, 93, 160, 263
    wet, 23, 25, 26, 50, 80, 89, 93, 263
Bermudagrass:
    hay, 22, 23, 24, 25, 26, 49, 50, 78, 80, 88, 89, 93, 160, 161, 193-4, 217, 263, 280, 283, 286, 287, 346-47, 354 *ff.*
    meal, dehydrated coastal, 22, 24, 26, 49, 79, 88, 92, 263
Beta, 125
Beta-carotene, 120
Biotin, 127
Blackstrap molasses, 203-204
Blood meal, 22, 24, 27, 49, 78, 88, 92, 178, 263
Bluegrass, 354 *ff.*
Bone composition, 81
Bone formation, 124, 133, 136
Bone meal, 112
    steamed, 21, 24, 27, 50, 79, 87, 92, 94-95, 114
Brewers dried grains, 22, 23, 26, 49, 78, 88, 92, 160, 181, 265
British Thermal Unit (BTU), 52
Brix, 205
Bromegrass, 345, 354 *ff.*
    forage, fresh, 23, 24, 26, 50, 80, 89, 93, 263
    hay, 22, 24, 25, 49, 79, 88, 93, 194-95, 263, 354 *ff.*
Butyric acids, 38
By-product concentrates, 23-24
By-products, animal, 27
By-products, wet, 208

Cabbage heads, 23, 25, 27, 50, 80, 89, 93, 263
Calcium, 3, 8, 82, 87-89, 90-91, 106, 112, 114, 133, 136, 138, 139, 162, 167, 267, 283
    in balancing rations, 257-58
    deficiency, 90, 96
California system for net energy evaluation of rations, 288-90
Calorie, 51
Calves, fattening, 307
Cane molasses, 25, 27, 48-49, 80, 88, 93, 203-204, 205, 266
Capacity calculation, 341
Carbohydrates, 4, 6, 11, 18, 53, 138
Carbon, 3, 4, 6, 7, 8, 53, 117
Carbon chain, 7
Carbon dioxide fixation, 127
Carotene, 16, 117, 120, 124, 125, 138, 139, 165, 166, 170, 215, 258, 267
Carrots, 23, 25, 27, 50, 80, 89, 93, 264
Cattle, 87
    beef:
        guides for feeting, 306-307
        nutrient requirements, 378
        pasture use for, 356-58
    dairy:
        guides for feeding, 306-307
        nutrient requirements, 379
        pasture use of, 358-59
Cellulose, 6, 17, 28, 30, 36, 39, 135
Cetyl trimethylammonium bromide, 30
Chlorine, 3, 8, 82, 84, 85, 122-23, 139
Chlorophyll, 16
Choline, 19, 128
Chromium, 4
Chyle, 37
Citrulline, 72
Citrus molasses, 25, 27, 49, 80, 88, 93, 204, 205, 264
Citrus pulp:
    dried, 22, 23, 26, 48-49, 79, 87, 88, 93, 169, 264
    fresh, 23, 25, 50, 80, 87, 88, 93, 264
    with molasses, dried, 22, 23, 26, 48, 79,

87, 88, 93, 264
Cobalt, 3, 8, 83, 103-104
Cobalt carbonate, 104
Cobalt chloride, 104
Cobalt sulfate, 104
Composition of feeds, 19-27, 210-15, 409-12, 416-47
Concentrated ferrous sulfate, 102
Concentrates:
    air-dry, 211-12
    energy, 212
    low-fiber air-dry, 26-27
    protein, 212
Copper, 3, 8, 83, 104-105, 108
Copper sulfate, 105
Copra meal, 22, 23, 26, 48, 78, 89, 92, 160, 161, 181-82, 264
Corn, 75, 106, 107, 122, 159, 161
Corn, ground ear, 22, 24, 26, 48, 79, 89, 92, 166, 213, 214, 264
Corn, ground shelled, 283
Corn, ground snapped, 22, 24, 26, 49, 79, 89, 92, 167, 213, 214, 264
Corn, shelled, 9, 79, 162, 259, 280, 286, 287
Corn, snap, 9, 79, 280
Corn, yellow shelled, 22, 23, 27, 48, 79, 89, 92, 215, 264
Corn bran, 22, 23, 26, 49, 79, 89, 93, 171, 264
Corn cob, ground, 22, 25, 49, 80, 89, 93, 264
Corn cobs, 199-200
Corn ear silage, 22, 24, 26, 50, 80, 89, 93, 264
Corn (fodder) silage, 23, 24, 26, 50, 80, 89, 93, 264
Corn gluten meal, 22, 24, 26, 48, 78, 89, 92, 180-81, 264
Corn husks, 199-200
Corn oil, 21, 23, 27, 48, 80, 89, 93, 174, 264
Corn starch, 22, 25, 27, 48, 80, 89, 93, 174, 264
Corn stover, 22, 24, 25, 49, 80, 88, 93, 199-200, 264
    silage, 22, 25, 26, 50, 80, 89, 93, 264
Cottonseed, 182-84
    ground, 21, 23, 26, 48, 78, 89, 92, 264
    hulls, 22, 24, 25, 49, 80, 89, 93, 201-202, 264
    meal, 21, 22, 23, 24, 26, 48, 49, 78, 88, 89, 92, 98, 160, 161, 178-89, 264
Cowpea hay, 22, 24, 26, 49, 79, 88, 92, 191-92, 264
Cows, 33, 35
    milking, guides to feeding, 307
    nursing, in dry-lot, 306

Crimson clover, 350, 354 *ff.*
    forage, fresh, 23, 25, 26, 50, 80, 88, 93, 264
Critical temperature, 131-32
Crude fat, 16-17, 19, 44-45, 46
Crude fiber, 17, 19, 44-45, 46
    extract fractions, 28, 29
    percentages, 25-27
Crude protein, 14-16, 18, 44-45, 46, 78-80, 165
Cutin, 31
Cyanocobalamin, 128
Cystine, 72

Dallisgrass, 348-49, 354 *ff.*
    forage, fresh, 23, 25, 26, 49, 50, 80, 89, 93, 264
    hay, 22, 24, 25, 79, 88, 93, 197, 264
    poisoning, 349
Decarboxylation, 127
Dehydrocholesterol, 125
Dextrin, 5
Dextrose, 5
Digester tankage, 21, 23, 27, 49, 78, 87, 92, 175-76, 263
Digestibility, 17, 41-45, 47
Digestible energy (DE), 53, 54
Digestible nutrients, total (TDN), 46-50, 53, 54
Digestion of nutrients, 36-40
Digestion trials, 41, 42, 43, 44-45
Digestive juices, 36, 37, (Table 6) 40
Digestive tract, 32-35
Diose, 4
Disaccharide, 4, 5
Distillers dried grains, 21, 23, 26, 48, 78, 89, 160, 181, 265
Dried feeds, 21-22, 47
Dry feed matter, 257
    digestion trial calculations, 44, 45, 267
    separation of, 29

Energy, 131, 133, 138, 141, 257, 267
    calculations in digestion trial, 44, 45
    concentrates, 212
    feeds, air-dry, 162-74
    proteins as source of, 75
    utilization, 51-54
    wool production, 141
Enzyme systems, 126, 127
Enzymes, 15, 36, 37, (table) 40, 82
Erepsin (intestinal peptidases), 37, 38
Ergosterol, 125
Ergot, 349
Ether extract, 12, 16, 23-24
Evaluation of forage, Van Soest method of, 28-31
Expeller extraction process, 77-78

Extraction of fat from oil seeds, 77-78
Extraction thimble, 16
Extractor, soxhlet, 16

Fat, 135
    animal, 21, 23
        feed grade, 27, 80, 89, 93, 172-74, 263
    crude, 12, 16-17, 19, 44-45
    digestion, 37, 39
    extraction of from oil seeds, 77-78
Fats, 6-7, 36, 53
    functions of, 11, 47
    pure, 27, 48, 80, 89, 93
    true, 16
Fattening, 134-36, 143
Fatty acids, 7, 39, 40, 133
Feather meal, 21, 24, 27, 49, 78, 88, 92, 177, 265
Feed composition, 210-15
Feed grains, 23-24, 89, 92-93
Feed groups, 159-61
Feed nutrients, 3-9
Feed requirements, estimation of, 335-38
Fescue:
    forage fresh, 23, 24, 26, 50, 80, 89, 93, 265
    hay, tall, 22, 24, 25, 49, 79, 88, 93, 196, 265, 347-48, 354 ff.
Fetal development, 139-40, 143
Fiber, crude, 12, 17, 19, 44-45
    extract fractions, 28, 29
    percentages, 25-27
Field cured grains, 21-22
Field cured hays, 21
Fish meal, 21, 23, 27, 49, 78, 87, 92, 160, 161, 176-77, 265
Fluorine, 3, 8, 83, 105-106
Folacin, 128
Folic acid, 128
Forage evaluation, Van Soest method, 28-31
Forages, 108, 114
    fresh, 22-23, 88, 208
    green, 8
Fructose, 5, 6, 36, 39, 40

Galactose, 5, 36, 39, 40
Gastric juice, 37, 40
Gastric lipase, 37, 39, 40
Glucose, 4, 5, 6, 36, 38, 40
Glutamic acid, 72
Gluten feed and meal, 160
Glycerol, 7, 36, 40
Glycine, 72
Glycogen, 6
Glycolysis, 126
Goose-stepping, 123, 127
Gossypol, 179

Grain by-products, 88
Grain sorghum grain, 22, 24, 26, 75, 79, 89, 92, 159, 214, 266
    silage, 22, 25, 26, 49, 50, 80, 89, 93, 265
Grains, 9, 24
    distillers dried, 92
Grass hay, 9
Grass staggers, 98
Grass tetany, 97, 98
Grease (yolk), 140
Greenchop, 208
Gross energy (GE), 52, 53
Growth of animals, 8, 106, 108, 111, 134, 135, 136, 143
Guides to feeding, 306-309

Hay, 216-23
    equivalents, Thaer's table of, 256
    grass, 9
    kudzu, 21 (see also individual types)
    making, 216-23
    meadow, 21, 24, 25, 49
    meals, 25-26, 49
    redtop, 21
Haylage, 207
Hays, 21-22, 24, 25-26, 49, 92-93
Heat increment (HI), 53
Hegari grain, 22, 24, 27, 48, 79, 89, 92, 163, 259, 266
Hemicelluloses, 17, 18, 28, 29, 30
Hemoglobin formation, 101, 104
Hexosans, 4
Hexose, 4, 5
High moisture feeds, 206-209, 211, 212, 213
Histidine, 72, 73
Hogs (see Swine)
Hominy feed, 22, 23, 26, 48, 79, 89, 92, 171, 264
Horses:
    guides to feeding, 308
    nutrient requirements, 379
    pasture use for, 360
Hulls, 24, 25, 49
Hydraulic extraction process, 77
Hydrogen, 3, 4, 6, 7, 8, 53, 117
Hydroxyproline, 72
Hypomagnesemia, 97
Hypomagnesemic tetany, 97

Ileum, 33
Indigestibles, 18, 28, 29, 30
Inositol, 128
Intermediate protein breakdown products (IPBP), 38, 40
Intermuscular deposition, 134-35
Intestinal juice, 37, 40
Intestinal maltase, 37, 38, 40
Intestinal peptidases (erepsin), 37, 40

Intestine, large, 33, 37, 38, 39
Intestine, small, sections of, 33, 37, 38, 39, 40
Intramuscular deposition, 135
Inulin, 6
Iodine, 3, 8, 83, 102-103
Iodine number, 7
Iodized salt, 103
Iron, 3, 7, 8, 82, 101-102
Iron absorption, 104
Iron dextran, 102
Isoleucine, 72, 73

Jerusalem artichokes, 6
Johnsongrass hay, 22, 24, 25, 49, 79, 88, 92, 197, 266, 354, 354 *ff*.

Kafir grain, 22, 24, 27, 49, 79, 89, 92, 163, 266
Kentucky bluegrass:
    forage, fresh, 23, 24, 26, 50, 80, 89, 93, 263
    hay, 22, 24, 26, 49, 79, 88, 92, 199, 263, 345-46
Keratin, 104
"Knuckling over," 107
Kudzu hay, 21, 24, 25, 49, 79, 88, 92, 265, 352, 354 *ff*.

Lactase, 37, 40
Lacteals, 37
Lactose, 5, 36, 39, 40
Ladino clover, 350-51, 354 *ff*.
    forage, fresh, 23, 25, 26, 50, 80, 88, 93, 264
Legume hays, 79, 188-89
Legume meals, 79, 188-89
Legumes, 211, 212, 342
Lespedeza, 351, 354 *ff*.
    hay, 21, 24, 26, 49, 79, 88, 92, 190, 265
Leucine, 72, 73
Lignin, 6, 7, 17, 18, 28, 30
Limestone, ground, 21, 25, 27, 50, 80, 87, 90, 93, 112, 114, 280, 283
Limiting amino acid, 72
Linoleic, 7, 133
Linolenic, 7, 133
Linseed meal, 22, 23, 26, 78, 88, 92, 265
Linseed oil meal, 160, 161, 179-80
Lipids, 29
Liquid supplements, 185-86
Liver, 33, 37, 40, 108
Lysine, 72, 73, 163

Macro minerals, 8, 84-100
Macromaterials, 14
Macronutrient, 7

Magnesium, 3, 8, 82, 96-98
Magnesium oxide, 98
Malt sugar, 5
Maltose, 5, 38, 40
Manganese, 3, 8, 83, 106-107
Marble dust, 90
Meadow hay, 21, 24, 25, 49, 79, 88, 93, 198, 265
Meat and bone meal, 21, 23, 27, 49, 78, 87, 92, 263
Meat and bone scrap, 176
Meat and meal, 176
Meat scrap, 21, 23, 26, 49, 78, 87, 92, 176, 263
Metabolism, 126, 127
Metabolism, abnormal bone, 105
Metabolism, basal, 131
Metabolizable energy (ME), 53
Methane, 38, 41
Methionine, 72, 73, 122, 123, 128, 163
Metric system in balancing steer ration, 284-87
Micro minerals, 8, 101-11
Microbes, 75
Microbial fermentation, 33, 38, 39
Micronutrients, vitamins as, 8
Milk:
    dried skimmed, 21, 25, 27, 48, 78-80, 87, 88, 92, 175, 264
    dried whole, 21, 22, 27, 48, 78-80, 88, 92, 177-78, 264
    fresh, 23, 27, 50, 80, 89, 91, 93, 209, 264
    fresh skimmed, 23, 25, 27, 50, 80, 87, 89, 91, 93, 213, 264
    production, 82, 136-39, 143
    products, 27
        dried, 88, 160, 161
    secretion, 137-38
    sugar, 5
Millet, 354 *ff*.
    hay, 22, 26, 49, 79, 88, 93, 198, 265, 353-54
Milo grain, 22, 24, 26, 49, 79, 89, 92, 163, 266
Mineral:
    elements, 8
    feeds, 21, 24-25, 50, 80, 112-14
    matter (ash), 12, 17-18, 19, 44-45, 47
    nutrition, 81-83
    products in feeds, 211
    supplements, 95
    wool production, for, 141
Minerals, 8, 11, 17, 36, 132, 133, 139, 258
Moisture content of feed, determination of, 13-14
Moisture feeds, 24-25, 50, 80
Moisture (high) feeds, 26-27, 93

Molasses, 9, 27, 80, 88, 159, 160, 203-5 (*see also individual molasseses*)
Molybdenum, 3, 8, 83, 104, 107-108
Monoglycerides, 39, 40
Monosaccharides, 4, 5
Mouth, 32, 38, 29
Mules, guides to feeding, 308

NDF (neutral detergent fiber), 30
NDS (neutral detergent solubles), 29
Net energy (NE), 53
Neutral detergent fiber (NDF), 30
Neutral detergent solubles (NDS), 29
NFE (nitrogen-free extract), 18-19, 28, 30, 44-45, 46
Niacin, 122
Nicotinamide, 126
Nicotinic acid, 126
Nitrogen, 3, 6, 7, 8, 14, 115, 116, 117
Nitrogen-free extract, 12, 18-19, 28, 44-45, 46
Nonlegumes, 79, 88, 89, 193-99, 211, 212, 342, 343
Nonmineral elements, 8
Nonprotein nitrogen, 75
Nonproteins, 15
Nonruminants, 28, 33, 37, 38, 39, 73, 75, 99
NPN, 75
Nutrient absorption, 36-40
Nutrient requirements, 378-87
Nutrient Requirements of Beef Cattle, 378, 380-87, 388, 410
Nutrient Requirements of Dairy Cattle, 379, 390-93, 410
Nutrient Requirements of Horses, 379, 394-97
Nutrient Requirements of Sheep, 379, 398-401
Nutrient Requirements of Swine, 379, 402-408, 410
Nutrient transport, 36-40
Nutrients, total digestible (TDN), 46-50, 53, 54
Nutrients of feed, 3-9, 10-11, 38, 39
Nutrition, protein, 70-80

Oat grain, 265
Oat groats, 22, 23, 26, 48, 79, 89, 92, 171, 265
Oat hay, 22, 24, 26, 49, 79, 88, 93, 198, 265
Oat hulls, 21, 24, 25, 49, 80, 89, 93, 201-202, 265
Oat straw, 22, 24, 25, 49, 80, 88, 93, 200-201, 265
Oats, 79, 159, 349, 354 *ff.*

Oats grain, 22, 23, 26, 49, 79, 89, 92, 165
Oil seed meals, 78-80, 160
Oil seeds, 78-80
    extraction of fat from, 77-78
Oils, 7
Oleic acid, 7
Olein, 7
Orchardgrass, 22, 23, 24, 25, 26, 49, 50, 79, 80, 88, 89, 92, 93, 195, 265, 345, 354 *ff.*
Osteomalacia, 90, 94, 125
Osteoporosis, 90, 94, 105, 125
Oxygen, 3, 4, 6, 7, 8, 117
Oyster shell flour, 21, 25, 27, 50, 80, 87, 90, 93, 112, 114

Palmitic acid, 7
Palmitin, 7
Pancreatic amylase, 37, 38, 40
Pancreatic juice, 37, 40
Pancreatic lipase, 37, 39, 40
Pantothenic acid, 122, 123, 127, 162
Para-aminobenzoic, 129
Parakeratosis, 109, 111
Pasture crops, 344-54
Pastures, 342-43, 255-61
Peanut hay, 22, 23, 26, 49, 79, 88, 93, 191, 265
Peanut hulls, 21, 24, 25, 50, 79, 93, 201-202, 265
Peanut kernels, 21, 23, 26, 48, 78, 89, 92, 265
Peanut oil meal, 21, 24, 26, 49, 78, 89, 92, 160, 161, 179, 265
Peanuts, 182-84
Pentosans, 4
Pentose, 4
Pepsin, 37, 38
Peptides, 38, 40
Permanganate lignin, 31
Petersen method of feed evaluation, 259, 260-66
Phenylalanine, 72, 73
Phosphate, defluorinated, 21, 25, 27, 50, 80, 87, 92, 95-96, 98, 112, 114
Phospholipids, 16
Phosphorus, 3, 7, 8, 82, 91-95, 96, 106, 112, 114, 133, 136, 138, 139, 162, 167, 258, 267
Pigs (*see* Swine)
Polypeptides, 38, 40
Polysaccharides, 4, 5, 6
Ponies, guides to feeding, 308
Potassium, 3, 8, 14, 82, 98-99, 141
Potassium iodide, 103
Potato meal, 21, 25, 27, 49, 79, 89, 93, 172, 265

Potato (sweet) meal, 22
Potatoes, sweet (see Sweet Potatoes)
Potatoes, white, 23, 25, 27, 50, 80, 89, 93, 265
Poultry by-product meal, 21, 23, 27, 48, 78, 87, 92, 177, 265
Prairie hay, 22, 24, 25, 49, 80, 88, 93, 198, 265
Proline, 72
Propionic acids, 38, 39
Proteins, 7-8, 11, 14, 17, 29, 36, 38, 39, 40, 73, 132, 133, 135, 136, 137, 138, 141, 163, 175-87, 257, 267
    concentrates, 212
    crude, 12, 14-16, 18, 44-45, 46, 78-80, 165
    feeds, mechanically extracted, 23-24
    of plant origin, 178-84
    (high) feeds, 77-78, 88, 91, 92
    nutrition, 70-80
    supplements, 184, 212
    true, 15
Proteoses, 38, 40
Pyridoxal, 127
Pyridoxamine, 127
Pyridoxine, 127

Red clover, 345, 354 ff.
    forage, fresh, 23, 25, 26, 50, 80, 88, 93, 264
    hay, 22, 24, 25, 49, 79, 88, 93, 189-90, 264
Redtop, 345, 354 ff.
    hay, 24, 25, 49, 79, 88, 93, 194, 265
Reed canarygrass hay, 22, 24, 25, 49, 79, 88, 92, 195, 263
Rennin, 37, 38
Resorptions, 107
Restaurant garbage, 23, 27, 50, 80, 89, 93, 265
Riboflavin, 121, 126, 162
Rice bran, 22, 23, 49, 79, 89, 92, 174, 265
Rice hulls, 21, 25, 50, 80, 89, 93, 202, 265
Rice millfeeds, 160
Rickets, 90, 94, 125
Root crops, 22-23, 89
Roots, 208-209
Roughages, 25, 53, 80, 87, 88, 89, 93, 106, 188-202, 211-12, 282-83
Rumen, 33, 37, 38, 39, 75, 99, 103, 107, 117
Ruminants, 33, 37, 38, 39, 42, 43, 75, 99, 103, 119-20
Rye, 79, 349, 354 ff.
    forage, fresh, 23, 25, 26, 50, 80, 89, 93, 265
    grain, 22, 24, 27, 49, 79, 89, 92, 172, 265

Ryegrass, 349, 354 ff.
    forage, fresh, 23, 25, 26, 50, 80, 89, 93, 265

Safflower meal without hulls, 22, 23, 26, 49, 78, 88, 92, 182, 265
Saliva, 37
Salivary amylase, 37, 38
Salivary maltase, 37, 38
Salt, 85-86, 112, 113, 139
    iodized, 103
    trace-mineralized, 98, 104, 105, 107, 111
Saponification, 7
Selenium, 3, 8, 83, 108-109, 120, 123, 125
Sericea, 354 ff.
    hay, 22, 24, 25, 49, 79, 88, 92, 190-91, 265, 351
Serine, 72
Serum alkaline phosphatase, 106
Sesame oil meal, 21, 23, 26, 48, 78, 88, 92, 182, 265
Sheep, 33, 87, 309, 359-60, 379
Silages, 9, 22-23, 89, 207-208, 224-33
Silica, 30, 31
Slurries, 239
Soap, 36, 39, 40
Sodium, 3, 8, 82, 84, 85, 139
Sodium hydroxide, 14, 17
Sodium lauryl sulfate, 29
Sodium selenate, 109
Sodium selenite, 109
Sodium sulfate, 14
Solvent-extracted feeds, 24-25
Solvent extraction process, 78
Sorghum grain, 163-65
Sorgo silage, 23, 25, 26, 50, 80, 89, 93, 266
Soybean:
    flakes, 266
    hay, 22, 24, 25, 49, 79, 88, 92, 191-99, 266
    hulls, 21, 24, 25, 49, 79, 88, 93, 200
    mill feed, 200
    oil meal, 21, 22, 25, 26, 48, 78, 88, 92, 107, 160, 161, 178, 266, 280, 283
    seed, 22, 23, 26, 48, 78, 88, 92, 266
Soybeans, 182-84
Square method, formulating feed mixtures by use of, 315-22
Starch, 4, 5, 6, 29, 36, 38, 40, 135
Starches, 17, 18, 29
Stearic acid, 7
Stearin, 7
Steer ration mixtures, fortifying with vitamin A, 323-24
Stepping syndrome, 97
Sterols, 16, 125

"Stiff-lamb," 125
Stocker steers in dry-lot, 306
Straws, 21-22, 24, 25, 49
Subcutaneous deposition, 134-35
Sucrase, 37, 39, 40
Sucrose, 4, 5, 36, 39, 40
Sudangrass, 354 ff.
    forage, fresh, 23, 25, 26, 50, 80, 89, 93,
        266, 353-54
    hay, 22, 24, 26, 49, 79, 88, 92, 197-98,
        266
Sugars, 5, 17, 18, 29, 135
Suint, 140, 141
Sulfur, 3, 8, 82, 99-100
Sulfuric acid, 14, 17, 30
Sunflower meal, 21, 24, 26, 49, 78, 88, 92,
    182, 266
Sweet clover poisoning, 191
Sweetclover hay, 21, 24, 26, 49, 79, 88, 92,
    191, 266
Sweet potato meal, 22, 25, 26, 48, 80, 89,
    93, 170, 266
Sweet potatoes, 22, 25, 26, 50, 80, 89, 93,
    170, 266
Swine, 33, 34, 120-23, 307-308, 360-61,
    379, 448-50, 451-53

Tankage, 87, 175-76
    with bone, 21, 23, 26, 49, 78, 87, 92,
        176
    digester, 175-76
    meat meal, and, 160, 161
    wet rendered, 175-76
TDN (Total digestible nutrients), 46-50, 53,
    54
Temperature, critical, 131-32
Tetrose, 4
Thermogenic effect, 53
Thiamine, 126
Thiamine hydrochloride, 126
Threonine, 72, 73
"Thumps," 102
Thyroxin, 102
Timothy, 345, 354 ff.
    hay, 22, 24, 25, 49, 79, 88, 93, 160,
        194, 266
Tocopherols, 125
Toluene, 14
Total digestible nutrients (TDN), 46-50, 53,
    54
Trace minerals, 8, 112, 114
Transmethylation, 128
Triose, 4
Trypsin, 37, 38, 40
Tryptophan, 72, 73, 163
Tubers, 208-209
Turnips, 23, 25, 27, 50, 80, 89, 93, 266
Tyrosine, 72

Uniform State Feed Bill, 240-54
Urea, 15, 75-77

Valine, 72, 73
Van Soest (method of forage evaluation),
    28-31
Vitamins, 8, 11, 15, 16, 19, 36, 37, 77, 96,
        103, 108, 115-18, 119-29,
        124-29, 132, 133, 138, 139, 141,
        162, 258, 323-24, 451-53
Volumetric distillation, 14

Water (importance of and animal need for),
    8-9, 11, 12, 13, 132, 133-34, 136,
    137
Weende Experiment Station (Germany), 12
Weende system of feed analysis, 28, 29
Wet brewers grain, 23, 24, 50, 80, 89, 93,
    265
Wet by-products, 22-23, 89
Wet rendered tankage, 175-76
Wheat, 79, 159, 349
    winter, 354 ff.
Wheat bran, 22, 23, 26, 49, 79, 89, 91, 92,
    167-68, 266
Wheat by-products, 79, 92
Wheat grain, 22, 24, 26, 48, 79, 89, 92, 166,
    214, 266
Wheat middlings, 22, 23, 26, 48, 79, 89, 91,
    92, 168, 266
Wheat millfeeds, 166
Wheat poisoning, 97
Wheat shorts, 22, 24, 26, 48, 79, 89, 92,
    168, 266
Wheat straw, 22, 24, 25, 49, 80, 89, 93,
    200-201, 266
Whey, dried, 21, 24, 27, 48, 79, 88, 92,
    174, 264
White clover, 350, 354 ff.
    forage, fresh, 23, 25, 26, 50, 80, 93, 264
"White muscle," 125
Wolff-Lehman standards, 256
Wood molasses, 25, 27, 49, 80, 88, 93, 204,
    266
Wool fat, 140
Wool fiber, 140
Wool growth, 105, 109, 111
Wool production, 140-41, 143
Work, 141-42, 143
Work of digestion, 53

Xanthine oxidase, 107

Yearlings, fattening, 307
Yolk (grease), 140

Zinc, 3, 8, 83, 109-11